Espresso Coffee

The Science of Quality

Espresso Coffee

The Science of Quality

Second edition

Edited by

ANDREA ILLY
and
RINANTONIO VIANI

with the assistance of Furio Suggi Liverani

ELSEVIER
ACADEMIC
PRESS

AMSTERDAM BOSTON HEIDELBERG LONDON NEW YORK OXFORD
PARIS SAN DIEGO SAN FRANCISCO SINGAPORE SYDNEY TOKYO

This book is printed on acid-free paper

First published 1995
Second edition 2005
Reprinted 2010 (twice), 2011 (twice), 2012, 2013 (twice)

Elsevier Academic Press
525 B Street, Suite 1900, San Diego, California 92101-4495, USA
http://www.elsevier.com

Elsevier Academic Press
84 Theobald's Road, London WC1X 8RR, UK
http://www.elsevier.com

British Library Cataloguing in Publication Data
A catalogue record for this book is available from the British Library

Library of Congress Catalog Number: 2004113615

ISBN 978-0-1237-0371-2

Contents

Contributors

Francisco Anzueto graduated in 1978 in Agronomy in Guatemala and obtained a doctorate in Coffee Plant Breeding in France in 1993. He was regional director of Anacafe from 1978 until 1986, where he is now research director. He holds a professorship at Landivar University in Guatemala. He has been co-manager or director of several regional research programs in Central America. He is the co-author of many scientific papers.

Thomas W. Baumann received his scientific education in Medicine and Biology at the University of Zurich, which he completed in 1969. After training in Molecular Biology at the University of Zurich he became Research Assistant at the Institute of Plant Biology of the same university, where he was 'infected' for the first time with coffee while chemically analysing and establishing tissue cultures of several species of *Coffea*. After postdoctoral studies at the University of Southern California (Allan Hancock Foundation) in Los Angeles in 1972 and 1973, he returned to the Institute of Plant Biology, where he continued his phytochemical work on coffee and other caffeine-containing plant species. In 1977 he became Docent, and 1995 Professor, of Plant Physiology. His research focused on the biosynthesis and ecological significance of caffeine and related purine alkaloids. He also studied the ecological chemistry of other plant genera, such as *Physalis*, *Echinacea* and *Catharanthus*, until retirement in 2001. His hobbies are scientific writing and gardening. In 2001 he published together with his wife a voluminous biography of Henri Pittier, the 'Humboldt of Switzerland' (Basle: Friedrich Reinhardt Verlag). At his home near Zurich he houses a living collection of almost all caffeine-containing species consumed in the world. Currently, he is working on a book dealing with the various facets of little-known fruits of the Neotropics.

Sarah Bee is a radiation physicist with a doctorate from University College London, UK. She is a Chartered Physicist and Member of the Institute of Physics. Since 1998 she has been Engineering Product

x Contributors

Manager within the R&D Department at Sortex. She was responsible for the technical description of the 'Niagara' vegetable sorting machine (colour and shape sorting at 40 000 objects/sec), which obtained Sortex's Queen's Award for Innovation application in 2001.

Bernd Bonnländer graduated in Food Chemistry at Erlangen University, Germany. After his doctorate on the characterization of aroma precursors in wine and spices, he joined AromaLab (R&D illycaffè), where he performs research on odour active and physiologically active compounds in green and roasted coffee as well as beverages.

Carlos Henrique Jorge Brando is an engineer with graduate work at doctoral level in Planning and Business at the Massachusetts Institute of Technology. He was the manager and director of leading Brazilian coffee machinery maker Pinhalense for over ten years. He is now director and partner of a coffee consulting, marketing and trading company, P&A International Marketing, which was a technical adviser to the Café do Brazil Marketing Program. He has been responsible for coffee processing and marketing projects in over 60 countries and written and spoken frequently about the two subjects.

Gianfranco Brumen graduated in Chemistry from the University of Trieste, and joined illycaffè in 1972 in charge of Quality Control and Research, becoming an expert in the raw product (characteristics of coffee types, production countries, raw coffee selection, blend formulation, etc.), controlling all the production cycle. As Quality Supervisor he is in charge of blend formulation, final product conformity, quality system application and its certification (ISO 9001), and all regulatory aspects.

Roberto Cappuccio graduated in Physics at the University of Trieste in 1997 with a thesis on Near Infrared Spectrometry. Immediately afterwards he entered the R&D Department at illycaffè as a researcher, where he dealt mainly with fluid dynamics. Since 2001 his interests have moved towards the exploration of the sensory and chemical aspects of taste, aroma and mouthfeel perception of coffee. Since 2002 he has coordinated illycaffè Sensorylab. He has written several scientific publications, mainly in coffee science and technology.

Nelson Carvalhaes is a lawyer who graduated in Agribusiness and Foreign Trade at Mackenzie University, Saõ Paulo in 1981. He has been a Director of Escritório Carvalhaes Café Ltda since 1980 and Porto de Santos C.E. Ltda since 1989.

Rudolf Eggers graduated and obtained his doctorate in Mechanical Engineering/Process Techniques at the Technical University in Hanover in 1972. He was research associate at the Technical University in Clausthal until 1977. From 1977 until 1984 he was head of Process Engineering at Thyssen Maschinenbau and Krupp Industrietechnik. Since 1984 he has been Professor for Heat and Mass Transfer at the Technical University in Hamburg–Harburg. He has conducted research and published in separation processes (extraction, adsorption, refining); coffee processing (roasting, extraction, decaffeination); interfacial phenomena (transport effects across phase boundaries); high pressure technique (supercritical fluids).

Ulrich H. Engelhardt graduated as a Food Chemist and obtained his doctorate with work on stimulant drinks at the Technical University of Braunschweig, where he is now a teacher in food analysis, food toxicology and chemistry and technology of foods containing polyphenols. He has published about 70 papers, mainly on coffee and tea chemistry. He is member of the working groups on tea and coffee of the German Institute of Standards (DIN). In 1996 he received the Young Scientist award of the German Chemical Society.

Giorgio Graziosi graduated in Natural Sciences in 1966 at the University of Trieste, where he was appointed Lecturer in Genetics in 1969. He obtained a position of Associate Professor in 1976 and of Full Professor in 1986. He was appointed Director of the Department of Biology in 1993 and President of the Faculty of Biological Sciences, University of Trieste, in 1998. He has been visiting scientist at the Developmental Genetics Unit, International Embryological Laboratory, 'Hubrecht', Utrecht, Holland, at the Genetics Department, University College London, UK and for some years at the Genetics Laboratory of Oxford University, UK. His research interests 1967–86 were developmental genetics of *Drosophila melanogaster* and since 1987 have focused on DNA polymorphism in several organisms including *Coffea* and genomics and gene expression in *Coffea*.

Andrea Illy graduated in Chemistry at the University of Trieste and completed the Master Executive programme at SDA Bocconi in Milan. He is the chief executive officer of illycaffè. His first working experience was in 1983, at the Nestlé Research and Development department. In 1990 he began his career at illycaffè as supervisor of the Quality Control department, where he instituted a total quality control programme. He has been president (1999) of the Association Scientifique Internationale du Café (ASIC) and organized the 19th International Colloquium on Coffee Science in Trieste; he is presently vice-president. Since 1999, he has been a member of the board of directors of the Italian Association of Premium Brand Industries. In 2003 he was appointed a member of the advisory board of SDA Bocconi in Milan.

Ernesto Illy graduated in Chemistry at the University of Bologna. He has been chairman of illycaffè since 1963 and an initiator active in international bodies devoted to the scientific and professional development of coffee. He was co-founder and a past president of Association Scientifique Internationale du Café (ASIC), of which he is now senior vice-president; co-founder, past president and an active member since its foundation of the Physiological Effects of Coffee Committee (PEC); co-founder and president of the Institute for Scientific Information on Coffee (ISIC); President of the Promotion Committee of the Private Sector Consultative Body (PSCB) of ICO; President of *Centromarca* (Italian Association of Trademark Industries). He has been honoured with important awards from the international coffee community and has received from the President of Italy the title of *Cavaliere del Lavoro*.

Isabelle Kölling-Speer graduated in Food Chemistry at the University of Hamburg. She obtained her PhD in food chemistry from the same university in 1993. Before joining the Institute for Food Chemistry at the Technical University of Dresden in 1993, where she is Scientific Researcher, she worked with the Hygiene Institute Hamburg in the Department of Chemistry and Food Chemistry. She is co-author of two books and has published over 20 scientific publications. Her research activities are related to the processing and quality control of foods of plant origin.

Hans Gerhard Maier studied Pharmacy and Food Chemistry at the Universities in Freiburg (Breisgau) and Frankfurt (Main), obtaining his doctorate in 1961. He also studied chemistry from 1963 to 1965 at the University of Frankfurt, obtaining the doctorate in 1969. After working from 1961 to 1963 in the Food Research Laboratory of Franck und Kathreiner in Ludwigsburg, he became in 1963 scientific assistant at the University of Frankfurt, lecturer in 1970, assistant professor in 1971. After two years spent at the University of Münster as Wissenschaftlicher Rat und Professor, he became full professor and director of the Institute of Food Chemistry at the Technical University in Braunschweig, from where he retired in 1998. He is the author of many important publications on coffee, and, in particular, of a book *Kaffee* (Berlin: Paul Parey, 1981).

Maria Cristina Nicoli graduated in Pharmaceutical Chemistry at the University of Bologna. After a decade of post-doctoral research activity at the University of Udine as assistant professor she moved to the University of Sassari, where she was appointed Professor in Food Technology. One year later she was called back to the University of Udine, where she continues research and teaching activity. Her research activity is mainly focused on chemical and physical factors affecting food functionality and shelf-life with particular attention to processed fruit and vegetables and roasted coffee. She is author of over 80 scientific papers published in international journals of food science and technology. She has been in charge of several Italian and European research projects.

Marino Petracco graduated in Chemical Engineering at the University of Trieste. After 10 years spent in industrial research working in a multinational petrochemical company, specializing on catalytic cracking, he joined illycaffè, working on topics ranging from plant botany to industrial transformation processes, beverage brewing dynamics and its effects on the consumer's body. He has published widely, including 30 research papers on sensory chemometrics, roasted coffee grinding dynamics, physicochemical characterization of espresso coffee, beverage brewing hydraulics and raw coffee sampling, and is co-author of several books. He is active in several scientific associations: in coffee science, ASIC and PEC, where he held the chair in 1991 and 1992; in coffee standardization, the Italian organization for standardization (UNI), the Association Française de Normalisation (AFNOR, commission V33A café) and the International Standardization Organization (ISO TC 34 SC 15 'Coffee'); in food hygiene and microbiology, the Association Africaine de Microbiologie et d'Hygiène Alimentaire (AAMHA) and the Sociedad Latino Americana de Micotoxicología (SLAM).

Carlos Roberto Piccin graduated in Agronomy at the Universidade Paulista, Jaboticabal SP, Brazil in 1981, and then attended the Graduate School in Mineral Nutrition at the Universidade Federal, Lavras, MG, Brazil. After a career as technical manager with various coffee companies, in 1999 he created Agropiccin, a technical consulting firm expert in coffee production systems, with emphasis on plant nutrition linked with integrated pest management seeking productivity, quality product and biological equilibrium. He is often an invited speaker at universities and in the private sector.

Oriana Savonitti obtained a Masters in Process Chemical Engineering from the University of Trieste and has been a researcher in the Research and Development department of illycaffè since 1998.

Maro R. Söndahl obtained in Brazil a BS degree in Agriculture Engineering in 1968 and MS in Plant Physiology in 1974. In the USA he obtained a PhD in Managing Technology and Innovation in 1968 and in Cell Biology in 1978. His professional career developed in government (13 years) and private organizations (20 years). Since 1994 he has been President of Fitolink Corporation in the USA, after working at an Institute of Agronomy in Brazil from 1970 to 1983, and at DNA Plant Corp. in the USA from 1983 until 1993. He is the author of five US patents and more than 40 scientific papers and review articles in the area of plant physiology and plant cell genetics of tropical plants with emphasis on coffee, cacao and oil palm. He acts as a consultant to several private coffee companies in the USA and Europe and to international organizations. He has been Visiting Professor and/or invited lecturer at several universities in the USA, Brazil, Argentina, Venezuela, Costa Rica, Colombia and Chile, and an invited speaker at scientific meetings in more than 20 countries.

Karl Speer completed his degree in food chemistry in 1985, and his doctorate at the University of Hamburg, Germany in 1993. He was affiliated with the Hygiene Institute Hamburg, Department of Chemistry and Food Chemistry, from 1984 until 1993. He has been with the Institute for Food Chemistry in Dresden since 1993. In 1997 and 1998 he was a visiting scholar at the Department of Food Science in West Lafayette, Purdue University, USA. He received the Josef Schormüller Award in 1992. He is co-author of two books and published over 50 scientific publications, especially in the area of coffee, honey and the analysis of pesticide residues and organic contaminants in foods. He is now professor of Food Chemistry and Food Production, Department of Food Chemistry, Technical University of Dresden.

Furio Suggi Liverani joined illycaffè in 1991 in the role of knowledge manager after having worked for 10 years as a computer science adviser. Since 1995 he has been managing director of the Department of Research and Development, directly under the CEO. He has participated in several projects of industrial and laboratory automation and in the development of illycaffè's production information system. He has developed technologies and innovative products such as an electronic coffee sorting machine. In 1993 he introduced the use of Internet and multimedia to the company, actualizing the www.illy.com website in 1996. In 1998 on behalf of illycaffè he founded AromaLab, a laboratory dedicated to the study of coffee chemistry, located in the Area Science Park of Trieste. He has been a council member of the AI*IA (Italian Association for Artificial Intelligence), co-ordinator of the workgroup AI*IA in industry, and is a member of several scientific societies. Currently he is member of the steering committee of the international conference 'Cellular Automata for Research and Industry' and of the 'Italian Congress of Artificial Intelligence'. He is also a board member of Qualicaf and adviser of the EDAMOK project. He is the author of 30 articles on the subjects of coffee technology, computer science and artificial intelligence, he has participated as spokesman in several courses, national and international conferences and conventions. He is the co-inventor of five industrial patents.

Aldir Alves Teixeira is an Engineer Agronomist, having graduated at Saõ Paulo University in 1959. In 1972 he gained a doctorate in Agronomy on Cup Tasting from Saõ Paulo University. He has been president of his own company, ASSICAFÉ, since 1992, president of the Scientific Research Association from 1990 to 1993, president of the Evaluation Board of Commission on the Brazilian Coffee Prize for 'Espresso', promoted by illycaffè, since 1991 and a member of the Brazilian Delegation to ISO TC 34 SC 15 'Coffee'. He has been a scientific consultant of illycaffè since 1991.

Ana Regina Rocha Teixeira is a Biologist who graduated at Saõ Paulo University in 1988. She worked at the Phytopatologic Biochemical Department of the Biologic Institute of Saõ Paulo from 1986 to 1992. Since 1992 she has worked with ASSICAFÉ, has been a member of the Evaluation Board of Commission on the Brazilian Coffee Prize for 'Espresso', promoted by illycaffè, since 1991, she contributes to the Brazilian Delegation to ISO TC 34 SC 15 'Coffee' and has collaborated with illycaffè since 1992.

Roberto Antônio Thomaziello is an Engineer Agronomist, having graduated at Saõ Paulo University in 1965. His positions have included: assistant technician of the Coffee Office, CATI, from 1968 to 1977 and from 1990/91 to 1998; Project Supervisor for the Coffee Agricultural Secretariat of Saõ Paulo from 1977 to 1986; head of the Department of Rural Assistance of the Agricultural Secretariat/SP from 1992 to 1995; director of group techniques at the Agricultural Secretariat from 1995 until 1997; and since 1999 at the Coffee Research Center 'Alcides Carvalho' of the Campinas Agronomic Institute.

Herbert A.M. van der Vossen graduated in 1964 with an Ir (MSc) degree in Tropical Agronomy and Plant Breeding and gained his doctorate in 1974 with a thesis on breeding and quantitative genetics in the oil palm from the Wageningen Agricultural University. He was research officer in charge of the Oil Palm Research Centre at Kade, Ghana, during 1964–1971 and then head of the Coffee Breeding Unit at the Coffee Research Station near Ruiru, Kenya, from 1971 to 1981. He was research manager then director of the breeding programmes for vegetable and flower seed crops with Sluis & Groot Seed Company at Enkhuizen in the Netherlands from 1981 until early retirement in 1993. From 1993 until 1996 he was seed policy adviser to the Ministry of Agriculture at Dhaka, Bangladesh and freelance consultant in plant breeding and seed production, including assignments in vegetable seeds, cereal crops, cocoa, oil palm and coffee. He has contributed more than 40 scientific papers and chapters. He is a member of the board of ASIC.

Rinantonio Viani, after graduating in Chemistry at the University of Pisa, did postgraduate work at CalTech in Pasadena, California, and Duke University in Durham, NC. From 1963 until 1974 he was a Research Chemist at the Research Laboratories of Nestlé in Vevey and Orbe. From 1975 until retirement in 1998 technical and scientific advisor on stimulant drinks in the Nestlé headquarters. He has been Chairman of PEC, Administrative Secretary and President of ASIC and Chairman of the Technical Commission of European Decaffeinators, of AFCASOLE and of ISO TC 34 SC 15 'Coffee' for many years. He has obtained several patents and contributed many scientific publications in food science and technology, particularly in the field of stimulant drinks. After retirement he has worked as consultant with the UN agencies UNIDO and FAO for coffee projects in producer countries.

Otto G. Vitzthum is a Chemist with a doctorate from the University of Erlangen and postgraduate studies in Madrid. He was Director of Scientific Research at HAG AG in Bremen from 1978 to 1986; Department Manager in Central Research at Jacobs Suchard from 1986 until 1994 and Director of Coffee Chemistry worldwide at Kraft Jacobs Suchard from 1994 until retirement in 1997; Lecturer at the Technical University Braunscheig since 1978 and Honorary Professor since 1987. He has been scientific secretary of ASIC and chairman of the technical commission of European Decaffeinators for many years, and twice Chairman of PEC. He is the author of many patents and scientific publications in the field of coffee.

Acknowledgements

The Editors thank Ms Elaine Dentan and Mr Fabio Silizio for the microscopy photos, Prospero S.r.l. for all figures and electronic drafts, and Ms Andrea Appelwick for her constant editorial assistance.

CHAPTER 1

Quality

A. Illy

The term 'quality' has been abused since the later 1980s, when Western industries realized that lack of quality accounted for their diminished competitiveness with respect to those of Japan. This tendency to speak, at times inappropriately, about quality seldom reflects the true meaning of the word itself, although this is perhaps understandable and justifiable if one considers that 'quality' can convey just as many meanings as the 'ability' an object has of producing effects, as will be seen below.

Quality is highly pervasive, and this prevents anyone who decides to 'create quality' from clearly defining a field of action. In companies recognized on the market for their quality, it is obvious this involves the entire staff, with no limits in time or space. In other words, quality is culture.

The following paragraphs will try to sketch at least an outline of this fascinating concept 'quality'.

1.1 ORIGINS AND MEANINGS OF QUALITY

The concept of quality is vast and cannot, therefore, be narrowed down to one meaning. Aristotle distinguished four different families of qualities, which were subsequently adopted by scholars:

1 Tendencies and aptitudes (examples of tendencies are temperance, science and virtue, while health, illness, heat, cold, etc. are examples of aptitudes).
2 Natural abilities and faculties, or active qualities.
3 Sentiments and passions, or passive qualities (sounds, colours, taste, etc.).

4   Shapes or geometric determination (square, circle, straight line, etc.).

These families can be classified as:

■   attitudinal qualities, which include aptitudes, habits, abilities, virtues and tendencies, or all the 'possibilities' of the object;
■   sensorial qualities, which can be perceived by our sense organs and include sounds, colours, smells, etc.;
■   measurable qualities, which can be measured objectively, such as speed, length, intensity, mass, etc.

The attitudinal qualities refer to the identity attributes of an object, or what distinguishes it from another, without value hierarchies, whereas the sensorial and measurable qualities refer to all the value attributes, which make one thing better than another.

Identity and value are the two poles of meaning of the word 'quality'. In a critical analysis, the two poles must be compared with the episteme – the cognitive dimension of the certain knowledge – and with the taste – the subjective orientation of individual knowledge. This is of particular importance for the understanding of quality as innate excellence.

In the modern industrial application, measurable quality prevails: 'You can't have quality if you can't measure it' wrote Juran (1951), based on the scientific method. Alternative and complementary approaches can be applied based on the product, the user or the production, and the prevailing dimensions are performance, reliability, conformity, life and functional efficiency, each with its own units of measurement.

Two ways of considering quality will therefore be compared: the first is philosophical, where aesthetics concerns itself with sensorial qualities, and the second is scientific, where everything is related to the characterization and quantification of measurable qualities.

1.2 DEFINITION OF QUALITY

The range of the term 'quality', with all its applications and facets, makes an exact and concise definition difficult. This explains why there are so many.

The 'official' definition is provided by the International Organization for Standardization (ISO):

The extent to which a group of intrinsic features (physical, sensorial, behavioural, temporal, ergonomic, functional, etc.) satisfies the requirements, where requirement means need or expectation which may be explicit, generally implicit or binding. (ISO, 2000)

There is, moreover, a series of definitions provided by specialists on the subject. Among the best known are:

Conformance to requirements (Crosby, 1979)

Fitness for use (Juran, 1951)

The efficient production of the quality that the market expects (Deming, 1982)

The total composite product and service characteristics of marketing, engineering, manufacturing, and maintenance through which the product and service in use will meet the expectations of the customer (Feigenbaum, 1964)

Meeting or exceeding customer expectations at a cost that represents value to them (Harrington, 1990)

Anything that can be improved (Imai, 1986)

Does not impart loss to society (Taguchi, 1987)

Degree of excellence (Webster's, 1984)

Although the official definition provides a desirable fusion of the most important meanings of the original definitions, it must be noted how the interpretations of these authors make a more or less marked reference to the various families of qualities described above. Focusing on a few of these definitions reveals the degree of subjectivity or objectivity of the qualitative characteristics they take into consideration.

1.3 COMMERCIAL QUALITY

Quality has always been of substantial importance in trade relations and this is understandable as the value of an exchanged good is the relationship between its quality and its price. Until a few years ago, commercial quality was product quality. Nowadays the term total quality is used to broaden the concept of quality to the service involved in the exchange.

1.3.1 Historic evolution

In the eighteenth and early nineteenth centuries when production was limited exclusively to cottage industries, quality, as we know it today, did not exist. The formal quality audits became necessary only with the arrival of mass production, mostly in the military field, to such an extent that, in 1941, the American War Department formed a committee aimed at enabling the army to procure arms and ammunition in large quantities without incurring problems of quality. In 1950, only one-third of the electronic military installations worked properly, and so a working group of the US Department of Defense was formed to deal specifically with reliability. This is how reliability came to be one of the 'chapters' of quality.

In the post-war period, with the publication in 1951 of the *Quality Control Handbook* by Juran, the concept of 'the economy of quality' took root and is the basis of all industrial quality management today.

In 1956, Feigenbaum introduced the concept of 'total quality control', saying that it must start with the design of the product and end only when the product has been placed in the hands of a customer who remains satisfied. So quality control was extended to research and development of methods and equipment, purchases, production, audit and acceptance testing, despatch, installation and service.

At the beginning of the 1960s the Japanese quality revolution marked the end of the previous era of 'quality control' and the beginning of the new era of 'quality management'. The 1980s saw the birth of a new concept, that of 'total quality'. Attention which had till then been focused on the product alone was shifted to the customer.

Today quality is read as total quality. The concept of 'fungibility to use' is replaced by 'suitability to needs' and the focal point is no longer merely the customer, but man, with all his (and her) physical, social and economical needs, be they explicit or otherwise.

1.3.2 Total quality

From the producer's point of view it can be said that total quality is offering products and services in conformity with customers' needs.

From a consumer's point of view it is, however, necessary to distinguish various types of quality:

1   *Expected quality* is the customer's expectations from a particular category of products.
2   *Promised quality* is the customer's expectations from a product of one particular brand. With respect to expected quality, promised quality is at a higher level, as the customer, who is initially unaware, gains awareness during his or her analysis of the offer. With the same product, the factors that differentiate the quality promised by one

producer from the other are the elements of communication such as design, appeal and brand image, price, sales point, advertising, etc.

3   *Effective quality* is the quality of the product in question, in other words, the measurable quality. If the promised quality is with regard to the form of the product, then here we are talking about pure substance. Effective quality is, in turn, made up of two components. The first, which is subjective, refers to the excellence of the characteristics of products and services. Juran's definition of 'fungibility to use' mainly refers to this component. Examples of product features relating to subjective quality are robustness, reliability, flexibility, precision, performance, beauty, goodness, respect for the environment, lifespan, service content, economy of use and so on. The second component, which is objective, refers, according to Crosby's definition (1979), to how the product conforms to the requirements specified of it in order to respect the above characteristics, and therefore to the absence of defects.

4   *Perceived quality* is the sum of the promised and effective qualities, and therefore both product and communication contribute to this. This is the most important quality, as the relationship between this and expected quality determines customer satisfaction.

Satisfaction = Perceived quality / Expected quality

5   *Potential quality* is how the product can be further improved.

Consumers assess all these qualities one by one. When they decide to approach a product, they have, or they create, depending on the information gathered, a 'map' corresponding to the expected quality for that particular category of products. When they must choose between one brand or another, they will choose the one they perceive as having the promised quality which is most in keeping with their needs for use and/or for price. Finally, once the product has been chosen and purchased, they will be faced with its effective quality, which is not always in line with the promised quality.

Psychological mechanisms also intervene in this increase in qualitative appraisal. For instance, it is more difficult for a consumer to realize that the effective quality of a product is poor if it has a high promised quality. This is due to the difference between form and substance, which is not something man is used to measure, as even in nature beauty represents good. On the other hand, man finds it difficult to recognize even an existing highly effective quality in a product with a poor image, and, indeed, even more difficult if one needs to be an expert to assess the quality of a product. How many people would be able to recognize a fine Bordeaux wine served in a bottle without a label? And vice versa, how many could recognize that a fairly good wine poured from a bottle of *premier cru classé* is not excellent?

1.3.3   Quality certification

Total quality is part of a large movement begun at the end of the post-war period in the West, aimed at improving the quality of life. This involved the introduction of more and more restrictive technical regulations regarding consumer safety and environmental protection, which gradually reduced the freedom of the producers of industrial goods. In order to maintain the free circulation of goods in such a restrictive market, it became necessary to be able to certify the quality of traded goods objectively and impartially. Therefore, especially in Europe under the pressure of the European unification, international legislation changed and quality certification was introduced.

There are two different types of quality certification: conformity certification and quality system certification.

Conformity certification means the certification of the conformity of the product to its declared characteristics. This type of certification can either be binding, as in the case of potentially dangerous products, or consensual, where the producers wish to attract attention to the qualitative contents of their product. The certificate is issued by a third party, often on the basis of tests carried out on the product in question. *System certification* seeks to have an official body certifying the 'quality assurance system', which includes all the structural and organizational factors put in place to obtain and maintain a particular standard of quality. Reference is made to the International Organization for Standardization ISO 9000 regulations, which are recognized nearly all over the world. Its latest revision, 'Vision 2000', has increased senior management involvement and has also introduced the concept of ethics in socio-economical activities. Conformity certification is direct in that the object certified is the effective quality of the product. System certification, on the other hand, is indirect, as it implies that a producer who complies with ISO 9000 regulations is *capable* of respecting the stipulated qualitative standard. In both cases, in order to be able to issue certificates, the certifying body must in turn be accredited by an appointed national authority. These bodies should actually be delegated by the public authorities, in compliance with a specific legislation on quality, the so-called 'Country Quality System'.

1.4 QUALITY OF FOOD PRODUCTS

Food products are consumed both owing to man's need to feed, but also from the search for pleasure provided by certain foods or drinks. In the case of unfinished products, one further aspect looked for in food products is the service level, i.e. how easy it is to use or quick to prepare. These three components of the expected quality for food products are listed below.

1.4.1 Nutritional quality

The nutritional quality of a food product depends on the nutritional content of the food product and how safe it is for the consumer's health.

The nutritional value implies both a quantitative aspect, in terms of the number of calories provided by the product per unit weight, and a qualitative aspect given by the composition in nutritive classes, or the relative percentages of carbohydrates, proteins, lipids, vitamins and mineral salts, plus other substances, such as fibres. Each nutritive class can, in turn, be analysed from a qualitative point of view, by taking into consideration the content of essential substances (such as amino acids and fatty acids) and its digestibility.

Wholesomeness, or hygienic quality, corresponds, on the other hand, to the complete absence of toxicity in food products. Leaving aside cases of uncontaminated food products, which become toxic due to improper use by the consumer, it can be said that food products only become toxic by contamination. This contamination may be chemical or bacteriological and can take place at any time during the life cycle of the product due to endogenous or exogenous factors.

Generally speaking, nutritional quality is deemed the acceptable minimum, that is, an indispensable condition for trading a product. This means that the nutritional content of a particular product must be consistent with its category and absolute wholesomeness. As the acceptable minimum, nutritional quality is at a lower level than expected quality therefore the consumer tends to take it for granted. Or consumers only take it into consideration when deciding whether or not to try out products in that particular category. This is why nutritional quality is normally standardized by law, especially regarding the aspects affecting its wholesomeness.

In some cases, however, nutritional quality is a condition for excellence. This is the case of special products, either of registered origin or, more generally, the products that, within their category, distinguish themselves by offering advantages for the consumer.

1.4.2   Sensorial quality

Sensorial quality is a secondary quality, in that it is the effect produced by certain primary qualities on our sense organs. The primary qualities in question (organoleptic attributes, according to ISO 5492) are to be tracked down to those chemical and/or physical properties of food – such as taste, aroma, texture, aspect (e.g. colour), maybe sound (e.g. crunchiness) – that impact on consumers' exteroception, namely on the senses of gustation, olfaction, haptic (tactile) sensitivity, vision and audition.

Sensorial quality can be defined as the ability of a product to satisfy the hedonic needs of consumers. It is, therefore, subjective as it is assessed through an interpretation by each consumer and cannot be measured in an absolute sense.

Despite this, products that most consumers like are considered quality products. As a consequence, the only means of defining the sensorial quality of a product is to bear in mind the tastes and views expressed by consumers. These opinions are influenced both by the individual characteristics of the sense organs of each person and each individual's ability to use them, and by the customs and traditions of the various regions in the world influenced by culture, ethnic groups, religion and social class.

Sensorial quality varies both in time and in space. Moreover, food products are very complex, which makes it difficult to attribute a particular sensorial quality to one or more easily analysed constituents. A way to solve this difficulty is to define accurately the ingredients used and the processes applied, as this is done with typical or registered products, and to proceed to the quality control of the finished product by means of sensorial analysis.

Consumers attach a great deal of importance to sensorial quality, indeed it can be said that this is what sells a product. This is due to the fact that, at least in the West, there is no longer a need to appease hunger; the Western diet is varied and 'complete' foods are no longer needed. Moreover, Western society, which takes care to guarantee consumer safety, ensures that only products of high nutritional quality are commercialized.

1.4.3   Service content

The lifespan of the packaged and/or opened product, the cooking time, the availability, how easy it is to transport, the quantity and quality of information that accompanies the product, the encumbrance of the packaging to be thrown away, the environmental impact of the packaging material, packaging safety etc. are all examples of the service content a product may or may not have.

At times the service content contrasts with other qualities. Just think of long-life milk, which has an increased lifespan at the expense of a reduced nutritional and sensorial quality. At other times, however, the service content strengthens other qualities; one lucky example of this is, indeed, coffee, which, if pressurized in inert gas, not only lasts three times as long, but also undergoes a considerable strengthening of its aroma content.

1.4.4   Conformity certifications

The fact that our society is more mature regarding producer responsibility in guaranteeing product safety has led to the development of both horizontal and vertical regulations in the food sector:

1   European Directive 85/374, 25 July 1985 (EC Official Gazette n. L210 – 7 August 1985) is concerned with the liability for faulty products (which applies also to all non-food sectors).
2   The cornerstone of the European regulations on the hygiene of food products is Regulation (EC) n. 852/2004, 29 April 2004 (EC Official Gazette n. 139 – 30 April 2004), which focuses on prevention by prescribing obligatory internal audits, assigning responsibility to manufacturers by imposing the analysis of critical control points – Hazard Analysis Critical Control Point (HACCP) – and requiring the drawing up of a manual of correct hygiene practices as a function of the type of products and processes and the characteristics of the manufacturer.
3   All the vertical regulations regarding disparate food products as a function of their critical state are coupled up with this horizontal set of regulations. Genetically Modified Organisms (GMOs), which so far do not include coffee product, are regulated by numerous directives and regulations; their presence in a food is among the situations most critically perceived by consumers.

1.4.5   Quality and the general food law

As a consequence of the recent European Regulation 178/2002/EC (General food law), 28 January 2002 (EC Official Gazette n. L31 – 1 February 2002), which states the principles and general requirements of the food law, provides the setting up of the European Authority for Food Safety (EFSA) and lays down procedures for food safety control, the European standardization organizations – which refer to the ISO – are drawing up guidelines on the systems of traceability in food companies. According to the new models of development based on the sustainability of the whole food chain, traceability is an instrument of health and hygiene guarantee, which, besides inspiring trust in the consumer, can help strengthen the identity and quality of the food product.

In addition to the safeguarding of fundamental health and hygiene requirements, protected by the regulations in force, more restrictive qualitative requirements, or requirements of typicality, are proposed through the guarantee of product certifications. These can refer to binding regulations (voluntary regulated product certifications), or to voluntary technical regulations defined by the producer companies with the control bodies. Organic farming production is an example.

Table 1.1 gives some examples of the validity of voluntary regulated product certifications laid down within the European Union.

The seal of quality is attributed to farm and food products for which a particular quality, reputation or other characteristics depend on their geographical origin, and whose production, transformation and/or processing take place in a certain area of production. At least one stage of the production process must therefore take place in a particular area.

As far as wines are concerned, they are regulated by national regulations in their respective production countries, which, in their turn, have been harmonized with the EC Council Regulation n. 823/87, 16 March 1987 (EC Official Gazette n. L084 – 27 March 1987 – Pg. 0059-0068 and subsequent amendments).

1.5 THE EXPERIENCE OF COFFEE CONSUMPTION

Ever since coffee was first consumed in the West, towards the middle of the seventeenth century, specific local habits and traditions have developed and overlapped, entailing great differences in coffee consumption. There is generally a remarkable ritual component, divided into two very distinct rites, preparation and tasting. Below is a cross-section of contemporary coffee cultures, which clearly shows the expected and perceived qualities in the various countries.

■ **Amsterdam:** Essential times for drinking coffee in Holland are in the morning, at breakfast and around 10 or 11 o'clock, often with guests.

Table 1.1 Regulations related to quality in the European Union

Product	Definition	Regulation
Organic product	A food product for which, throughout the production cycle, the use of chemicals (pesticides and fertilizers) is excluded, and only the use of environmentally friendly techniques of cultivation and stock farming is foreseen. Land is made fertile by crop rotation and the use of organic manure and natural minerals, while environmentally friendly products and techniques are employed to defend the crops from parasites. Coffee can be certified as organic	EC Regulation n. 2092/91 – Official Gazette n. L198 – 22 July 1991 and subsequent amendments
Protected designation of origin (PDO)	Seal of quality attributed to food products whose particular characteristics essentially depend on the territory in which they are produced. The geographical environment includes natural and human factors, which make it possible to obtain a product, which cannot be imitated outside its defined area of production. All stages of production, transformation and processing must take place in a delimited geographical area	EC Regulation n. 2081/92 – Official Gazette n. L208 – 24 July 1992 – pp. 1–8 and subsequent amendments
Protected geographical indication (PGI)	Seal of quality attributed to farm and food products for which a particular quality, reputation or other characteristics depend on their geographical origin, and whose production, transformation and/or processing take place in a certain area of production. At least one stage of the production process must therefore take place in a particular area.	EC Regulation n. 2081/92 – Official Gazette n. L208 – 24 July 1992 – pp. 1–8 and subsequent amendments

It is mostly prepared with a filter machine, even though a growing number of people now have an espresso machine in their homes. Much care is taken in the preparation, the presentation and how the coffee is served. The 'koffie verkeerd' is their typical coffee, drunk with plenty of milk in mugs and often served with sweets, such as apple pie. Socializing and comfort are two aspects particularly associated with coffee consumption in Holland and, in public, young people want to drink coffee 'in the right way'. Espresso has become the ideal after-dinner coffee beverage, and possession of an espresso machine with all its accessories has become a status symbol.

■ **Hamburg:** Coffee in Germany is associated, first of all, with well-being and euphoria, relaxation and fun, and, despite the fact that it is consumed throughout the day, it is seldom drunk after a meal. The day starts off with a steaming cup of coffee and a large breakfast. The careful preparation of the brew takes place in a relaxed atmosphere, in which the coffee is served on a perfectly laid table, every day of the year, and drunk in quantity, followed by fresh bread rolls. Throughout the rest of the day, plenty of coffee is drunk, both at work and out and about, in the 'EisKaffe', places where beverages are served with cakes, or in the 'Steh-kaffee', bars with high counters looking out onto the street, where one can drink coffee standing up. The traditional coffee is mostly prepared with a filter, but for a more modern consumer espresso coffee consumption is growing. Milk is gradually replacing cream, which traditionally used to be served with coffee. Germans pay particular attention to coffee being at the right temperature, and so they tend to keep it handy in a thermos. In the afternoon, around 5 o'clock, it is time for 'Kaffeeklatsch', an important moment for socializing at home, which is the ideal formula for inviting someone to the home.

■ **London:** The main reason British people decide to drink a cup of coffee on a weekday is for its stimulating effect to improve their performances. Taste and aroma are of secondary importance, so much so that some take a thermos of coffee with them from home. Lack of time and space have contributed to the spread of extremely functional methods of preparing coffee, such as instant coffee prepared with the kettle that is also used for tea, which still has a strong hold in British homes. At weekends, when people have more time, habits change and coffee is then drunk with real pleasure, because it no longer means only 'getting on with something quickly'. Some bars have become popular because they offer coffee in real settings consumers can relate to. Coffee consumption in the UK is therefore a modern and cosmopolitan experience, most certainly not a substitute for the traditional rite of tea.

■ **Naples:** Naples probably represents the city with the richest daily idiosyncratic coffee culture in the world, having elaborated both a philosophy and a theory on its use and consumption. It is no accident if all expressive activities that distinguish this city, from the theatre of de Filippo to the songs of Pino Daniele, have dealt with coffee by creating and offering particularly original visions and interpretations of the experience. At home, coffee is prepared with the Neapolitan coffeemaker and served in the '*tazzulella*'. At the coffee bar – an authentic omnipresent coffee temple (approximately one for every 450 people) – a very concentrated espresso of no more than 20 ml (called *ristretto*) is served, prepared by the expert hand of a Neapolitan barman, operating on an old-fashioned lever-espresso machine, with sugar mixed with coffee during percolation.

■ **New York:** The first coffee of the morning, generally prepared with a filter machine or by the infusion system, is drunk in a hurry in the traditional mugs accompanied by the traditional muffins or bagels, and seen as a habit, with no special rite. A common scene in New York is that of coffee drunk in the street in a paper cup, 'in transit', like street fuel, very basic and always available on every street corner. There is an enormously wide supply of blends for the home – besides the various brands, you can find pure origin, flavoured coffee, organic coffee, etc. Coffee is less commonly taken after main meals, whereas, especially on holidays, people often take a break during the day to prepare a coffee for themselves or for friends. There are no fixed rules about occasions for drinking coffee as there are in other countries. Espresso, stronger tasting than the traditional American coffee, and the beverages associated with it, the so-called ESBAD (espresso-based), such as *latte* and *cappuccino*, are consumed more frequently in the coffee shop chains or in Italian restaurants and are therefore mainly considered 'special' beverages.

■ **Oslo:** The basic Norwegian coffee is what is known as ordinary coffee, which is black, with no milk or sugar, mostly associated with home consumption. Preparation with a filter is gradually taking over from the use of a kettle. Consumption out of the home, which mainly takes place in coffee bars, is becoming more and more widespread and this has opened the way for new ways of drinking coffee, such as espresso and *cappuccino*. Sometimes these are even prepared at home, especially when people have guests for the weekend or in the evening. Consumption is very high – an average of five cups a day – which makes Norwegians, together with the rest of the Nordic people, the highest coffee consumers per head in the world.

■ **Paris:** Coffee is the heart of breakfast; most French people have a real yearning for it. It is normally prepared using a filter and is drunk with other typically French products, such as croissants and baguettes. Another cup is drunk after lunch, without milk. Real coffee lovers still go to '*brûleries*', if they can find one, where one can enjoy coffee after choosing the desired strength and aroma. 'Café', in its two accepted meanings, as a beverage and as a place of encounter, represents an important social and collective rite that reached its climax, thanks to the likes of Voltaire, Balzac and all the way to Sartre. In bars and restaurants, especially in the north of the country, coffee is generally prepared with an espresso machine, although it is not served as concentrated as in Italy.

■ **San Francisco:** Their first desire in this city is to start the day off with the rite of a good cup of coffee, that is, to begin the day with an experience of quality where its careful preparation is something to be proud of. There are roasted coffee dealers, shops and bars that serve coffee almost anywhere in town, and they are proud of roasting their own coffee. San Francisco is an important centre for coffee in the United States and played a leading role in the 1980s in the revolution of gourmet coffee. The wide availability of different types of coffee tempts people to try out all kinds of novelties. Milk and espresso-based beverages are like meals, so at lunch lighter drip coffee is preferred, while 'fun' coffees are consumed after lunch, as a dessert. The bars are like an extension of the sitting room at home where you can drink a coffee as you relax, reading the newspaper or surfing on the Internet.

■ **Tel Aviv:** In the morning, instant coffee, quick and without much ceremony, is a necessary provision for the working day ahead; no great interest is taken in preparing the coffee and certainly no time is dedicated to this. At other times of the day drinking coffee is a way of spending time together. People in the city love sitting for hours in roomy, high-tech cafes, where they can go to see and be seen, especially on Friday, savouring coffee at length with little sips, often accompanied by dry cakes, biscuits or fruits.

■ **Tokyo:** The pace of life is very fast and people do not have much time for themselves. For this reason, most people drink instant coffee, of which there is a wide choice. The most widespread coffee drunk, especially out-of-home, is American coffee. The coffee is consumed in bars or from vending machines, mostly as an energizing drink. The vending machines, to be found everywhere, even in city streets and on country roads, sell various kinds of soft drinks, but above all, cans of liquid coffee, for which Japan is at the top of the world market. Coffee is on the menu in Western-style restaurants, which, together with the coffee shop chains, are the main places where coffee is drunk, especially among coffee lovers. At home, this beverage is indeed little used, people almost only drinking instant coffee once in the morning. Espresso coffee is very successful, even if it is thought by many to be too bitter.

1.6 THE QUALITY OF ESPRESSO COFFEE

Three coffee experiences can be lived throughout the day:

1   coffee for waking up is still preferred as a hot, dilute beverage;
2   espresso, as has been seen, takes the lion's share of coffees at breaks, especially in Latin countries and above all in Italy;
3   in Anglo-Saxon or northern European countries, espresso is considered a speciality suitable for relaxing, where feeling and care are involved in its preparation.

The most appreciated characteristics of espresso are its creaminess, body and the strength of its aroma, associated with stimulating properties, despite the fact that espresso contains less caffeine than the more dilute coffees do (see Chapter 8). Espresso therefore represents a benchmark universally recognized as being a great pleasure and a symbol of Italian culture, so much so that the highest expression of coffee is that served in bars in Italy, in restaurants in France, and restaurants and coffee shops in the rest of the world, precisely because it is coffee prepared with a professional machine.

As regards the nutritional content (see Chapter 10), the only expectations one can have from the consumption of an espresso are the intake of a moderate dose of caffeine, known above all as a stimulating pleasure, boosting intellectual activity, improving memory and concentration, quickening reflexes, making it easier to stay awake, improving one's mood, etc., and the fact of consuming a product devoid *per se* of calories.

The real consumer's expectations, however, are of a hedonistic nature (see Chapter 10), because drinking a good espresso is genuinely wonderful. Espresso is a true elixir, a concentrate of exquisite aromas lasting long after it has been drunk. Even sight and touch are satisfied, thanks to the striped hazelnut colour of the foam and to its full body. In moments of relaxation at home this tasting rite is preceded by the eagerly awaited rite of preparation, when it is enjoyable to play out the actions of the barman, enriching it with your own secrets, as you anticipate the pleasure of its taste.

As far as service is concerned, consumers require the roasted and ground coffee to retain the fragrance of the aromas developed during roasting until the time the package is opened, and they want to know the characteristics of the product they are consuming (see Chapter 6). It should also go a long way, in that only a small amount is needed to make a good cup of coffee and, once open, the aroma should last until the contents of the package run out.

Grinding deserves a whole chapter (see Chapter 5). Connoisseurs believe that in order to be excellent, the beans should only be ground immediately prior to preparation; unfortunately, it is difficult to find and take care of top quality household grinders. This is why 90% of packaged coffee sold in retail shops is ground. The most promising solution for the future is servings – ground and pressed doses of coffee, sealed in a wrapper and ready for use. Servings offer the fastest, cleanest way of preparing espressos, and, most importantly, they guarantee a consistency of quality that neither coffee beans nor ground coffee can in any way guarantee due to the serious problems in keeping all the parameters involved in preparing a perfect espresso cup under control.

Espresso consumption is an aesthetic experience, like tasting a vintage wine or admiring a painting. It is a search for beauty and goodness for improving the quality of our life. As it offers such subjectively ineffable 'goodness', devoid of defects, the only adequate reaction to it is astonishment – astonishment that can give birth to enthusiasm, and therefore intellectual and spiritual enrichment.

The predominance of the experience aspect means that the official definition of quality may be too limited for espresso. It may be more appropriate to speak of 'degree of excellence'. The elements characterizing effective quality are the subject of this book and will be revealed in detail as you read. From an organoleptic point of view, we have already seen the importance of the aroma and the full-body – to a certain extent represented by the visual component of the foam – while, as regards the taste, consumers look for a slight bitterness in southern countries or a slight acidity in northern countries, in both cases accompanied by the characteristic sweetness of the coffee. On the opposite side, the most common, serious defects penalizing consumption are the extreme bitterness and foul flavour of poor quality beans.

1.7 DEFINITION OF ESPRESSO

Everyone in Italy has a clear mental picture of a cup of espresso: a small heavy china cup with a capacity just over 50 ml, half full with a dark brew topped by a thick layer of a reddish-brown foam of tiny bubbles, also known by the Italian term 'crema'. More than 50 million cups of espresso are consumed every day in the world: its fragrance and flavour are the first stimuli in the morning, they crown an excellent meal later in the day, and act as frequent revivers during lengthy working sessions.

1.7.1 Espresso as a lifestyle: brewed on the spur of the moment

One of the meanings of the word *espresso* (express) is that it is made for a special purpose, on the moment, on order (Marzullo, 1965; Hazon, 1981); therefore it is made for the occasion on express request, extemporaneously rather than fast. This concept is clarified by the saying 'the consumer, not the espresso must wait'. As a direct consequence, once brewed, espresso cannot be kept and must be drunk immediately, before the foam shrinks and collapses breaking into patches on the surface. After a while, the surface of the liquid is completely free from foam, which has dried out on the walls of the cup above the liquid.

If an espresso is kept waiting, smoothness of taste is lost and perceived acidity increases with time regardless of cooling. Furthermore, if the cup cools down, an unbalanced saltiness becomes noticeable.

■ Freshness of preparation must be an integral part of the definition of this very special brew.

1.7.2 Espresso as a brewing technique: it requires pressure

At the beginning of the twentieth century the need for preparing a cup of coffee within seconds of a customer's request led to an increase in the pressure of the extraction water. Water was heated up to its boiling point in a sealed kettle, so that the steam in equilibrium created pressure, accelerating extraction. A drawback of this technique was that brewing with boiling water provokes over-extraction of astringent and bitter, usually less soluble, substances, which give a *burnt* taste to the brew.

Brewing was first improved by separating the water used for brewing, best hot but not boiling, from the heating water. Pressures as high as 10 bars could be created by a lever, multiplying the force of the arm of the bartender, producing a thick layer of foam on the cup. The lever has now been replaced by an electric pump, simpler and more regular to operate.

A pressure field applied within a fluid produces potential energy – what is known as Bernoulli's piezometric energy – which can be easily transformed into kinetic energy, and further transformed into surface potential energy and heat.

Pressure is important for the definition of espresso, making it different from other brews. During espresso percolation (see Chapter 7), a small amount of hot water under pressure is applied to a layer of ground roasted coffee, the *coffee cake*, and this very efficiently produces a concentrated brew, containing not only soluble solids, but also lipophilic substances, lacking in filter and instant coffees. The foam on the top and the opaque brew are unique to espresso, owing to the presence of a disperse phase formed by very small oil droplets in emulsion (Petracco, 1989) (see 8.1.1), which are perceived in the mouth as a special creamy sensation, the *body*. Furthermore, the oil droplets preserve many volatile aromatic components, which would otherwise either escape into the atmosphere or be destroyed by contact with water as in other brewing techniques, so that the rich coffee taste lingers in the mouth for several minutes. If coffee were percolated under high static pressure only, the pressure would be lost downstream and no work could be performed on the cake; while, if kinetic energy from stirring propellers, choke nozzles, sprayers, etc., were applied downstream from the cake, a smooth layer of foam could be produced, but it would lack body.

The Latin etymology of the word espresso, literally meaning *pressed out* (Campanini and Carboni, 1993), clearly points out the importance of pressure in espresso brewing, making the technique an integral part of the definition:

■ Espresso is a brew obtained by percolation of hot water under pressure through tamped/compacted roasted ground coffee, where the energy of the water pressure is spent within the cake.

1.7.3   Italian espresso: it must be rapidly brewed

Another important feature of espresso, especially as traditionally drunk in Italy, is the length of percolation (see 7.5.8). The diversified energy input in espresso pressure-brewing efficiently brings into the cup both hydrophilic and lipophilic substances. A best mix is reached within 30 seconds; if the extraction is shorter than 15 seconds a weak and exceedingly acid unbalanced and under-extracted cup is obtained. Conversely, if extraction lasts longer than 30 seconds, over-extraction of substances with poor flavour will produce an ordinary harsh-tasting cup, as can be easily seen by separately tasting the liquid fraction percolated after the prescribed 30 seconds.

A quantitative definition can now be given:

Italian espresso is a small cup of concentrated brew prepared on request by extraction of ground roasted coffee beans, with hot water under pressure for a defined short time.

The range of the parameters is:

Ground coffee portion 6.5 ± 1.5 g
Water temperature $90 \pm 5\,°C$
Inlet water pressure 9 ± 2 bar
Percolation time 30 ± 5 seconds

The requisite conditions to make a good cup of espresso will be reviewed in detail in the following chapters.

1.8 CONCLUSIONS

We have seen how varied and complex the concept of quality is and how difficult it is to use it as an objective measure, particularly in the case of the espresso. We have, however, understood that the espresso lives almost solely on the pleasure it gives consumers. Therefore, if the number of lovers is to be maintained and increased, it is necessary to seek the means of continually improving its quality, meaning its degree of excellence. In the following chapters we will try to outline this road to improvement, setting out everything that is known so far about the quality of the espresso.

This becomes a stage on the road to improving the quality of the production of coffee brewed worldwide, which, as the resounding success of gourmet coffee has shown in the Anglo-Saxon world, can lead to a significant increase in consumption. This may result in an important contribution to the rebalancing of the supply and demand of coffee, thus improving the precarious financial and social situation the producer countries find themselves in.

Wine has travelled a similar road very successfully and, besides providing pleasure to consumers, has achieved a significant rise in the value of overall production. This has resulted in a segmentation of the market where no one in the world is any longer surprised by the fact that the price of a bottle can range over several orders of magnitude. Consequently, there has been a general increase in the well-being and satisfaction of those along the whole chain. This road was not easy and involved an almost 'manic' search for excellence, the specialization of the distribution networks and education of the consumers.

Espresso can play the lead role in a similar situation in the coffee field, thanks to the image it enjoys as a symbolic beverage and thanks to the commitment of the whole industry to continuous improvement.

REFERENCES

Campanini G. and Carboni G. (1993) *Vocabolario latino-italiano*. Milan: Paravia, p. 524.
Crosby P. (1979) *Quality Is Free*. New York: McGraw–Hill.
Deming W.E. (1982) *Out of Crisis*. Cambridge, MA: MIT, Center for Advanced Engineering Study.
Feigenbaum A.V. (1964) *Total Quality Control*. New York: McGraw–Hill.
Harrington H.J. (1990) *Il costo della non-qualità*. Milan: Editoriale Itaca.
Hazon M. (1981) Grande dizionario italiano–inglese e inglese–italiano. Milan: Garzanti, p. 1348.
Imai M.K. (1986) *The Key to Japan's Competitive Success*. New York: McGraw–Hill.
ISO (1992) *Sensory Analysis – Vocabulary*. ISO 5492:1992. Geneva: International Organization for Standardization.
ISO (2000) *Quality Management Systems, Principles and Terminology*. ISO 9000:2000. Geneva: International Organization for Standardization.
Juran J.M. (1951) *Quality Control Handbook*. New York: McGraw–Hill.
Marzullo A. (1965) *Dizionario della lingua italiana*. Milan: Fratelli Fabbri, p. 417.
Petracco M. (1989) 'Physico-chemical characterisation of espresso coffee brew', *Proc. 13th ASIC Coll.*, pp. 246–261.
Taguchi G. (1987) 'The system of experimental design: engineering methods to optimize quality and minimize cost'. Unpublished, White Plains, New York.
Webster's (1984) *New Riverside University Dictionary*.

CHAPTER 2

The plant

F. Anzueto, T.W. Baumann, G. Graziosi,
C.R. Piccin, M.R. Söndahl and
H.A.M. van der Vossen

2.1 ORIGIN, PRODUCTION AND BOTANY

M.R. Söndahl and H.A.M. van der Vossen

2.1.1 Origin and geographic distribution

Commercial coffee production is based on two plant species, *Coffea arabica* L. (arabica coffee) and *C. canephora* Pierre ex Froehn. (robusta coffee). A third species, *C. liberica* Bull ex Hiern (liberica or excelsa coffee) contributes less than 1% to world coffee production. All species within the genus *Coffea* are of tropical African origin (Bridson and Verdcourt, 1988). Many forms of wild *C. canephora* can be found in the equatorial lowland forests from Guinea to Uganda, but natural populations of *C. arabica* are restricted to the montane forests of southwestern Ethiopia (Berthaud and Charrier, 1988).

The date of first domestication of arabica coffee in Ethiopia is uncertain and so is the claim that Ethiopian invaders brought coffee to Arabia in the sixth century. However, there is written evidence for extensive cultivation of coffee in the Arabian Peninsula (Yemen) by the twelfth century. During the following 300 years, the stimulating hot beverage prepared from roasted and ground coffee beans, called *qawha* (a general word for wine and other stimulants) by the Arabs and *cahveh* by the Turks, became immensely popular in the Islamic world and from 1600 onwards also in Europe, with Mocha as the exclusive centre of coffee trade (Pendergrast, 1999). In Uganda and other central African countries, there has been a centuries-old tradition of growing coffee (mainly *C. canephora*) near homesteads for the purpose of chewing the dried fruits or beans for their stimulating effect (Wrigley, 1988).

Commercial production of arabica coffee outside Yemen started in Sri Lanka (formerly Ceylon) in the 1660s, and subsequently in Java around 1700 with coffee plants introduced by the Dutch East India Company from Sri Lanka or the Malabar Coast of southwestern India. The arabica coffee in India and Sri Lanka originated probably from seeds brought directly from Yemen by Moslem pilgrims during the first decade of the seventeenth century. A few plants taken from Java to the Botanic Garden in Amsterdam in 1706 formed the basis of practically all the cultivars of arabica coffee planted in the Western Hemisphere. These early introductions of coffee plants from Yemen to Asia and Latin America are of the variety Typica, *C. arabica* var. *arabica*. Coffee plants taken by the French around 1715 from Yemen to the Réunion island (formerly Bourbon) and subsequently to Latin America and East Africa are *C. arabica* var. *bourbon*. Varieties of the Bourbon type have generally a more compact and upright growth habit than Typica varieties, are higher yielding and produce coffee of better quality (Carvalho, 1988).

By 1860 world coffee trade involved some 4 million bags (60 kg, the standard unit of trade), mostly from Brazil, Indonesia and Sri Lanka. On account of its mild and aromatic cup quality, *C. arabica* would have certainly continued to be the only cultivated species, had it not been so vulnerable to diseases, particularly to coffee leaf rust (*Hemileia vastatrix*). Coffee leaf rust (CLR) was first reported in 1869 in Sri Lanka and within 20 years it had virtually wiped out coffee cultivation in Asia. Sri Lanka switched to tea cultivation, but Indonesia has continued to be a major coffee producer based on *C. canephora*, which turned out to have resistance to CLR. Robusta was first introduced into Indonesia (Java) from Congo in 1900 (via Belgium) and a few decades later, high-yielding plant materials had been developed by selection, which became known as robusta coffee and have until today remained the basis of most robusta coffee production in the world.

Coffee cultivation is now widespread in tropical and subtropical regions, with the bulk of arabica coffee concentrated in Latin America and robusta coffee predominant in South-East Asia and Africa (Figure 2.1).

2.1.2 Production

World coffee production increased from 86 million bags in 1980 to 112 million bags in 2000/01 (from 10.5 million ha). However, production graphs show large annual fluctuations, from a very low 80 million bags in 1986/87 to a record crop of some 115 million bags in 1999/2000, mainly caused by the leading producer Brazil (ICO, 2002). The Brazilian coffee

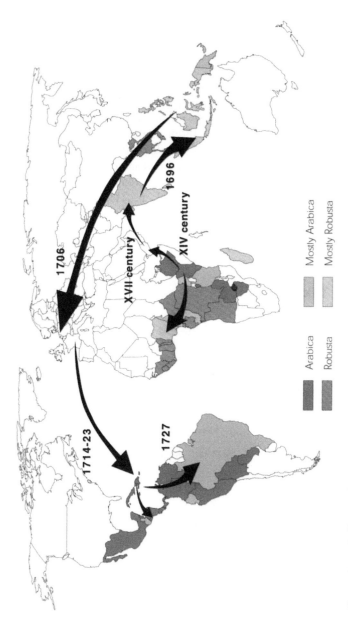

Figure 2.1   Coffee producing regions of the world

crop may vary from 19% to 33% of annual world production as result of biennial fluctuations, recurrent frosts and droughts. The tremendous expansion of robusta coffee cultivation during the past decade, in Vietnam in particular, but also in Indonesia and Brazil, has further exacerbated the excess in supply over demand on the world coffee market. Consequently, the share of arabicas in the world coffee supply has also dropped from the erstwhile 75% to 68–70%. Production data for the 10 largest coffee-producing countries are presented in Table 2.1. About 59% of the world's coffee in 2000/01 was produced in Latin America, 21% in Africa and 20% in Asia. Brazil (80% arabica) alone produced 26%, Vietnam (97% robusta) 12%, and Colombia (100% arabica) 11% of the 2000/01 coffee crop. Coffee is produced in some 60 countries, including the coffee-exporting members of the International Coffee Organization (40 at 30 September 2001: ITC, 2002, p. 21). In most countries, smallholders are the main coffee producers, but medium to large coffee plantations (30–3000 ha) can be found in countries like Brazil, El Salvador, Guatemala, India, Indonesia, Vietnam, Kenya and Ivory Coast.

2.1.3 Taxonomy

The genus *Coffea* belongs to the family *Rubiaceae*, subfamily *Ixoroideae* and tribe *Coffeae*. Linnaeus described the first coffee species (*Coffea arabica*)

Table 2.1 Coffee production in the 10 largest producing countries and world total (millions of 60 kg bags)

Country	Year 2000/01		Annual average	
	Arabica	Robusta	Total	1996–2001
Brazil	23.7	6.4	30.1	28.9
Vietnam	0.4	14.8	15.2	8.6
Colombia	10.5	–	10.5	11.1
Indonesia	0.7	5.7	6.4	7.1
Mexico	5.1	0.1	5.2	5.5
India	2.4	2.6	5.0	4.2
Ivory Coast		4.0	4.0	3.9
Guatemala	4.7	0.05	4.7	4.7
Ethiopia	2.9	–	2.9	3.2
Uganda	0.4	2.8	3.2	3.3
Total (60 countries)	**68.8**	**42.9**	**111.7**	**106.6**

Sources: ICO, 2002; ITC, 2002

in 1753. The taxonomy of coffee has been under recent review and may not yet be completely finalized yet. Bridson and Verdcourt (1988) have grouped coffee species in two genera: *Coffea* (subgenera *Coffea* and *Baracoffea*) and *Psilanthus* (subgenera *Psilanthus* and *Afrocoffea*). The subgenus *Coffea* includes 80 species, 25 from continental Africa and 55 unique to Madagascar and the Mascarene Islands. The subgenus *Baracoffea* was assigned only seven species (Bridson, 1994). Conventional and molecular genetic studies confirm that five clusters of related species can be distinguished corresponding to geographical regions (West, Central and East Africa, Ethiopia and Madagascar), but that the process of differentiation between these groups has not yet progressed into a stage of strong crossing barriers (Berthaud and Charrier, 1988; Lashermes *et al.*, 1997). The earlier taxonomic classifications into sections and subsections (Chevalier, 1947) or subgenera (Leroy, 1980) do not appear to be justified any longer. Instead, a monophyletic origin of all species within the genus *Coffea* has been generally accepted. Molecular evidence (Lashermes *et al.*, 1997; Cros *et al.*, 1998) for fairly close genetic relations between *Coffea* and the genus *Psilanthus* (Bridson and Verdcourt, 1988), as well as the successful hybridization between *C. arabica* and a *Psilanthus* species (Couturon *et al.*, 1998) may lead to a taxonomic revision into one genus *Coffea* with more than 100 species.

All *Coffea* species are diploid (2n = 22 chromosomes) except for the allotetraploid *C. arabica* (2n = 44), which probably originated as a spontaneous interspecific cross between two diploid species, followed by duplication (Carvalho, 1946; Sondahl and Sharp, 1979). Recent cytogenetic and molecular studies have indicated that indeed *C. eugenioides* and *C. canephora* (or *C. congensis*) are the most likely progenitors of *C. arabica* (Raina *et al.*, 1998; Lashermes *et al.*, 1999). The analysis of segregating molecular markers (Lashermes *et al.*, 2000a) has confirmed earlier genetic and cytogenetic evidence that *C. arabica* is a functional diploid. The observed disomic meiotic behaviour is probably under the control of genes regulating pairing of chromosomes.

2.1.4 Growth, flowering and fruiting

The coffee plant is an evergreen shrub or small tree, which under free growth may become 4–6 m tall for *C. arabica* and 8–12 m for *C. canephora*. In cultivation, both species are pruned to manageable heights of less than 2 m (less than 3 m in mechanically harvested plantations in Brazil) with one or more stems. The strictly dimorphic branching habit of coffee, as is schematically presented in Figure 2.2, determines plant growth habit and

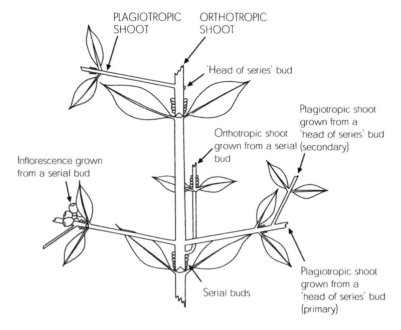

Figure 2.2   Dimorphic branching habit of coffee (Cannell, 1985, p. 113)

has important consequences for agronomic practices, such as pruning for maintenance or rejuvenation and vegetative propagation (Cannell, 1985). Continuously growing orthotropic (vertical) stems produce a pair of plagiotropic (horizontal) branches at each new node from the 'head of series' buds. The other buds on the same node remain dormant but will grow out into orthotropic shoots called suckers, when forced by capping of the main stem. Young seedlings produce the first pair of plagiotropic branches on the ninth to eleventh node. Branches bear two opposite leaves per node, with glossy dark green upper surfaces, 10–15 × 5–10 cm in size, and slightly undulating in arabica. In robusta plants, leaves are much larger and strongly corrugated. The serial buds present in the leaf axils on plagiotropic branches may develop into inflorescences or secondary lateral branches, but never into orthotropic shoots. In consequence of this strictly dimorphic growth habit in coffee, only cuttings prepared from orthotropic shoots (suckers) will, after rooting, grow into trees with a normal upright growth habit.

At the end of the rain period, differentiation of flower buds takes place, and this phenomenon is frequently associated with lower temperatures and shorter daylengths. Soon after differentiation, flower buds will grow for about 8 weeks until reaching 5 mm in length, and then become quiescent for about 2 months. At the end of this period, they will complete their differentiation process to form a glomerule (short inflorescence) bearing a cluster of 2–19 flower buds in arabica (Carvalho *et al.*, 1969) and up to 80 buds in robusta. Later, flower buds resume growth until they reach 8–10 mm in length, and then enter into a second quiescent period. At this point, they are quite visible, with a pale-yellow coloration. Arrested flower buds will grow again when exposed to first showers of the new rainy period. Flower buds are then released from dormancy, undergo a rapid expansion during 8–10 days, changing colour from pale-yellow to white. This last growth phase will end with the opening of the flowers (anthesis), resulting in massive flushes of blossoming (Alvim, 1973; Cannell, 1985). Depending on the prevailing dry season, two to five blossoming events will occur per flowering season, leading to sequential ripening events and differentiated pickings. The longer the dry season, the more synchronized flowering and reduced number of blossoming events.

In equatorial regions with bimodal rainfall patterns, flowerings occur at the beginning of both rainy seasons, and this will result in overlapping stages of fruit development and fruit ripening extending over several months. In unimodal rainfall regions further from the equator, there is a single flowering season, and a much shorter harvesting period.

The white and fragrant bisexual flowers have a corolla tube 8–10 mm in length, with five lobes (five to seven in C. *canephora*). The five stamens are fused to the corolla and the bilocular anthers open lengthwise. The pistil has an inferior, bilocular ovary and a filiform style with two stigmatic branches (Van der Vossen *et al.*, 2000) (Figure 2.3).

Flowers wither within 2–3 days after anthesis and pollination. C. *arabica* is self-fertile and less than 10% of the flowers are naturally cross-pollinated, mainly by visiting bees (Carvalho, 1988). On the other hand C. *canephora* is strictly allogamous with a gametophytic system of self-incompatibility preventing self-pollination (Berthaud, 1980; Lashermes *et al.*, 1996a). Pollen tubes take about 24 hours to reach the ovary followed by fertilization, but the first cell division of the zygote takes place not until 60–70 days later (Dedecca, 1957). Three stages of fruit development can be distinguished (Cannell, 1985). After an initial period of dormancy with little growth during the first 6–8 weeks after flowering (pinhead stage) fruitlets expand in volume almost to the final size over a period of 8–10 weeks (soft-green stage) followed by a period of bean (endosperm) filling for another 10–15 weeks (hard-green stage). Fruits mature in 7–9 (C. *arabica*) to 9–11 months (C. *canephora*) from date of flowering. The green fruits will turn yellow or red, depending on the variety, during ripening (Figure 2.4).

Figure 2.3 Coffee flowering

The fruit is a drupe, usually called a berry or cherry, containing two seeds embedded in a fleshy pericarp (fruit skin) and a sweet tasting mucilage layer. Sometimes the fruit contains only one round seed called a *peaberry*. Fruits of C. *arabica* are 12–18 mm and those of C. *canephora* 8–16 mm long. The seeds (coffee beans) are plano-convex in shape and grooved (centre cut) on the flat side. They consist mostly of endosperm with a small embryo at the base of the seed and are enveloped in a rudimentary integument ('silverskin') and fibrous endocarp (parchment). Beans of C. *canephora* are generally smaller, rounder and with a tighter centre cut than those of C. *arabica*.

Figure 2.4 Green, ripe and overripe cherries of C. *arabica*: (left) yellow variety; (right) red variety

Coffee fruits are strong assimilate-accepting sinks and the tree is unable to regulate the crop load effectively by shedding fruits. The key issue in coffee growing is therefore the prevention of excessive cropping that leads to biennial bearing and also shoot and root dieback. Crop husbandry is, therefore, aimed at maintaining enough foliage to sustain the crop (about 20 cm leaf area per fruit) as well as new shoot growth by pruning and other agronomic practices (Cannell, 1985). Copper and some other types of fungicides improve leaf retention, particularly in ecosystems with pronounced periods of water stress. This 'tonic' effect of fungicides may result in considerably higher yields, irrespective of disease control *per se* (Van der Vossen and Browning, 1978).

2.2 VARIETY DEVELOPMENT

M.R. Söndahl and H.A.M. van der Vossen

2.2.1 Breeding strategies

Breeding programmes implemented to the two coffee species have basically the same main objective of developing new cultivars with the potential of yielding optimum economic returns to coffee growers. Yield, plant vigour and quality are the main selection criteria, but particularly in arabica disease resistance and compact growth habit have also been given much attention. Variation in climate and soil, incidence of diseases and pests, cropping systems, socio-economic factors, market dynamics and consumer preferences further define priorities given to selection criteria applied in specific programmes. Breeding methods depend primarily on the mating system and plant improvement goals (Van der Vossen, 1985, 2001; Charrier and Eskes, 1997). For the self-pollinating C. *arabica* the methods applied include, in increasing order of complexity (Carvalho, 1988):

■ introduction of coffee germplasm and pure line selection;
■ pedigree selection after intervarietal crossing and often also back-crossing;
■ F1 hybrids between genetically divergent breeding lines;
■ interspecific hybridization followed by backcrossing to arabica lines and pedigree selection.

In case of the out-crossing C. *canephora*, these are (Montagnon *et al.*, 1998 a,b):

■   introduction of coffee germplasm and mass selection;
■   family and clonal selection;
■   reciprocal recurrent selection;
■   interspecific hybridization followed by family and clonal selection.

2.2.2 Traditional and modern cultivars

2.2.2.1 Arabica

World coffee production is still based to a large extent on traditional cultivars. In C. *arabica* cultivars were developed long ago by line selection within the Typica and Bourbon varieties, or in offspring of crosses between these two types. Besides, cultivars like Caturra, Pacas, San Ramon, Sumatra and Maragogipe have their origin as single-gene mutants found in C. *arabica* populations (Krug and Carvalho, 1951). The traditional cultivars are often renowned for their excellent cup quality, but most are very susceptible to the major coffee diseases, which makes them increasingly obsolete for economic (costs of chemical disease control) and ecological (chemical pollution) reasons in many coffee regions (Table 2.2).

However, as a result of considerable breeding efforts during the past 50 years, new arabica cultivars with higher yield potential and resistance to important diseases have now started to replace traditional varieties in several countries. These include Catimor and Sarchimor type of cultivars in Colombia, Brazil, Central American countries and India, Icatu in Brazil, as well as the F1 hybrid cultivars Ruiru II, generally referred as Ruiru 11 in Kenya, and Ababuna in Ethiopia. Some of the recently developed cultivars of arabica coffee are described in Table 2.3.

Efforts to obtain resistance to CLR, already started in India around 1920, have had a long history of initial successes followed by disappointments because of repeated appearance of new virulent races of the rust fungus. However, some lines of the cultivar Catimor, which was developed from crosses between Caturra and Hibrido de Timor (a natural interspecific hybrid between C. *arabica* and C. *canephora* with arabica phenotype), have shown complete resistance to all physiological races of the CLR pathogen in most countries. These results were obtained by breeding plans normally applied to self-pollinating crops, including recombination crosses followed by backcrossing, inbreeding and pedigree selection (Carvalho, 1988; Bettencourt and Rodrigues, 1988). The breeding programme in Kenya demonstrated the advantages of F1 hybrid cultivars also for arabica coffee, especially the simultaneous combination of compact plant type, high yield, good quality and resistance to coffee berry disease (CBD) and CLR in one cultivar (Van der Vossen and Walyaro, 1981). Subsequently, breeding

Table 2.2 Commercially important traditional arabica cultivars (disease susceptible)

Name	Country	Description
Typica	Worldwide	The original type of arabica coffee introduced from Yemen into Asia in the early sixteenth century and after 1720 into the Caribbean and Latin America
Bourbon	Worldwide	Introduced around 1715 from Yemen into Réunion Island (formerly Bourbon) and subsequently into Latin America and East Africa; more compact and upright growth, higher yielding and better bean and cup quality than Typica; red and yellow fruited Bourbon types
Tekesik	C. America	High-yielding selection from Red Bourbon made in El Salvador
Kona	Hawaii	Selection from Typica with large beans, good quality, but low yield
Mundo Novo	Brazil	Selection from a natural cross between Sumatra (Typica) and red Bourbon found in Saõ Paulo in 1931; replacing much of Typica after 1960; vigorous growth and high yielding
Blue Mountain	Jamaica	A Typica selection grown in the Blue Mountains and famous for its cup quality; susceptible to CLR, moderately resistant to CBD
Kent	India	Selection from Typica in 1911; some resistance to CLR (race II); good cup quality; still grown at high altitudes
K7	Kenya	Selection from Kent in 1936; some resistance to CLR (race II) and CBD; vigorous growth and high yielding, but lower cup quality than SL28
SL28	Kenya	Selection from Bourbon 'Tanganyika Drought Resistant II' in 1935; high yielding and some drought resistance; excellent bean and cup quality
SL34	Kenya	Selected from 'French Mission' (Bourbon-type); good yield and excellent bean size and liquor quality; adapted to highest altitudes
KP423	Tanzania	Selection from Kent similar in yield and quality to K7 in Kenya
N39	Tanzania	Selection from Bourbon-type coffee; excellent bean and cup quality, but moderate yielder
Jimma	Ethiopia	Dry-processed forest and garden arabica coffees of many origins and very variable quality
Harar, Gimbi	Ethiopia	Distinct varieties grown in Hararge and Welega provinces resp.; dry-processed coffees highly appreciated for their 'mocha' flavour
Yirga Chefe, Limu	Ethiopia	Washed coffees of very good quality; names refer to place of origin (Sidamo and Kefa provinces resp.) rather than being distinct cultivars
Caturra	Worldwide	Mutant discovered in Bourbon field in Brazil in 1935; compact growth (short internodes); medium-sized beans and moderate cup quality
Catuai	Brazil	Selection from a cross (Caturra x Mundo Novo) made in 1949; compact growth; in Brazil higher yielding and better quality than Caturra
Villa Sarchi	Costa Rica	Mutant with short internodes found in Bourbon coffee; syn. Villa Lobos; similar to Caturra in plant stature and yield, but smaller beans
Pacas	El Salvador	Caturra-like mutant found in Red Bourbon coffee in 1909

CLR, coffee leaf rust.

Table 2.3 Commercially important modern arabica cultivars
(cultivars with disease resistance)

Name	Country	Description
Catimor	Worldwide	Selections from crosses (Caturra/Catuai x Hibrido de Timor); compact growth like Caturra; resistant to CLR; Oeiras (Brazil), Cauvery (India), IHCAFE90 and CR95 (C. America) are Catimor-like cultivars
Sarchimor	Worldwide	Selections from cross (Villa Sarchi x H. de Timor); similar to Catimor in compact growth and resistance to CLR; Tupi, Obata and IAPAR59 (Brazil) are Sarchimor-like cultivars
Colombia	Colombia	Synthetic variety composed of a number of Catimor lines; large bean and good liquor quality; resistant to CLR; some lines with resistance to CBD
S795	India	Selection from a natural interspecific hybrid (C. arabica x C. liberica) backcrossed to C. arabica; resistant to some CLR races; high yield and good liquor quality; most important arabica cultivar in India
Ruiru II (Ruiru 11)	Kenya	F1 hybrids (seed by hand-pollination) between selected Catimor lines and selected clones from multiple crosses (tall plants); resistant to CBD and CLR; slightly less compact than Catimor; early and high yielding; most F1 combinations have very good bean size and cup quality
Ababuna	Ethiopia	F1 hybrid (seed by hand-pollination) between selected lines of Ethiopian germplasm with normal (tall) plant habit; resistance to CBD
Icatu	Brazil	Selections developed by crossing (tetraploid) C. canephora x C. arabica (Bourbon) followed by backcrossing to Mundo Novo; resistance to CLR; tall growth habit; susceptible to drought and cold; high yielding and good cup quality. IAC3282 is an early maturing Icatu with yellow berries
S2828	India	Developed by interspecific crossing and backcrossing similar to Icatu; tall plants; resistance to all (?) races of CLR; high yielding and good cup quality

1. CLR, coffee leaf rust; CBD, coffee berry disease.
2. Important progenitors for disease resistance:
Hibrido de Timor: arabica-type variety cultivated in Timor; assumed to have developed from a natural cross between C. arabica and C. canephora; CIFC (Portugal) clones 832/1, 832/2 and 1343 have resistance to all or most races of CLR; CIFC clone 1343 also carries a gene for resistance to CBD. According to Bertrand et al. (In Press) it should be possible to find lines with resistance genes and good beverage quality. Selection can avoid accompanying the introgression of resistance genes with a drop in beverage quality (Bertrand et al., 2004).
Rume Sudan: semi-wild arabica variety collected in 1942 from the Boma Plateau in southeastern Sudan; some individual plants have very high resistance to CBD.

strategies aiming at F1 hybrid cultivars in arabica coffee have been adopted elsewhere, e.g. in Ethiopia (Bellachew, 1997) and in Central American countries (Bertrand et al., 1997).

Selection for bean size and cup quality has received much attention in arabica coffee breeding programmes in Kenya and Colombia in particular, because the quality of new disease-resistant cultivars should be at least equal to that of the traditional cultivars in order to uphold the country's high reputation and special position in the world coffee market. This was obviously achieved in Kenya with the CBD- and CLR-resistant hybrid cultivar Ruiru 11 (Njoroge et al., 1990) and in Colombia with the CLR-resistant cultivar Colombia (Moreno et al., 1995), as judged by international coffee-tasting panels. Rigorous standardization of pre- and post-harvest practices, bean grading and cup tasting applied in these two breeding programmes contributed to increased selection progress and helped to overcome the initially negative effects on quality due to introgression of disease resistances from exotic germoplasm (cupping characteristics are, however, different from those of the parent material and still considered by trade less suitable for espresso-type blends). Bean size improvement is particularly noticeable in the Pacamara variety (cross between Pacas × Maragogype), developed in El Salvador, with screen size bean values of 19 andmire.

2.2.2.2   Robusta

In case of the cross-pollinating C. canephora, most cultivars are also propagated by seed, often from polyclonal seed gardens. Commercially important robusta cultivars include the BP and SA series in Indonesia, S274 and BR series in India, the IF series in the Ivory Coast and the cultivar Apoata in Brazil (Table 2.4). Some breeding programmes in robusta coffee (e.g. in the Ivory Coast) have recently adopted methods of reciprocal recurrent selection with distinct sub-populations to increase chances of producing genotypes superior in yield, quality and other important traits (Leroy et al., 1997b).

2.2.2.3   Interspecific hybrids

Interspecific hybridization has played a significant role in coffee, such as crosses between arabica and robusta coffee with the objective of introgressing disease resistance into arabica, e.g. the cultivars Icatu in Brazil (Carvalho, 1988) and S2828 in India (Srinivasan et al., 2000), or improved liquor quality into robusta coffee, such as the variety Arabusta in the Ivory Coast (Capot, 1972). Other examples of interspecific hybridization leading to successful robusta cultivars are Congusta in Indonesia and the CxR variety in India.

Table 2.4   Some robusta cultivars

Name	Country	Description
BP and SA series	Indonesia	Important robusta cultivars developed on two research stations in Java in the 1920s; propagated by seed or grafting
S274, BR series	India	Major robusta cultivars in India; selected from plant material of Java origin and released in the 1950s; seed propagated
IF series	Ivory Coast	Selected from robusta plant material from Java and the DR Congo; propagated by seed and cuttings
Kouilou (Conilon)	Brazil	Named after the Kioulou river in the DR Congo; the main robusta variety (95%) in Brazil; seed propagated
Apoata	Brazil	Initially used as nematode-resistant rootstock for arabica cultivars; now also used as robusta cultivar; seed propagated
Arabusta	Ivory Coast	Developed from a (*C. canephora* x *C. arabica*) cross in the 1960s: cup quality better than robusta: limited commercialization due to persistent low fertility; clonal propagation
C x R variety	India	Developed from a (*C. canephora* x *C. congensis*) cross in 1942, followed by backcrossing to S274 and selection; released in 1976; bolder beans and better cup quality than robusta

2.3 AGRONOMY

M.R. Söndahl, F. Anzueto, C.R. Piccin and H.A.M. van der Vossen

2.3.1 Climate and soil

Arabica coffee is cultivated at medium to high altitudes (1000–2100 m a.s.l.) in equatorial regions, or at 400–1200 m altitudes further from the equator (9–24 °N and S latitudes), where average daily temperatures are about 18–22 °C. In general, the cooler the climate the better will be the cup quality of arabica coffees. However, temperatures near 0 °C will kill the leaves immediately. On the other hand, long periods of hot weather will cause abnormal flowering (star flowers) and shoot dieback. Depending on local evapotranspiration, it requires a minimum of 1200–1500 mm annual rainfall with no more than 3 months of less than 70 mm for good growth and production. In contrast, robusta coffee requires warm and humid climates of tropical lowlands and foothills (100–1000 m altitude) with average daily temperatures of 22–26 °C, absolute minimum temperatures not below 10 °C and well-distributed annual rainfall of at least 2000 mm.

The parent material of major coffee soils include lava and tuff (e.g. Kenya), volcanic ash (e.g. Indonesia, Central America), basalt and granite (e.g. Brazil, West Africa, India). Soils for coffee cultivation should be deep (at least 2 m), free-draining loams with a good water-holding capacity, fertile and slightly acid (pH 5–6) and contain at least 2% organic matter in the upper horizon.

2.3.2 Propagation and crop husbandry

2.3.2.1 Propagation and planting

Cultivars of arabica and robusta coffee are generally propagated by seed. Seeds germinate within 4–5 weeks after sowing in wet sand. Removal of parchment prior to sowing reduces coffee seed germination time by half. Vegetative propagation by rooted cuttings or grafting has found limited application in the past to multiply high-yielding robusta clones (e.g., in Java and Ivory Coast). Grafting of arabica cultivars on robusta rootstock is very common in certain areas of Brazil and Central America where nematodes are a serious problem. Recently, more efficient *in-vitro* methods of clonal propagation by somatic embryogenesis have been developed for robusta as well as arabica coffees (Sondahl and Baumann, 2001; Etienne *et al.*, 2002).

Seedlings or clonal plants are raised in shaded nurseries, in polythene bags filled with a soil-compost mixture, for 6–12 months before transplanting to the field at the start of the rainy season. Seedlings are planted in previously dug holes filled with top soil supplemented with organic and inorganic fertilizers, or alternatively in trenches dug along the rows by a tractor-mounted implement and refilled with a soil–compost mixture (Mitchell, 1988). Under optimum field conditions young plants may flower 12–15 months after planting, producing a first light crop 2.5 years after planting. Conventional plant densities vary from 1300–2600 plants/ha for tall-stature arabica varieties to 1100–1400 plants/ha for robusta coffee. Higher densities of 3333–10 000 plants/ha are applied with compact growing arabica cultivars like Caturra, Catuai and Catimor in Latin America or Ruiru 11 in Kenya.

2.3.2.2 Shade

In South America and East Africa coffee is mostly grown in pure stands without permanent overstorey shade, except for widely spaced rows of shade trees acting as windbreaks. In other countries (e.g. Central America, Mexico, Cameroon, Indonesia, India), coffee is grown either with temporary or permanent shade trees, or in association with perennial crops (coconut palms, rubber, clove, fruit trees), or in home gardens associated with food crops, bananas (e.g. northern Tanzania, Uganda) and tree crops. In India and Indonesia the stems of shade trees are often utilized to support the vines of black pepper, and so provide an additional source of revenue to coffee growers. With intensive crop management including high inputs (fertilizers and pest/disease control), unshaded coffee will usually produce higher yields than coffee grown under shade. Without high inputs or under sub-optimal ecological conditions coffee usually produces better results under shade. Shade trees have multiple functions, such as: reducing extreme temperatures, breaking the force of monsoon rain and winds, controlling erosion on steep slopes, reducing the incidence of certain pests (e.g. white stem borer), and generally preventing overbearing and shoot dieback.

Common shade trees are *Inga* spp. and *Grevillea robusta* (Latin America), *Albizzia* spp. (Africa), *Erythrina indica* (India) and *Leucaena* sp. and *Casuarina* sp. (Indonesia and Papua New Guinea). In Central America the density of shade trees varies from 156–204 trees/ha at lower altitude to 83–100 trees/ha at higher elevations. As a general recommendation, overstorey shade should not reduce more than 50% of total irradiances.

2.3.2.3 Pruning

Pruning is essential in coffee production to achieve the desired plant shape, to maximize the amount of new wood for the next season crop, to maintain a correct balance between leaf area and fruits, to prevent overbearing and thus reduce biennial production, stimulation of root growth, enhancement of light penetration, and also to facilitate disease and pest control. The following main pruning systems can be distinguished (Wrigley, 1988):

■ single- or multiple-stem, capped at about 1.8 m height, accompanied by regular pruning of lateral branches and removal of suckers;

■ multiple-stem with three to four uncapped stems with little maintenance pruning; rotational system of replacement of old stems (after three to four cropping years) by new orthotropic shoots;

■ 'agobiado', which is a multiple-stem system on one main stem that has been bent over at an early age;

■ complete stumping to 30 cm above ground level (with or without a temporary lung branch) to encourage re-growth of suckers, which, after thinning to two to three vertical shoots, will be brought up to form the vertical stems for the next cycle of production.

This last pruning system is applied on a rotational basis every fourth or fifth year in closely spaced arabica coffee blocks, usually planted with compact-type cultivars (e.g. Caturra, Catuai), or to rejuvenate an old block of conventionally spaced coffee after 8–12 years of production.

2.3.2.4 Weed control

Weeds should be controlled, as they compete with the coffee trees for moisture and nutrients. Broad-leaved weeds are generally less harmful than perennial grasses (e.g. couch, star grass and lalang) and sedges (e.g. nut- and water-grass), because the roots of the latter also produce exudates toxic to coffee and they are more difficult to eradicate. Tillage, herbicides, mulching or interplanting with cover crops can suppress weeds. Tillage should be done carefully to prevent damage of the superficial feeder roots of the coffee and also to avoid soil erosion. Herbicide applications in combination with zero cultivation has the added advantages of not disturbing the superficial feeder roots of the coffee trees, which can therefore develop in the layer of decomposing organic matter from dead weeds and so increase uptake of nutrients. Mulching produces significant yield responses in low rainfall areas, such as in the arabica coffee regions of East Africa and Brazil. The interplanting of leguminous cover crops is less common with arabica than with robusta coffee, even where rainfall is not a limiting factor. Arabica coffee is more sensitive to competition from weeds and cover crops. During the first three years after planting, when the young coffee trees occupy a small portion of the soil surface, between-row interplanting with annual food (e.g. beans, maize, rice) and commercial (e.g. pineapple, cassava) crops serves as an alternative manner of controlling weeds, while producing also an additional source of food and income (Mitchell, 1988).

2.3.2.5 Fertilization

Fertilizer requirements depend on yield level and natural soil fertility. Nutrients are removed by harvested fruits, but additional nutrients are also required for sustaining the vegetative growth. Part of the total nutrients is recycled back to soil by leaf fall, prunings and decaying feeder roots. The annual nutrient uptake per hectare of a vigorously growing block of arabica coffee yielding 1.0 t/ha green coffee beans is estimated at 135 kg N, 34 kg P_2O_5 and 145 kg K_2O. Soil and leaf analyses are important for monitoring the balance of nutrients in soil and plant tissues, and so provide good guidance for fertilizer application. Generally, nitrogen fertilizers at rates of 50–400 kg N/ha/year give considerable increases in yield, especially when applied in split applications. Responses to potassium fertilizers (up to 400 kg K_2O/ha/year) vary from nil in mulched coffee grown on volcanic soils rich in K (e.g. Kenya) to highly significant in soils with a low K status (e.g. Brazil). Very high K applications may induce Mg deficiency. Phosphate is often applied as compound fertilizer (NPK: 2–1–2), but its effect is faster in foliar applications. Calcium in the form of lime is used to correct soil acidity. Deficiencies in Mg and minor nutrients such as boron and manganese are also best corrected by foliar applications. Organic manures (stable manure, cover crops, mulch and decaying coffee pulp) are alternatives to chemical fertilizers and are often the only type available to smallholders. Organic matter also improves physical (water retention and aeration) and biological (microflora and soil fauna) properties of the soil (Willson, 1985; Mitchell, 1988).

2.3.2.6   Irrigation

Irrigation is essential in nurseries and during the first 1–2 years after field planting to enable the young plants to establish well and to survive the dry seasons. Irrigation of unshaded adult coffee can produce economic yield responses, when annual rainfall is less than 1200 mm and the dry season(s) are extended over more than 2–3 months. Critical periods when irrigation should be applied are at flowering time, at berry expansion stage and at bean filling stage. Especially moisture stress at the latter (hard green berry) stage will result in light and poor quality beans (Mitchell, 1988). However, irrigation is only applied on large estates, as it is expensive, e.g. overhead sprinkler or drip irrigation systems. The latter is most efficient in water use, but requires a very costly initial investment of equipment.

2.3.2.7   Organic and sustainable coffee

In several coffee-producing countries there is increasing awareness of the need to produce sustainable and environment-friendly coffee. This is achieved by improved land management, increased organic instead of synthetic fertilizers, reduced pesticide use by integrated disease and pest management (disease resistant cultivars and biological control of insect pests) and by reduced water use and pollution during post-harvest handling. Some of these coffees may be certified as organically or sustainably produced coffees and so attract premium prices in a gradually developing niche market of health- and environment-conscious consumers, particularly in the USA and Europe.

2.3.3   Diseases and pests

2.3.3.1   Diseases

Coffee leaf rust (CLR), caused by the fungus *Hemileia vastatrix*, is the most important disease of arabica. It reached Brazil in 1970, and is now present worldwide in all coffee-producing countries (except Hawaii and Australia). The use of copper and systemic fungicides and resistant cultivars has reduced the CLR threat. Nevertheless, CLR continues to cause considerable economic damage to arabica on a global scale due to increased production costs by fungicide sprays (10%) and crop losses (20% and more), because most coffee is still produced on CLR-susceptible cultivars. There is also a problem in some countries, India in particular, of repeated breakdowns of host resistance on CLR-resistant cultivars by the development of new pathogenic races of the leaf rust fungus (Eskes, 1989).

The very destructive coffee berry disease (CBD) is still restricted to arabica in Africa, although climatic conditions in certain high-altitude coffee areas of Latin America and Asia appear to be favourable to the fungus *Colletotrichum* kahawae (*coffeanum*). CBD epidemics can quickly destroy 50–80% of the developing crop on susceptible arabica cultivars during prolonged wet and cool weather conditions. Preventive control by frequent fungicide sprays may account for 30–40% of total production costs. New cultivars with high levels of durable CBD resistance have been successfully developed (e.g. in Kenya), but have not yet sufficiently replaced old susceptible cultivars (Van der Vossen, 1997). Most robusta cultivars are resistant to both diseases.

Of the several other fungal and bacterial diseases affecting both coffee species (Wrigley, 1988), the following four have drawn attention in recent years. In Central America the leaf spot disease 'Ojo de Gallo' (*Mycena citricolor*) can be a serious problem in coffee grown under shade at high altitudes, particularly affecting CLR-resistant cultivars like Catimor and Sarchimor. Black rot (*Koleroga noxia*) is now the second most important coffee disease in India after coffee leaf rust (Bhat *et al.*, 1995). The bacterium *Xylella fastidiosa* has become a major problem in some regions in Brazil (Beretta *et al.*, 1996). Fusarium wilt disease or tracheomycosis (*Fusarium xylarioides*) has been seriously affecting *C. canephora* in DR Congo and Uganda since the 1980s (Flood and Brayford, 1997). The traditional Nganda types are very susceptible, but some robusta clones appear to be resistant to Fusarium wilt.

2.3.3.2   Nematodes and insect pests

Nematodes (mainly *Meloidogyne* spp. and *Pratylenchus* spp.) can cause considerable problems on arabica in Central America in particular, but also in Brazil, India and Indonesia. Much progress has been made in developing arabica cultivars with resistance to certain nematode species (Anzueto *et al.*, 2001) as well as in utilizing nematode-resistant robusta clones as rootstock for arabica cultivars (Bertrand *et al.*, 2000).

Among the most damaging and worldwide coffee pests are the coffee berry borer (*Hypothenemus hampei*), stem borers (e.g. *Xylotrechus quadripes*), scale insects (e.g. green scale, *Coccus viridis*) and leaf miners (e.g. *Perileucoptera coffeella*). Host resistance to insect pests in natural coffee germplasm is extremely rare, but transgenic coffee plants with resistance to leaf miners and the coffee berry borer may become available in the near future (Guerreiro Filho *et al.*, 1998; Leroy *et al.*, 2000). Integrated pest management (IPM) has proved to be much more effective in reducing damage to the coffee trees and crops than frequent applications of insecticides (Bardner, 1985). IPM includes early warning systems (monitoring of insect populations) in combination with cultural (pruning and shading), biological (insect traps; introducing insects or micro-organisms parasitic to the pest) and chemical (carefully timed applications with non-persistent insecticides) methods of control.

2.3.4   Harvesting and yields

In equatorial regions where the harvesting season extends over several months, selective picking of ripe berries at 7–10 days intervals is common. Where the harvest season is short, e.g. in Brazil, or the cost of hired labour high, whole branches are stripped when the majority of the berries are ripe. Harvesting costs are two to three times higher for selective hand picking, but quality of the product is usually better (but, see also 3.2.4). The use of self-propelled mechanical harvesters has found application in plantation coffee of Brazil and a few other countries (see 3.2.3).

Yields may vary from 200 kg green coffee beans per ha from low-input smallholder plots to 2 t/ha for arabica and 3.5 t/ha for robusta coffees in well-managed plantations at conventional plant density and without shade. Yields of 5 t/ha or higher have been obtained in some closely spaced and unshaded coffee blocks planted with compact-type arabica cultivars, e.g. in Brazil, Colombia and Kenya.

2.3.5 Agronomic practices in selected countries

There is a great variation in production systems among the coffee-producing countries. Comprehensive overviews of all or most important countries have been presented by Krug and de Poerck (1968), Wrigley (1988) and the ITC report (2002). The four examples presented here are Brazil, as the largest producer of natural arabica and leader in mechanization of coffee cultivation, the Central American region, as largest producer of washed arabicas, Kenya in East Africa, with a tradition of producing mild arabicas of the highest quality, and India, as leader in technology development in arabica and robusta coffees in Asia. See Chapter 3 for details of the production techniques mentioned below.

2.3.5.1 Brazil

Coffee is grown on some 2.5 million ha (5.5 billion trees) between latitudes 10 and 24°S, mostly on gently sloping land, often with high-tech cultivation practices. Brazil is the largest arabica producer in the world, and its coffee plantations are composed of 73% arabica and 27% robusta trees. Its share of robusta in the total annual crop has increased to more than 21% in recent years (ITC, 2002). Domestic consumption accounted for about 50% of total production of 30.1 million bags in 2000/01. In recent years new coffee plantings have moved further north up to the states of Bahia (arabica) and Rondonia (robusta). Recently, incidental severe droughts have caused more impact on the Brazilian coffee crop than frost. It is estimated that about 85% of all planted arabica coffee consists of traditional cultivars, Mundo Novo and Catuai in particular, and the rest are CLR-resistant cultivars like Tupi, Iapar-95, Icatu and others.

Coffee plantations in Brazil vary from relatively small (5–50 ha) and medium-sized farms (50–200 ha) to very mechanized large plantations (200–3000 ha). Plant densities have evolved from 1000–1900 plants/ha, often with two plants per hole with free growth, to 3500–5700 plants/ha in hedgerow planting with periodic pruning. The availability of more compact cultivars has led to the adoption of even higher plant densities in combination with regular pruning.

Preparation for planting is often entirely mechanized. Tractor-mounted planting implements have been developed that first draw deep furrows along the planting lines, then mix the soil with fertilizers and lime and close the furrows all in one single passage. The bulk of planting material utilized in Brazil consists of 6-month old seedlings produced in small plastic bags, but more recently the 'tubette' seedlings are becoming popular since they allow for semi-mechanized planting methods with consequently 40% reduction on planting costs. The tubette seedlings are of particular use for arabica cultivars grafted on nematode-resistant robusta rootstock.

Almost all the coffee in Brazil is grown without shade. Weeds are controlled with a combination of pre-emergent herbicides and intercropping with native leguminous plants and monocots. A green cover of spontaneous species is maintained by mechanized trimming and diluted solutions of herbicides plus liquid fertilizer. Plantations under full sun produce more, and so require higher levels of fertilizers. Recommended fertilizer rates for each 10 bags of green coffee are (in kg/ha): N (100), P_2O_5 (30), K_2O (120), Mg (20), S (5), B (2) and Mn (3). Fertilization is split in three to four applications and monitored by leaf analysis: if N is above 3.2% and K above 3.1%, the remaining N–K dosages are suspended. Coffee leaf rust is being effectively controlled in susceptible cultivars by three to four rounds of sprays with copper-based and systemic fungicides during the main growing season. Coffee berry borer, leaf miners and in some areas also nematodes are the main coffee pests.

Yields vary from 0.6 t/ha green coffee in low-input coffee farms to 2–3 t/ha in well-maintained plantations. Very high annual yields of 4–7 t/ha have been reported from technology-driven coffee plantations established in the last decade in northwest Bahia with short-stature cultivars.

The method of pruning both vertical stems and lateral branches is becoming very popular among farmers, and special machines have been developed for these operations.

Harvest season begins in mid-April (at low-altitude farms) and extends until August (in high lands). Stripping all cherries from branches at once is the common method of harvesting. To maximize the percentage of ripe cherries during the single harvesting round, it is common to grow coffee cultivars with early, medium and late fruit ripening characteristics. Harvesting is still predominantly a manual operation, accounting for 35–40% of production costs. Brazil is the world leader in the testing and development of mechanical implements to reduce costs of coffee harvesting (see Figure 3.3). Equipment includes hand-held vibrating machines similar to olive harvesters, tractor-pulled harvesters equipped with one or two cylinders of vibrating rods to harvest one or both sides of a row of coffee trees. More advanced harvesting machines include the self-propelled models equipped with fruit collecting and cleaning systems. Harvesting is accomplished by driving over the coffee plants and the berries are placed in bags or directly loaded onto trucks in bulk. These vibrating harvesting machines usually operate at speeds between 0.5–1.0 km/hour. Self-propelled grape harvesters from Europe with a system of oscillating arms are also being tested for 'selective picking' of ripe coffee berries. They can operate at three to four times the speed of current models, but still require some adaptations to make them suitable for coffee harvesting.

In Brazil, arabica coffee is traditionally harvested by the stripping method and dry-processed to produce 'natural' coffees. Fruits from stripping harvest include several ripening stages (green, green-ripe, ripe, overripe and dry). By water flotation, dry and overripe fruits are isolated for sun drying in separated lots. Certain farmers depulp ripe cherries and dry the parchment without removing the mucilage. This semi-natural preparation method is increasing in acceptance, since it reduces the total patio drying time and offers reduced risk of undesirable fermentation. Presently, it is estimated that Brazil produces about 3.5 million bags of pulped natural coffee (see 3.3.3). The Brazilian arabica coffee is characterized by mild acidity, aromatic notes and heavy body, but qualitative variations do exist due to different microclimates, varieties and processing methods (see 3.8 for details on sensorial analysis). Limited amounts of Brazilian coffee are also processed by the fully wet method.

2.3.5.2 Central America

Central America (Guatemala, El Salvador, Honduras, Nicaragua and Costa Rica) produced altogether some 13 million bags of washed arabica and only very little robusta coffee in 2000/01 (ITC, 2002). Coffee is grown in mountainous regions with volcanic soils at altitudes ranging from 400–1700 m, but the majority of fields are within 1200–1500 m altitude. In general, the rainy season extends from April/May to October, but there is a tremendous variation in total annual rainfall (1000–4500 mm) between regions due to topographical differences. Cultivation under shade is common in more than 95% of the coffee areas in Central America, except in Costa Rica, which has 25% of full-sun plantings. The most common shade trees are *Inga* spp., *Erythrina* spp. and *Grevillea* spp., at densities ranging from 100 to 156 trees/ha. While in Costa Rica, Nicaragua and Honduras most of the coffee is produced by smallholders, medium to large coffee plantations have an important share of the total crop in Guatemala and El Salvador.

The main cultivars grown at present are Caturra, Pacas, Catuai and Red Bourbon. These have largely replaced the traditional Typica or 'Arabigo' variety, which was introduced from the Caribbean in the eighteenth century. The cultivation of compact cultivars like Caturra has stimulated important technological changes, such as high plant densities, reduced shading and inputs of fertilizers and pesticides, increasing productivity. Actually, three coffee production systems can be distinguished in Central America:

- ■ technical, with compact cultivars (Caturra, Catuai) grown at high plant densities (5000–7000 plants/ha);
- ■ technical, with tall cultivars (Bourbon lines) at close spacing (3400–3700 plants/ha);
- ■ traditional, with tall Bourbon cultivars at conventional spacing and low inputs.

In Costa Rica, the technical farming system with Caturra and Catuai is most common, whereas in Guatemala and El Salvador, Honduras and Nicaragua one can see all three systems of cultivation. The adoption rate of compact growing CLR-resistant cultivars like Catimor and Sarchimor is still rather modest (about 5%). National average yields of the Central American countries vary from 800 to 1500 kg/ha of green coffee. Technical-operation coffee farms can reach up to 3000 kg/ha, whereas low-input traditional farms produce about 500 kg/ha.

Modern cultivation practices include periodic pruning by rows or by blocs, while in traditional farming systems selective pruning is often applied, i.e. plant-by-plant as the need arises. It is also common to bend orthotropic stems (agobio) to induce multiple vertical branches forming a typical plant architecture called 'parra'. The amount and type of fertilizers applied varies with region and level of production, but is usually given in NPK dosages of 50–75, 100–150 and 150–300 kg/ha, in two to three split applications during the rainy season. The economically most important diseases are coffee leaf rust and 'Ojo de Gallo', whereas coffee berry borer and nematodes are the main pests. In highly affected nematode regions in Guatemala the control is achieved by grafting arabica cultivars on resistant robusta rootstock.

Harvesting in low-altitude coffee farms takes place in the period August–October, at mid-altitude during October–December and at high altitudes, January–April. Ripe berries are picked in three to five passages and all coffee is processed by the wet method. Mucilage is mainly removed by fermentation, but some farms are adopting mechanical removal of the mucilage after depulping. On-farm wet processing is common in Guatemala, Honduras and Nicaragua, but in Costa Rica and El Salvador most harvested cherry is processed in central coffee factories. Wet parchment coffee is dried on patios under the sun, but mechanical dryers are used for final drying in mid- to large processing factories. In Central America green coffee is commercialized by growing altitude and, sometimes, location in relation to the Pacific or Atlantic Ocean. For instance, in Guatemala the type 'strictly hard bean' originates from altitudes above 1350 m. In Costa Rica 'strictly hard bean North' coffee comes from altitudes of 1200–1600 m. In Honduras and El Salvador 'strictly high grown' coffee is derived from altitudes above 1200 m.

2.3.5.3 Kenya

Arabica coffee is grown in the highlands east and west of the Great Rift Valley, just below and above the equator at 1450 to 2000 m altitudes. Kenya produced 1.1 million bags of mild arabica coffee from 150 000 ha in 2000/01, which is almost entirely exported (local consumption 4.5%). About 60% of total Kenya coffee is produced by smallholders on 75% of the coffee land area, and the remaining by medium to large plantations. Coffee production in Kenya increased from 560 000 bags (30 000 ha) in 1963/64 to a peak of 2.2 million bags in 1987/88, but has been declining since that time due to economic and socio-political factors.

The coffee flowers in March/April and again in November, under influence of the bimodal rainfall pattern typical of equatorial climates. Consequently, the harvest season extends over several months, with the main crop usually maturing in October–December and an early crop in June–August. Average total rainfall per year in the coffee zone varies from 900 mm at low to 2000 mm at high altitudes, but the East African climate is notorious for its recurring cycles of wet and dry years. Larger plantations are situated in the lower range of the coffee zone, and the water deficit is compensated by (overhead sprinkler) irrigation to sustain economic yield levels. Smallholders in the upper altitudes benefit from higher rainfall, but with no irrigation facilities available their coffee suffers in years of severe droughts. Coffee is grown without shade trees, except in very high altitude areas for the protection against wind and extremely low temperatures.

Much of the existing coffee on old plantations is a direct descendant of the Bourbon-type 'French Mission' or 'Mocha' coffee plot established at the St Austin's Catholic Mission near Nairobi around 1900 with seeds originating from Yemen. The planting material of farms established after 1950, and all smallholder coffee fields, consists mostly of the cultivars SL34 and SL28. These two varieties are derived from single plant selection from French Mission and other Bourbon coffees, respectively, and growing at high and medium altitudes. The variety K7 is a selection from the Indian cultivar Kent, and is cultivated in the lowest coffee zones. The cup quality of SL34, SL28 and also of French Mission is significantly better than that of K7, but the latter became popular because it has some resistance to CLR, which is the predominant coffee disease at lower altitudes. Blue Mountain was recommended in the 1950s to smallholders in Kissii, a high-altitude and wet coffee zone west of the Great Rift Valley, on account of its lower susceptibility to CBD than SL34 or SL28, but its cup quality is rather disappointing according to Kenyan standards. In 1986, the Kenyan Coffee Research Foundation (CRF) released Ruiru 11, a seed-propagated (F1 hybrid seeds) compact growing and productive arabica cultivar with high resistance to CBD and also to CLR. Under good management Ruiru 11 is capable of producing a cup quality similar to cultivars SL34 or SL28; it is, however, not looked for in top espresso coffee blends. An estimated 20 000 ha have been planted with Ruiru 11.

Coffee is conventionally planted at a density of about 1320 trees/ha and maintained as three-stemmed capped (plantations) or free-growth (smallholders) trees. Plantations with irrigation equipment have often converted to 2640 trees/ha by doubling within-row density. The disease-resistant and compact Ruiru 11 is best planted at 3000–5000 trees/ha, maintained on one or two uncapped stems per mature tree and after three full crops clean-stumped to start a new cycle of production. Mulching (grasses, etc.) is general practice on larger plantations to control weeds and conserve moisture, but smallholders are often unable to follow recommended practices because the little crop residues available on the farm may be required to feed livestock. During recent years of low coffee prices, regular fertilizer applications are mainly restricted to coffee plantations, as smallholders lack the financial resources for external inputs. This also applies to chemical control of diseases (in CBD- and CLR-susceptible coffee cultivars) and pests. Yields of small coffee farms have declined considerably and are now often not more than 300–400 kg/ha. Well-maintained large coffee plantations produce 1–2.5 t/ha per year. Yields of 3 t/ha and more have been obtained in high-density coffee blocks with Ruiru 11.

All Kenyan coffee is harvested by several rounds of selective picking of ripe berries and processed according to the wet method (see 3.3.2). A small fraction, consisting of overripe and black cherry, is dry-processed and sold as 'mbuni' on the local trade. Plantations operate their own coffee factories, while smallholders bring harvested cherry to cooperative factories for depulping, fermenting, washing and sun drying the parchment coffee, usually done on raised tables with wire-mesh tops.

Some factories, mainly on large plantations, also have mechanical dryers (see 3.4.3) to speed up the drying process during wet weather. Coffee processing stations are situated near rivers, as large quantities of water are required for the wet process. Regulations are now in place to prevent coffee pulp and dirty wastewater from polluting the rivers. Dry parchment is delivered to one of the four existing coffee mills for curing, which includes hulling, cleaning and grading by size and density. Cup quality of each lot of green coffee is then determined by liquoring before being offered for sale at the weekly auction in Nairobi. There are seven grades of green coffee (from AA to T), and cup classification goes from excellent mild arabica with high notes of pointed acidity, rich flavour and light body (class 1) and FAQ (class 4) to a low cup type (class 7). In addition, there are another three under-grades, including 'mbuni', to cup class 10. A rare Kenyan AA coffee of class 1–2 is unmatched in quality on the European coffee market, except occasionally by mild arabica coffee (Bourbon cultivar N39) from the higher slopes of Mount Kilimanjaro in Tanzania.

2.3.5.4   India

Coffee is cultivated mostly in the mountain ranges in the states of Karnataka, Kerala and Tamil Nadu in southern India, between latitudes 9° and 14° N, robusta at 400–1000 m and arabica areas at 700–1600 m altitudes. Small pockets of coffee cultivation also exist in east Andhra Pradesh, Orissa (18° N) and seven northeastern states (27° N). Arabica and robusta coffees are generally grown under shade. The upper tier consists of natural forest trees (*Ficus*, *Artocarpus*, *Terminalia* and other species) or of planted *Grevillea robusta*, whereas the lower tier has *Erythrina indica* and orange trees. Coffee flowering occurs in March/April at the start of the main rainy season. Arabica is harvested during the period November to February and robusta from January to March. Harvesting is entirely manual, by picking ripe berries in several harvesting rounds.

Coffee production increased from 1.1 million bags (73% arabica) from 92 000 ha in 1960/61 to almost 5 million bags (48% arabica) from 310 000 ha in 2000/01. The national average yield is, therefore, almost 1.0 t green coffee/ha. Robusta coffee generally produces 10–15% more than arabica plants and large plantations yield about 25% more than small farms. Smallholders cover about two-third of the coffee area and contribute with 60% of the total annual coffee crop, and the remaining 40% comes from plantations. Arabica and robusta coffees are processed according to the wet (washed) or dry methods, and subsequently cleaned and graded in coffee curing centres. An estimated 80% of all arabica and 30% of the robusta coffees are washed. Coffee produced in large plantations is practically all processed by the wet method. About 80% of all the coffee produced in India is exported.

Resistance to CLR has remained the most important objective for arabica coffee breeding in India since the release of cultivar Kent in the 1920s. The humid and relatively warm climate appears very favourable to the pathogen and Kent (resistance to CLR race II) was soon attacked by

CLR. There is no other coffee-producing country in the world where CLR has developed so many new physiological races, the latest ones apparently capable of overcoming host resistance of some of the most advanced cultivars. Kent is still grown at very high altitude, where CLR incidence is relatively low, but almost 70% of all arabica grown in India at present is S795, a cultivar released in 1946 with resistance to CLR races II and I only, but high yielding and producing good quality Indian arabica coffee. CLR can be controlled fairly effectively by prophylactic sprays (Bordeaux mixture 0.5%) and by systemic fungicides (Plantvax 0.03% or Bayleton 0.02%), but that adds considerably to production costs. Even Cauvery, a cultivar selected from Catimor (ex CIFC Portugal), showed CLR infection within a few years after its first release in 1985 (Srinivasan et al., 2000). There are now about 22 000 ha planted with Cauvery and it is performing well at higher altitudes, where leaf rust epidemics are less severe. Among the other arabica cultivars released by the Central Coffee Research Institute (CCRI) near Chikmagalur, S2828 is most promising for its complete resistance to all CLR races in combination with good yield, bean size and cup quality. This cultivar was developed from an interspecific cross between S274 (robusta) and Kent made in 1934, followed by two back-crosses to Kent and intensive pedigree selection over several generations.

The first robusta coffee planted in India was in the Wayanad district of Kerala State around 1905 and much of the present robusta plantings is directly derived from those early introductions. Family selection for yield and bean size resulted in a number of selections, of which eventually S274 and the BR series were released in the 1950s. S274 in particular has become an important robusta cultivar. Probably inspired by favourable reports from Java of a spontaneous hybrid between C. congensis and C. canephora named 'congusta', breeders at the CCRI made interspecific crosses between C. congensis and S274 around 1942. After backcrossing to S274 and a number of generations of sib-mating and selection, the new cultivar C x R was first released to some estates in 1973–6 in Kerala, but further selection for uniformity is being pursued. The C x R cultivar has characteristics very similar to those reported earlier for Congusta on Java, such as bolder beans and better cup quality, somewhat lower caffeine content, more compact growth habit, good yield and adaptation to cooler climates. It appears to have the possibility of considerably improving the quality of robusta coffees produced in India. Some 10 000 ha have been planted so far, new planting as well as conversion of old plantations by top-working. There is also in development a compact growing version of the C x R variety, analogous to Caturra in arabica.

In general, the capped single-stem pruning system is recommended in India for both species. Plant densities applied are 3000 plants/ha for tall arabica cultivars, 4400 plants/ha for Catimor arabica and 1100 plants/ha for robusta plantations. Regular applications of chemical fertilizers are common practice in large plantations to support economic yield levels complemented with farmyard manure or composted organic wastes (e.g. coffee pulp). Weed control is accomplished by manual slashing or herbicides. Recently, the Indian coffee sector has started actively to promote organic coffee farming, in which no chemical fertilizers or herbicides are used, and with minimum use of pesticides.

Black rot is considered the second most important disease after CLR and this disease affects arabica as well as robusta. So far, no sources of resistance have been identified and control depends on a number of cultural measures (e.g. thinning of shade and regular pruning) and fungicide spray applications. The most devastating insect pests in India are the white stem borer and coffee berry borer (since 1990). There are no sources of genetically controlled host resistances available to these two pests, excepting that generally the white stem borer does not attack robusta. The CCRI has developed instead a number of effective IPM systems, including the use of pheromone traps for white stem borer, simple traps with alcohol mixture as attractant for coffee berry borer, and parasitoids and natural predators against coffee berry borer, shot-hole borer, mealy bugs and scales.

Indian coffees are classified according to a rather complex system of grading, region of origin and specialty coffees. In addition to the regular coffee grades (PB, AA, A, AB, B and C) for washed/unwashed arabicas and robustas, the Indian system also distinguishes grades for 'blacks/ browns', removed from the regular grades by electronic and manual colour sorting (see 3.5.4 and 3.5.5), 'bits' (beans smaller than C grade and broken beans) and 'bulk' (ungraded and not sorted for off-types). Coffees are also called after their region of origin, such as Mysore, Chikmagalur, Coorg (Kodagu), Malabar, Wayanad, Biligiris, Nilgiris, Bababudan, Shevaroys and Pulneys. Speciality coffees include the Mysore Nuggets EB (extra bold washed arabicas), Robusta Kaapi Royale (A-grade washed robustas) and Monsooned coffees. The latter are prepared from cherry arabica and robusta coffees of AA and A grades, mainly for export to Scandinavian countries.

2.3.6   Factors determining green bean quality

The following main factors determine the quality of green coffee: environment, cultural practices, genotype of the variety/cultivar and post-harvest handling (processing). The first three factors determine the potential quality of the green coffee, which can still be easily degraded by poor processing practices. Post-harvest handling of the cherries is carried out by the wet, semi-wet, or dry (natural) method of processing. The wet process favours a light body and acid beverage, whereas semi-wet and dry methods lead to enhanced body with reduced acidity. The effect of environmental, agronomic and genetics factors on coffee quality will be discussed below, while coffee processing and its relation to quality is described in 3.3 to 3.5.

2.3.6.1 Environment

The influence of altitude and temperature on the quality of coffee has been well documented. Arabica originated from the highlands of Ethiopia and its mild and pleasant beverage is best preserved under similar growing conditions. High altitude is critical for cultivation of arabica near the equator. Moderate temperatures will favour a slow and uniform matura-tion process of coffee berries and especially the wide amplitude between day and night temperatures (thermoperiod) will increase flavours and aromatic precursors in the beans. Coffee fruits developed at higher altitudes produce more mucilage and they are richer in sugars and other soluble solids.

The minimum annual rainfall for arabica production is about 1200 mm per year and the maximum should not exceed 2000–2400 mm. Coffee plants grow and yield better if exposed to alternated cycles of wet and dry seasons, and, moreover, a period of water deficit is important to synchronize flower bud differentiation. Areas with excess precipitations, especially during crop maturation, have a tendency to produce lower quality coffee due to irregular cherry ripening and poor conditions for drying the crop after harvesting. High air humidity may cause beans with off-flavours like fermented and hardish, phenolic, medicinal and musty 'rioy' notes, especially in dry-processed (natural) coffees (see 3.7.2 for a definition of these terms). Fermentation notes can be avoided if fruits are harvested at the ripe cherry stage and wet-processed immediately. The rioy taste seems to be associated with a biological transformation of 2,4,6-trichlorophenol to 2,4,6-trichloroanisole, which is facilitated by excess of moisture. In years of excessively long dry seasons, due to changes in climate (e.g. caused by El Niño), shoot dieback and premature ripening of the berries will result in green beans producing liquor with immature and astringent notes. For example, a fishy off-flavour in immature beans is attributed to 4-heptenal, an oxidation product of linolenic acid (Full *et al.*, 1999).

2.3.6.2 Cultural practices

The correct amount of nutrients, the equilibrium of elements and the application of the right fertilizers are of prime importance for production of high-quality coffee beans. Among the essential elements, nitrogen potassium, calcium, zinc and boron are considered the most important. The endogenous level of potassium will influence total sugar and citric acid content and is a key element during bean filling. Potassium sulphate is a better source than potassium chloride. Nitrogen is important for amino acid and protein build-up and influences caffeine content. N-fertilization increases the total nitrogen in beans, with a weak negative correlation to cup quality (Amorim et al., 1973). Calcium is an essential element for cell wall formation (Ca-pectate) providing more compact beans and improving the resistance to pathogen attacks. Zinc influences protein and carbohydrate metabolism, and also affects the synthesis of auxins, which promote cell elongation during fruit formation in the 'soft green bean stage' (Ferreira and Cruz, 1988). Boron influences flowering and fruit set, consequently affecting yield.

The effective control of pests and diseases is essential for the production of quality coffee. The coffee berry borer is a real threat to green bean quality. Perforated beans lose weight and infected fruits are susceptible to attack by air-borne fungi facilitating undesirable fermentations, degrading the final quality of the whole lot. Several off-flavours have been associated with uncontrolled fermentation in coffee, as illustrated in the normal versus elevated values for the following compounds: ethyl 2-methylbutyrate (2.4 in normal vs 37 µg/kg in fermented beans), ethyl 3-methylbutyrate (22 vs 345 µg/kg) and cyclo-hexanoic acid ethyl ester (Bade-Wegner et al., 1997). The infestation of coffee berries by the fruit fly (Ceratitis capitata) is commonly associated with a fungus infection (Fusarium concolor), which gives a pink colour to the bean and strongly affects cup quality (Bergamin, 1963). The presence of virus, transmitted by mites, also affects negatively the final green bean quality through precocious fruit ripening, irregular bean formation, and in several instances the presence of dry and immature fruits (beans with black colour and pungent/bitter taste at cupping).

2.3.6.3 Genetic aspects

The qualitative differences in cup quality among different arabica varieties have received special attention during past decade, especially after the release of new disease-resistant cultivars, such as Catimor, Sarchimor, Ruiru 11 and Icatu. Comparing cup quality of arabica cultivars requires rigorous standardization to eliminate confounding effects from differences in climate, cultural practices and processing methods. New cultivars are usually compared with traditional varieties renowned for their excellent liquor characteristics, such as Bourbon, Blue Mountain, Kent and SL28. Cup quality is evaluated on beverage characteristics like aroma (flowery, 'peasy'), taste (bitter, acid, sweet, woody-earthy), flavour (chocolate, caramel), and body (light, medium, heavy). Data from analytical tests of the chemical composition of green beans are poorly correlated with the perceived cup quality as determined organoleptically. This is an additional handicap for objectively establishing the varietal differences, unless a panel of expert liquorers is available for judgement. Recently, catimor and sarchimor progenies of var. Colombia have been submitted to biochemical evaluations (caffeine, chlorogenic acids, fat, trigonelline, sucrose) in parallel with cup tasting with respect to a standard (Caturra): sucrose content and beverage acidity could be correlated; no significant correlation could, however, be found between degree of introgression and the biochemical parameters (Bertrand et al., unpublished results).

Differences between arabica and robusta coffee are generally very pronounced and well documented. For instance, the musty smell of robusta is attributed to the presence of 2-methylsoborneol (MIB) (Vitzthum et al., 1990), and MIB ranges from 120–430 ng/kg in robusta versus a level lower than 20 ng/kg in arabica (Bade-Wegner et al., 1993). This knowledge led to a practical application, whereby steam treatment of green robusta beans denatures MIB, and so is claimed to allow increased percentage of robusta in commercial coffee blends. Other major differences in green bean composition between arabica and robusta coffee are summarized in Table 2.5.

Arabica has lower levels of caffeine, amino acids and chlorogenic acids in comparison to robusta, but 60% more total oils. Chlorogenic acids contribute to astringent notes, so the reduced amounts in arabica favours its final cup quality (see also 3.11.6). It is known that many aromatic volatile compounds are dissolved (trapped) in oil droplets and released during brewing, so the oil fraction may explain some differences in cup quality between arabica and robusta, particularly in espresso (see Chapters 8 and 9). The influence of the relative composition of the lipid fractions in coffee on final cup quality is not yet well understood. The major lipid fractions in arabica include (Fonseca and Gutierrez, 1971) linoleic acid (C18:1, 47%), palmitic acid (C16:0, 41%), oleic acid (C18:2, 6.4%) and stearic acid (C18:0, 6%) (see also 3.11.9). Most chemical data in green arabica coffee are presented by 'origin of sample' (Santos, Colombia, Kenya, etc.) without indicating the specific variety. Analytical data based

Table 2.5 Main chemical differences between
arabica and robusta green coffees

Compound/fraction	Arabica (%)	Robusta (%)	Reference
Caffeine	1.2	2.4 (> 4)	1
Trigonelline	1.0	0.7	1
Amino-acids	0.5	0.8	1
Chlorogenic acids	7.1	10.3	1
Lipids: total	16 (13–17)	10 (7–11)	1, 2
Oleic acid	6.7–8.2	9.7–14.2	2
Dipertenes: cafestol	0.5–0.9	0.2	1
Kahweol	0.3	–	1
16-0-methylcafestol	–	0.07–0.15	1

Sources: [1]Illy and Viani, 1995; [2]Speer and Kolling-Speer, 2001

on variety will provide additional information on interactions of genotype
with environment and type of processing. Sampling over successive crops
helps to understand the year-to-year fluctuations for key compounds, as
exemplified below for % total oils from Brazilian trees growing in the
same farm and under identical cultural practices (Table 2.6).

Besides fluctuations in total oil content, there are genotypic differences
in other non-volatile and volatile compounds, but these differences have
not yet been correlated to particular beverage types. It is thought that
reducing sugars contribute to bitterness upon caramelization during
roasting, so the ideal sugar content and relative fractions must be
determined. The acidity notes depend on the total acid content as well as

Table 2.6 Yearly fluctuation in total oil content of
Brazilian arabica coffees

Cultivar	1989 crop	1990 crop	1991 crop	Average
Caturra	15.9	11.5	13.8	13.7
Red Catuai	15.8	13.6	14.3	14.6
Catimor	15.5	13.5	14.1	14.4
Mundo Novo	15.7	13.8	14.4	14.6
Laurina	12.3	13.5	14.9	13.6
Icatu	17.2	15.1	15.5	15.9
Crop Average	15.4	13.5	14.4	14.5

Source: Illy and Viani, 1995

on the balance among acid compounds present in green bean (Illy and Viani, 1995). More than 30 acids are present in green coffee, the most important being the chlorogenic acids, quinic acid, malic acid, citric acid and phosphoric acid. Some variability seems to occur in different coffee genotypes (Table 2.7).

The main acid fractions listed in Table 2.7 should not be considered absolute since there are fluctuations from year to year, and differences among varieties (Balzer, 2001). In these data, the arabica-Santos sample is mainly represented by Mundo Novo and Catuai varieties and the arabica-Kenya sample is an indication of Bourbon SL28 material. It will be of interest to know how much of the acid fraction is controlled by specific varieties, and how much is due to environmental and cultural practices.

The coffee aroma profile is mainly controlled by genotype (species, varieties), environmental origin and degree of roasting (see 4.4). Mayer *et al.* (1999) made a comparative study among genotypes and found that variety Typica (from Colombia) had a 52% higher content of a sulphur/roasty compound (2-furfurylthiol) than an arabica sample from Brazil. Holscher (1996) found similar data, reporting at least five-fold differences of 2-furfurylthiol between different samples.

The genetic component of coffee aroma profiles may be more a quantitative rather than a qualitative effect, i.e. the levels present in green beans at harvest will be modulated by the interaction genotype/environment. In other words, once an environment is fixed, it will be always possible to select a variety that produces greater amounts of particular aromas under a specific location. With further improvement of analytical tools, it should become possible to differentiate specific coffee varieties and origins by their profile of volatile compounds, which would help tremendously in blending standards and quality control.

Table 2.7 Variability in acid content among genotypes

Genotype/origin	Quinic acid	Malic acid	Citric acid	Phosphoric acid
Arabica/Santos	5.6	6.1	13.8	1.1
Arabica/Kenya	4.7	6.6	11.6	1.4
Robusta/Togo	3.1	2.5	6.7	1.6

Source: Balzer, 2001

2.4 BIOCHEMICAL ECOLOGY

T.W. Baumann

2.4.1 Introduction

Imagining the squadrons of bacteria, moulds, insects and grazing mammals (not to speak of human beings and their activities), which attack, feed on and destroy plants, we must wonder why the landscapes are not bare but still covered with a lush vegetation. Biochemical ecology (or chemical ecology) emerged from the discovery of the chemicals involved in the above-mentioned interactions between plants, animals and microorganisms. Since Fraenkel's remarkable article (1959) regarding the *raison d'être* of the so-called secondary plant substances (Hartmann, 1996), scientists are aware that plants (and to a smaller extent also animals and bacteria) produce a vast array of substances not only to defend themselves against pathogens and predators but also to attract organisms for their own benefit. Today, biochemical ecology is an established field of natural sciences (for a comprehensive review see Harborne, 2001). It is conceivable that a better knowledge of such chemically mediated interactions will stimulate biocontrolled farming and assist also coffee growers in their fight against pests in the plantations, and thus – in the long term – will undoubtedly improve coffee quality, to which all, farmers, manufacturers and scientists are committed (Illy, 1997).

Before passing over specifically to biochemical ecology in coffee, two well-studied examples will serve to illustrate this rapidly emerging discipline of science. Many flowering species attract the pollinator, such as an insect (e.g. bee, butterfly), a bird (e.g. humming bird), or a mammal (e.g. bat), by odours and/or pigmentation. Additionally and very frequently, the morphology of the flower complies with the needs of a specific pollinator, for example by providing a landing platform for bees and bumblebees or a hanging thin throat for hummingbirds, etc. Generally, the pollinator is rewarded by nectar and by a more or less welcome dust of pollen. However, about one-third of all orchids do not invest energy into nectar production but have evolved flowers simulating various kinds of rewards. For example, several species mimic by odours, colours, shape and texture the female of a distinct solitary bee. The delusion is so accurate that the male tries to copulate with the flower whereby the entire pollinia are positioned, for example, on his head to carry them to the place of a subsequent frustrating pseudocopulation, finally resulting in the pollination of that flower (Schiestl et al., 1999). And moreover, after pollination the flower produces a volatile substance usually emitted by non-receptive female insects to inhibit copulation (Schiestl and Ayasse, 2001). This example nicely illustrates that biochemistry (odours and pigmentation) of biochemical ecology cannot be separated from physical factors including morphology, texture, mechanical strength, nutritional energy and, last but not least, appearance in space and time.

A second example demonstrates how the plant calls for help when attacked by a predator. The larvae of the beet army worm, *Spodoptera exigua*, are a pest of the maize plant. In a series of brilliantly designed experiments, it has been shown that the saliva of the larvae contains a signal compound (perhaps partly originating from the destroyed plant membranes) that, locally, and later also systemically, induces the formation of leaf volatiles specifically attracting parasitic wasps (*Cotesia marginiventris*) which in turn lay their eggs into the larvae, and thus eliminate the predating insect (Turlings *et al.*, 1990; Alborn *et al.*, 1997). Additionally, the volatiles may protect the plant from attack by other herbivores (Kessler and Baldwin, 2001). Today, various examples of such interactions within – at least – three levels (tritrophic interaction: plant/predator/parasite) are known and have been successfully studied by coupling highly sensitive electrophysiological techniques with GC-MS (for reviews see: Dicke and Van Loon, 2000; Pichersky and Gershenzon, 2002).

Yet, secondary compounds govern, besides the sophisticated interactions described above, another key process of plant survival, which is named chemical defence (reviewed in Edwards, 1992; Harborne, 2001). From an evolutionary point of view, general strategies for optimal defence against predation were formulated, for example that plants are expected to accumulate protective phytochemicals in a tissue or organ in direct proportion to the risk of predation of that unit (Rhoades, 1979). Therefore, organs and tissues with a high nutritional value (seeds, young leaves, pollen) have a particularly high risk of predation.

2.4.2 The main secondary metabolites in coffee (see also 3.11)

The phytochemical catalogue of coffee is very large (for an overview of physiologically active substances see Viani, 1988; Baumann and Seitz, 1992; Clarke and Vitzthum, 2001); however, with respect to chemical ecology, only a few most prominent coffee compounds have been intensely studied so far, namely caffeine, chlorogenic acids and trigonelline. Caffeine is a purine alkaloid (PuA). PuA are divided into methylxanthines (caffeine, theobromine, theophylline etc.) and methylated uric acids (theacrine, liberine etc). Their ecological functions in relation to plant development will be discussed in detail below. 5-Caffeoylquinic acid (5-CQA) is the main component among the coffee's chlorogenic acids. It is a metabolite of the phenylpropanoid pathway and often induced by biotic (pathogens, herbivores) and abiotic (UV, temperature, nutrient, light) stress. Thus, it is thought to exert specific functions in plant protection such as defending against microbial infection and herbivores, acting like a screen against harmful UV radiation, and scavenging free radicals and other oxidative species (reviewed in Grace and Logan, 2000). As will be outlined below, chlorogenic acid is the ally of caffeine, as it associates with it in a physico-chemical complex (Sondheimer et al., 1961; Horman and Viani, 1972; Martin et al., 1987). Lastly, trigonelline is a derivative of nicotinic acid and influences many events in plant life (reviewed in Minorsky, 2002): it is a hormone-like regulator of the cell cycle and arrests the cells in the phase prior to mitosis (G2); it has a not yet fully cleared function in nodulation of the legume alfalfa by rhizobial symbionts (interacting individuals of different species); it is believed to be a signal transmitter in the response to oxidative stress; it may act as an osmoregulator; and, finally, it has been reported to induce leaf closure in various species showing sleep movements, the so-called nyctinasty. Additionally, trigonelline was shown to act as a reserve for the synthesis of coenzymes (NAD) during early coffee seed germination (Shimizu and Mazzafera, 2000). To our knowledge no report exists regarding the chemical defence properties of trigonelline, though its considerable concentration of around 1% (dry weight) in both seeds and leaves, and even higher in the youngest internode (>2%), this indicating a transport from old to young leaves (Kende, 1960).

Clearly, the chemical defence by caffeine and chlorogenic acids works because of (a) their high concentration in the related coffee organ or tissue, (b) the generally low body weight of the herbivore (a phytophagous insect may be deterred after one cautious trial, while a large naive animal, such as a mammal, may swallow one or several plant organs at a time and thus be intoxicated, but later will avoid that organ and feed selectively on the rest of the plant) and (c) the need of the herbivore to ingest a large quantity of leaves due to their overall low protein content. Very convincingly, the coffee seed, furnished with comparatively high protein has, in addition to the chemical protection, a strong mechanical defence: the endosperm is extremely hard (see 2.4.5) and the inner fruit wall, the endocarp, called 'parchment', is tough.

If we relate chemical defence of the coffee bean to the human being foraging on it, we can state the following: the coffee drinker's body weight, which nota bene does not result from the protein content of the bean(!), is high compared to the biomass of ground endosperm (roast coffee) used for the coffee brew. And still, some coffee drinkers are only pleased with a plant extract smoothed by sugar, cream or milk. The protecting layers around the bean (see 2.4.3 and 2.4.5) already have been removed by the lengthy and labour-intense processing of the coffee fruit. Finally, the target molecule (receptor) of pharmacological action in the coffee drinker is different from that in, for example, an insect (Table 2.8).

2.4.3   From seed to plantlet

As mentioned in the introduction, biochemical ecology is also a matter of space and time and therefore has to deal with questions such as 'how and to where is coffee dispersed?', 'how long does it take for a seed (= coffee bean) to germinate?', or 'what happens during the transitions from a seed to a whole plant?'. Most of the results cited below were obtained by studies on arabica but, slightly modified, also apply to robusta.

The hobby gardener may seed a so-called 'green' (unroasted) coffee bean possibly available at the local coffee roasting company. By doing so, he or

Table 2.8   Organism-related effects of caffeine (multifunctionality)

Organism	Effect	Underlying mechanism, target	References
Bacteria and fungi (yeasts)	Bacterio- and fungistatic	Inhibits UV dark repair in DNA	Kihlman, 1977, McCready et al., 2000
Fungi	Fungistatic, reduces mycotoxin production	Unknown	Buchanan et al., 1981
Plants	Inhibits germination, reduces growth	Not known; inhibits the formation of the cell plate, calcium?	Rizvi et al., 1981, Gimenez-Martin et al., 1968
Molluscs (snails and slugs)	Molluscicidal, reduces heart rate	Calcium release, increases the duration of action-potential plateaus?	Hollingsworth et al., 2002
Insects	Disturbs developmental processes	Inhibits cyclic AMP phosphodiesterase	Nathanson, 1984
Mammals	Activates CNS, constricts cerebral blood vessels, increases lipolysis, positive inotropic	Binds to adenosine receptors	Nehlig, 1999

she has to bear in mind two things. First, the bean rapidly loses its germinating power when harvested and processed, depending on the bean humidity and on the temperature during storage. Generally, within 3 months after harvest and processing, the germination rate drops to zero, but is extended to one full year or longer at 18 °C, when the bean water content never falls below 40%, achieved by storing the beans vacuum-packed in a polyethylene bag (Valio, 1976; Couturon, 1980; Van der Vossen, 1980). Second, under natural conditions the coffee bean will not or very rarely get directly in touch with soil, because the seed-dispersing animals (monkeys, elephants, large birds) attracted by the red, fleshy and sweet coffee berries will, after having ingested them, excrete or regurgitate seeds still covered by the hard endocarp, so to speak a parchment coffee, as it also results technically from wet processing (see 3.3.2). Occasionally, the entire fruit is dispersed or drops beneath the coffee tree. Then, the berry will dry and shrink very slowly, looking like dry-processed coffee (see 3.3.1) before hulling. The seeds within the shrunk fruit are even better protected and lose less water than in the parchment coffee. In any case, in nature under favourable conditions (humidity) the seed will take up water and germinate, breaking through the outer shell made up either of the parchment or the complete husk. Finally, going back to the hobby gardener, we have to add the following: the naked bean will germinate earlier (by around 2 weeks) than the bean in parch, but has a higher risk of being attacked by pests and pathogens in the soil during imbibition (swelling) and germination. Even though caffeine is described as an effective agent (Table 2.8) against all kinds of organism (generalists), a few specialists have overcome the defence mechanism and not only detoxify but possibly also metabolize the PuA for their own needs. Here, a striking example is the coffee berry borer, *Hypothenemus caffeicola*, a serious worldwide pest, which develops within the caffeine-rich endosperm.

The generalist/specialist rule is pertinent to all secondary compounds so far studied in this respect. Additionally, one can recognize that these phytochemicals are multifunctional, meaning that they have not only multiple functions in the plant producing them but exert also various and specific effects towards the target organism.

Let us return to germination: Most interestingly, during water uptake (imbibition) caffeine remains almost completely fixed within the bean, very likely due to an efficient caffeine barrier at its surface made out of chlorogenic acids (Dentan, 1985). As will be outlined below, these phenolic compounds are of crucial importance during the entire life of the coffee plant, since they prevent autotoxicity by caffeine! Only when the radicle (the thick primary young root) grows into the substrate, is caffeine released at high concentrations into the environment (Baumann and

Gabriel, 1984). Obviously, under natural conditions the imbibed bean is safely encapsulated – there is no need to excrete the defence compound caffeine – whereas during the later stages of germination about one-third of the caffeine of the young seedling, now exposed to pathogens and predators, leaks via the root surface into the substrate, where it may inhibit the competition by other plant species and prepare, along with other compounds, the way for specific root colonization by microorganisms.

As nicely illustrated in Figure 2.5 (upper row), the emerging primary root is the first visible sign of germination. However, dramatic changes occur even earlier inside the bean: the tiny cotyledons invade the nutrient-rich storage tissue of the bean, the so-called endosperm, and eventually occupy the entire cavity within the bean. Simultaneously, the hypocotyl (the region of the stem beneath the stalks of the cotyledons) stretches and carries the head above the ground. Finally, the cotyledons unfold and shed the seed coat. Since during invasion into the endosperm they suck up all its constituents, cotyledons are, in simplified terms, a blot of the original coffee bean: they contain and conserve all the caffeine of the endosperm, and hence are most suitable for a screening to detect and select caffeine mutants (Baumann et al., 1998). Similarly, the chlorogenic acids are also transported into the cotyledons where, however, a large fraction is possibly used for lignin synthesis in order to mechanically stabilize the leaf tissue (Aerts and Baumann, 1994). Now the seedling is ready to develop into a plantlet.

2.4.4   From bud to leaf

The leaves are arranged in pairs, which alternate along the stem but are adjusted almost in one plane along the side branches (see also Figure 2.2), this in order to optimize light harvest. The individual leaf pair is born from the terminal bud of the stem or of a side branch of first and second order. In the bud, the leaf primordia (tiny pre-formed installations) are covered by a resinous layer followed by two tough scale-like bracts (stipules). The life of the leaf pair starts with its emergence from the bud with the tiny leaf blades still attached to each other. Thereafter, they separate and the individual leaf expands considerably to achieve its final size and shape about 4–5 weeks after emergence. It is still soft, light green and glossy. During the next 2–3 weeks, the leaf texture gets tough and, coincidentally, the upper surface of the lamina turns from glossy to dull dark-green, possibly resulting from a chemical change of the epicuticular waxy coatings: long chain fatty acids are transformed into the corresponding alkanes (Stocker, 1976). Now, 50–60 days after emergence, the leaf is fully developed and optimally gathers solar light energy for the

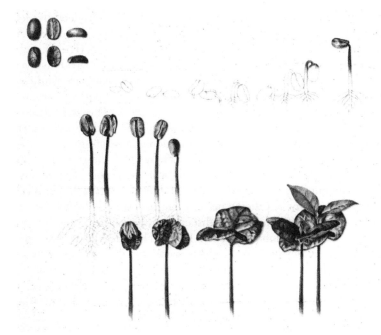

Figure 2.5 Seed germination and seedling development. (Drawing by Yvonne Boitel-Baur©, Zurich, Switzerland). This illustration shows the various stages from the bean in parch up to the emergence of the first foliage leaf pair

First row from left to right: Germination, the primary root, and later, the hypocotyl emerge from the seed. Small lateral roots are formed and the hypocotyl extends to carry the 'head' above the ground (c.3 weeks)

Middle row from right to left: The apical hook straightens and the cotyledons (not visible) completely invade in and dissolve the endosperm (7 weeks)

Lower row left: The cotyledons start unfolding, the remaining unresorbed layer of the endosperm is shed, and subsequently the cotyledons fully expand (10 weeks). The first foliage leaves appear later. In the upper left corner, coffee in parchment (upper row) and 'naked' beans are shown with one of the latter with the silverskin partially removed

formation of sugars from carbon dioxide and water, or, in other words, the net photosynthesis has reached its maximum where it remains for a long time (Frischknecht *et al.*, 1982; Mösli Waldhauser *et al.*, 1997). Under natural and favourable conditions, the coffee leaf's lifetime lasts for 10–15 months (Van der Vossen and Browning, 1978). Thereafter, senescence starts and finally the leaf is shed.

If we throw a glance on the concomitant course of the key secondary compounds, we find exceptionally high foliar concentrations of PuA and chlorogenic acids as soon as the very tender, nutrient-rich leaflets have left the mechanically protective bud, what can easily be explained by their high risk of predation by, for example, phytophagous insects (Frischknecht *et al.*, 1986). At this stage, both the enzymes (methyltransferases) involved in caffeine biosynthesis and the key enzyme of phenylpropane synthesis, phenylalanine ammonia lyase (PAL), show very high activities (Aerts and Baumann, 1994; Mösli Waldhauser *et al.*, 1997). The leaf alkaloid concentration increases almost to 0.1 M (as related to the tissue water) and thus has about 10 times the concentration of an espresso!

The velocity of synthesis of both caffeine and chlorogenic acids decreases sharply during the subsequent leaf expansion. The relative caffeine content drops as a consequence of dilution by growth. However, the absolute amount of caffeine per leaf increases steadily because of low enzyme activities still persisting throughout the entire period of leaf expansion (Mösli Waldhauser *et al.*, 1997). The fully developed coffee leaf has, on a dry weight basis, a caffeine content in the range of the bean. Amazingly, the chlorogenic acids, even though formed in parallel to caffeine, still continue to increase during the next six weeks (Kappeler, 1988). Shedding leaves have been reported to be caffeine-free, indicating that the caffeine nitrogen is re-used by the plant.

The concerted formation of both the alkaloids (mainly caffeine) and chlorogenic acids (mainly 5-CQA) has a physiological significance: caffeine easily permeates through all kinds of biological barriers, except a few installed by the caffeine-containing plants themselves (for example, the coffee bean's surface mentioned above). In order to avoid auto-toxicity, caffeine is physico-chemically complexed by 5-CQA, and compartmented in the cell vacuole (Mösli Waldhauser and Baumann, 1996). Since chlorogenic acid is encaged in the cells where it was synthesized, one has to assume that caffeine, due to its hydro- and lipophilic nature, and to its complexing ability, slowly migrates within the coffee plant towards the sites of highly accumulated chlorogenic acids. In other words, caffeine is passively dislocated and, so to speak, collected within the plant in proportion to the tissue concentration of chlorogenic acid. Apparently, the coffee plant controls caffeine distribution by the allocation of chlorogenic acids. This is nicely illustrated by the uneven distribution of caffeine within the lamina of the coffee leaf: it is highly accumulated at the margins and sharply decreases in concentration towards the mid-rib (Wenger and Baumann, unpublished). Needless to say, the chlorogenic acids show the same distribution pattern. In terms of biochemical ecology it is important to note that this phytochemical leaf architecture is significant: the leaf margin, a preferential site of insect attack, is particularly well furnished with the key defence compounds.

Whether phytochemical leaf architecture is genetically controlled and can be influenced by breeding remains to be investigated.

Thus, caffeine is an ideal defence compound: the organism feeding on caffeine-containing tissues is unable to hinder it from rapid distribution within and action on its body. However, from the plant's view the problem of autotoxicity had to be solved. The solution was, as already mentioned, complexation by phenolics: all PuA-containing plants allocate high concentrations of either chlorogenic acids (coffee, maté), catechins (cocoa, cola, guaraná), or both (tea). By the use of suspension-cultured coffee cells as a model system, it has been shown that up to 77.4% of the caffeine is fixed in the complex at 25 °C (Mösli Waldhauser and Baumann, 1996). Amazingly, in 1964 Sondheimer calculated a complexation degree of 78% for caffeine in the coffee bean at the same temperature (1964). Obviously, the concentration achieved by the remaining fraction (approximately 20–25%) is beyond the autotoxicity level. A temperature increase lowers the degree of complexation and *vice versa*. In this context, several intriguing questions regarding the interdependence between phenolics and PuA in terms of metabolic regulation, their final concentrations and organ growth rate are not yet fully answered but are well considered (e.g. Ky *et al.*, 1999) in the work on interspecific crosses by the group of Michel Noirot in Montpellier, France (www.coffee-genomics.com).

2.4.5   From flower to fruit

The time period between flower opening (anthesis) and the fully ripe fruit is species-specific and varies considerably among the coffee species. It depends further on the genotype and on climatic and cultural conditions. The economically important species C. *arabica* and C. *canephora* require 6–8 and 9–11 months for maturation, respectively (Guerreiro Filho, 1992).

As might be expected, all flower organs contain the purine alkaloid caffeine, with highest concentrations in the stamens. Amazingly, the latter accumulate, besides traces of theobromine, easily detectable amounts of theophylline, indicating an alternative biosynthetic pathway in the male part of the flower with theophylline as the direct precursor of caffeine. In leaves and seeds, caffeine biosynthesis was shown to proceed via theobromine. In analogy to citrus plants, where the highest caffeine concentration has been found in the protein-rich pollen (Kretschmar and Baumann, 1999), we may assume a preferential PuA allocation also to coffee pollen grains. However, the related analyses have not yet been done. Bees are, in contrast to many insects, not only amazingly tolerant against caffeine and other phytochemicals (Detzel and Wink, 1993), but rather, after caffeine uptake, have an improved performance such as a boost in ovipostion by the young queen, an enhanced activity of the bees outside the hive, and an improved defence by bees against hornets at the hive entrance (reviewed in Kretschmar and Baumann, 1999).

When the blossom falls from the coffee tree the persisting ovary develops into the young green coffee fruit (Figure 2.6 and Figure 2.7). Fruits always signify high investment costs and, therefore, to defend them against predators the coffee plant pursues several linked strategies.

First, the very young and green fruits are not showy but rather inapparently arranged in clusters in the leaf axil (Figure 2.6). Secondly, both chlorogenic acids and purine alkaloids are highly concentrated, and, thirdly, the development of the real endosperm is postponed until mechanical protection works. This last point is a most remarkable feature of the coffee fruit development. Within 3–4 months after anthesis the still green fruit reaches a size suggesting readiness for maturation. When cut across, two greenish beans, already typically rolled up, can be recognized. However, the appearances are deceptive: the fruit is far from being mature, the (generally) two beans are false perisperm beans made up of mother tissue (Carvalho et al., 1969). At the adaxial pole (towards the fruit stalk) of each bean one can see the beginnings of endosperm development: a whitish, very soft tissue (also called liquid endosperm) starts to invade into and resorb the perisperm bean. Recent studies show that perisperm metabolites such as sugars and organic acids are most likely acquired by the endosperm (Rogers et al., 1999b). We may assume that this process is similar to the invasion of the cotyledons into the endosperm during germination described above: the metabolites shift from one tissue to the other, whereby they have to pass through the so-called apoplast, i.e. the extracellular space between peri- and endosperm during seed development, and between endosperm and cotyledons during germination. However, the endosperm is more than a simple blot of the perisperm, since it owns high biosynthetic activities. In conclusion, and philosophically speaking, in coffee the way to the next generation is characterized by transitions in which the metabolites are shuffled around twice.

During this invasion the inner layer (endocarp) of the fruit wall (pericarp) noticeably and increasingly solidifies and later results in the parch layer described above (see 2.4.2 and 2.4.3). The mechanical defence of the endosperm itself is remarkably increased by the formation of thick cell walls containing, besides cellulose, the so-called hemi-celluloses, i.e. arabinogalactan and galactomannan (Bradbury, 2001). Hemicelluloses are highly complex polysaccharides (see 3.11.3) primarily renowned for giving an amazing degree of hardness to palm seeds (cf. date,

Figure 2.6   Coffee plant (*C. arabica*) flowering and fruiting. (Drawing by Beatrice Häsler©, Uster, Switzerland). The drawing shows a side branch with flowers at various stages: wilting and falling off, in fresh bloom, or buds (from the base to the apex). Above, flowers are illustrated in detail and enlarged. The yellow stamens are inserted in the throat of the corolla. Similarly, various fruit stages are shown: like the flowers they are arranged in composite clusters at the leaf axil

Phoenix dactylifera; vegetable ivory, *Phytelephas macrocarpa*). It remains to be mentioned that the coffee perisperm finally atrophies into the thin seed coat, the silverskin, that falls off during roasting. Very soon after anthesis the pericarp contains an absolute amount of caffeine kept unchanged until ripeness. However, the initially high (>2%) caffeine

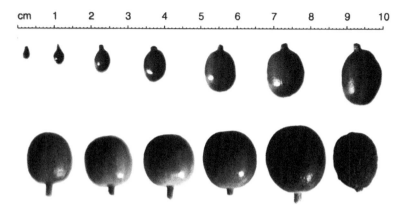

Figure 2.7   Fruit development (*C. arabica*)
Above: Within about 4 months the green fruit grows to a considerable size. Comparatively late, i.e. between stage 5 and 7, the endosperm starts to develop (2–3 months)
Below: When the fruit turns olive, the endosperm is already hard (4–5 months). Now the mesocarp gets fleshy and the exocarp partially red (5–6 months). After 6 months the exocarp is bright red and the mesocarp very fleshy. Later (7–8 months), the fruit colour turns to dark (dull) red and the mesocarp starts to dry out. Finally, the fruit starts to shrink and the exocarp gets dark and darker

concentration drops by dilution to around 0.2% during the further growth and maturation processes, culminating in the transformation of the fruit wall (pericarp) into three distinct layers which serve for fruit dispersal: the tough endocarp protects the seed from digesting enzyme activities in the gut of the frugivores such as birds or mammals; the fleshy, sugar-containing (Urbaneja *et al.*, 1996) middle layer (mesocarp) softened by enzymes (Golden *et al.*, 1993) acts as a reward, while the vivid coloration by anthocyanins (Barboza and Ramirez-Martinez, 1991) of the outermost layer (exocarp) is to attract the dispersing animal.

We should not close this section without relating our thoughts about biochemical ecology to a practical question of our daily life: how does the espresso bean get its caffeine? Though numerous publications on caffeine biosynthesis exist (for a comprehensive review see Ashihara and Crozier, 1999), this problem has never been addressed and therefore we can only speculate about it. Clearly, the endosperm has a certain biosynthetic capacity for caffeine. But is this all? Are there contributions of other sources? The perisperm provides around one-third of the seed caffeine as estimated from the caffeine content of the perisperm bean (see Figure 10.5 in Sondahl and Baumann, 2001). The leaves are not directly contributing to seed caffeine, but the pericarp may be a valuable source, as studies with labelled caffeine have shown (Keller et al., 1972; Sondahl and Baumann, 2001). Obviously, caffeine migrates from the fruit wall into the developing seed, most likely due to the high concentration of chlorogenic acids allocated to the perisperm/endosperm. Unfortunately, the extent of this caffeine transport is unknown. Conceivably, this fraction depends on both the fruit developmental time and the chlorogenic acids allocations, and is correspondingly larger in a slow-ripening species with a high ratio of seed to pericarp chlorogenic acids. Again, synthesis, transport and accumulation of chlorogenic acids eventually determine where and how much caffeine is to be allocated in the seed. In conclusion, perisperm and pericarp are certainly important sources of the seed caffeine, whereas the leaves, the pericarp and perhaps also the greenish perisperm may provide most of the chlorogenic acids crucial to gather and firmly fix the caffeine to the coffee bean. However, the degree of contribution from each side (maternal tissues versus endosperm and embryo) is not yet known. Additional studies on the developmental biology of the coffee seed (Marraccini et al., 2001a, 2001b) as well as reciprocal crosses between coffee species differing in their caffeine and chlorogenic acids content (see 2.4.4, From bud to leaf) will cast some light into the espresso's darkness!

2.5 MOLECULAR GENETICS OF COFFEE

G. Graziosi*

2.5.1 Introduction

Research on the main *arabica* cultivars has two main complications: one historical and the other biological. As reported by Berthaud and Charrier (1988), most of the cultivars were derived from a small number of plants that survived transport from Yemen to Southern Asia and to Central and South America via Europe. Consequently, the genetic base of the main

*I am indebted to Maro Sondahl, Philippe Lashermes, Francois Anthony and Alexander de Kochko for providing samples of beans and leaves on many occasions. I wish to thank Maro Sondahl for critical reading of the manuscript and Alberto Pallavicini, Barbara de Nardi, Paola Rovelli and Elisa Asquini for preparing the figures. The results reported here were supported in part by the European Community (Microsatellites: Contract ERBIC18CT970181 and ICA4-CT-2001-10070) and in part by illycaffè SpA (ESTs).

cultivars is rather restricted and the overall diversity is poor. To overcome this problem and to introduce new valuable genes, a number of interspecific crosses has been carried out; Arabusta, Icatù and Catimor are well known hybrids (Capot, 1972; Carvalho, 1988). The biological complication is brought about by the self-fertilizing capability of C. *arabica*, which enhances genetic homogeneity.

The relatively small amount of fundamental research on coffee has been somewhat expanded by recent technical developments and by a moderate increase in scientific studies. New momentum has been gained through the advent of genomics. DNA sequencing techniques are now commonly available in most laboratories, and the *Coffea arabica* genome (see Glossary at the end of this section for technical words) has been considered for large-scale analysis.

2.5.2   The genome

The genome of C. *arabica* consists of 44 small chromosomes ($2n = 4x$), twice the number of C. *canephora* chromosomes and all other *Coffea* species (Krug and Mendes, 1940; Bouharmount, 1959; Kammacher and Capot, 1972; Charrier, 1978; de Kochko *et al.*, 2001). C. *arabica* is an allotetraploid plant formed by the spontaneous fusion between the genome of C. *canephora*, as a male parent, and the genome of a plant of the C. *eugenioides* group as a female parent (Lashermes *et al.*, 1999). Indeed the total amount of DNA per nucleus in arabica (2.6 pg, about 2.4×10^9 base pairs) is twice as much as any other *Coffea* (Cros *et al.*, 1995). The interspecific cross occurred considerably less than one million years ago and therefore this species should be considered, in evolutionary terms, as recent.

If C. *arabica* is a relatively young species, then the two ancestral genomes should have maintained their individuality with little redistribution of genes among the chromosomes of different origin (similar chromosomes between these two genomes are said to be homoeologous). Evidence to this effect has been provided by Lashermes *et al.* (2000a), who studied the inheritance of single genes in this allotetraploid organism. They found a normal disomic inheritance, as in any other diploid species. Nevertheless, *arabica* must have at least four copies of each gene, two derived from the *canephora* ancestor and two derived from the *eugenioides* ancestor. It is also reasonable to assume that the pairs of orthologous loci (corresponding genes on the homoeologous chromosomes) could be somewhat different. Thus, the homozygosity generated by the self-fertilizing property of this organism is presumably compensated by the constitutive heterozygosity brought about by the two ancestral genomes.

As a final note, it is important to mention that a genetic map of C. *arabica* does not exist. Nevertheless, low-density maps of C. *canephora* and of interspecific crosses are available (Paillard *et al.*, 1996; Ky *et al.*, 2000; Lashermes *et al.*, 2001; Herrera *et al.*, 2002). Apart from the relatively low level of research mentioned above, the lack of an *arabica* map is due to a number of reasons: the complexity of a tetraploid genome of recent origin, the high level of homozygosity and last, but not least, the low number of available polymorphisms.

2.5.3   DNA polymorphisms and molecular diversity

DNA polymorphisms are an important tool for studying genomes and they have allowed for the identification and mapping of expressed genes in many organisms. Whilst it is difficult to identify genes responsible for a given trait, polymorphisms associated with an interesting gene can be easily followed through various generations and crosses to assist breeders in their introgressive and selective protocols. In addition, in the past few years the C. *arabica* genome has undergone screening projects aimed to identify polymorphisms. Some of these studies and the main techniques are reported here.

Cros *et al.* (1993) and Lashermes *et al.* (1996b) reported the identification of many restriction fragment length polymorphisms (RFLPs) in various *Coffea* species. This approach detects sequence variations, frequently single base mutations, and involves the digestion of DNA by a specific restriction enzyme. Unfortunately, this classical technique requires Southern blots, which are rather complex and costly. Moreover, these polymorphisms are frequently not very informative because they do not have more than two alleles. A new generation of RFLP is required, possibly based on the enzymatic amplification (PCR: polymerase chain reaction) of expressed genes.

Several publications have reported a random amplified polymorphic DNA (RAPD) approach (Cros *et al.*, 1993; Lashermes *et al.*, 1993, 1996c; Orozco-Castillo *et al.*, 1994; Zezlina *et al.*, 1999; Ruas *et al.*, 2000). This technique is very useful since it allows for the identification of DNA polymorphisms in almost any organism even when there is no other information available on the DNA under study. It is based on the enzymatic amplification of short DNA stretches (100–1000 base pairs) lying between two inverted repeats. Agwanda *et al.* (1997) reported the first case of association between RAPD polymorphisms and the T gene for resistance to coffee berry disease (CBD) in *Coffea*.

The AFLP technique (amplified fragment length polymorphism) has proved to be very informative: many electrophoretic bands (30–90 bands) can be obtained in a single PCR and the probability of finding a polymorphic band is relatively high. This technique is based on the digestion of the DNA by two restriction enzymes, adapters are added at both ends of the DNA fragments and some of the DNA fragments are PCR amplified. AFLP has been successfully used for characterizing varieties (Anthony et al., 2002) as well as in evaluating the introgression of genes in hybrids (Lashermes et al., 2000b). Nevertheless, AFLP is relatively complex and it does not allow for the identification of heterozygotes.

Undoubtedly, microsatellites are the most useful source of polymorphisms and they are widely used in plant, animal and human research. Microsatellites are DNA sequences repeated in tandem and the number of repeats can vary in different organism of the same species. Usually, they are very informative because one locus can show many alleles (high heterozygosity) and because they display codominance, thus allowing for the identification of heterozygotes. Moreover, analyses are easily carried out by PCR and a large number of samples can be analysed at low cost and in a short time. However, the development of assays for these polymorphisms is sometimes difficult and expensive. The first description of microsatellites in C. arabica was reported by Mettulio et al. (1999), Rovelli et al. (2000) and Combes et al. (2000). Since then, microsatellites have been used in a variety of studies, as reported by Lashermes et al. (2000b), Prakash et al. (2001) and Anthony et al. (2002). As expected, microsatellites can show different alleles as well as individual differences in C. canephora (Taylor et al., 2002) but they can show some complications in C. arabica. In the cross reported in Figure 2.8, the microsatellite GTG52 shows normal disomic inheritance, but it is not unusual to find individual plants displaying more than two alleles (Figure 2.9). Most probably, the primers used in these experiments recognized more than one locus, as would be expected for a polyploid organism.

There are other classes of polymorphic sequences, for instance the Internally Transcribed Sequence (ITS) of the ribosomal region, but their use is limited to specific purposes since they can display sequence variations within the same plant (Zezlina et al., 1999).

The general picture that can be drawn from polymorphic DNA studies is that C. arabica is a peculiar crop with a very low level of heterozygosity. This is not surprising because of the restricted genetic base and the high incidence of autofertility. There is little doubt that this low heterozygosity will hamper both the development of a genetic map and the use of molecular genetics in breeding programmes until the development of a reasonable number of informative polymorphisms.

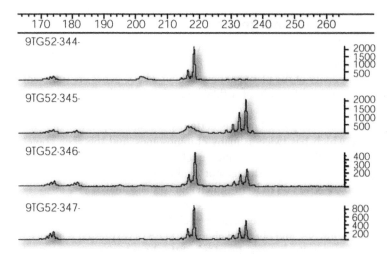

Figure 2.8 Genescan analysis of the cross Sarcimor x ET6 with the 9TG52 microsatellite
   Line 1: The Sarcimor parent was homozygote for the 219 bp allele
   Line 2: The ET6 parent was homozygote for the 235 bp allele
   Lines 3 and 4: The two F1 samples presented both alleles (heterozygote) as the other 15 plants of the same progeny. (Cross performed by F. Anthony at CATIE)

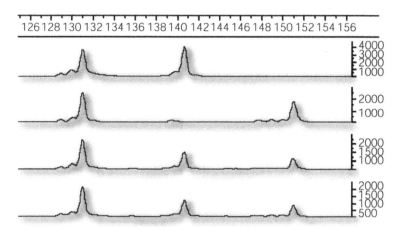

Figure 2.9 Samples as in Figure 2.8 analysed with the 7TG46 microsatellite. The two F1 samples showed three alleles: 131 bp, 140 bp and 151 bp. Most probably the primers amplified two loci, one of which was homozygote for the 131 bp allele in both parents

2.5.4 Expressed genes

Completion of the model plant *Arabidopsis thaliana* genome sequence (The *Arabidopsis* Genome Initiative, 2000) has given new insights into the complexity of the plant genome. About 25 000 genes are sufficient for supporting the life cycle of a flowering plant and there is no reason to believe that the coffee plant requires more genes. This estimate could be valid for C. *canephora* but it is certainly an underrepresentation for C. *arabica*, which has twice as many chromosomes, and the gene copies on the homoeologous chromosomes might have significantly diverged in sequence and function throughout evolution. None the less, the size of the arabica genome renders it amenable to mass analysis.

The first catalogue of expressed genes was reported by Pallavicini *et al.* (2001). They partially sequenced a cDNA library prepared from root meristem mRNA and obtained about 1200 ESTs (expressed sequence tag) which eventually were reduced to 901 clusters. Sequence homology analysis identified a provisional function for about 70% of the clusters (Figure 2.10). The complete list of clusters can be obtained from the home page of the Department of Biology, University of Trieste.

Recently, the public DNA sequence databases reported new C. *arabica* ESTs provided by M.A. Cristancho and S.R. McCouch (Genebank Accession Numbers: BQ4488720-BQ449166) but a reference publication is unavailable.

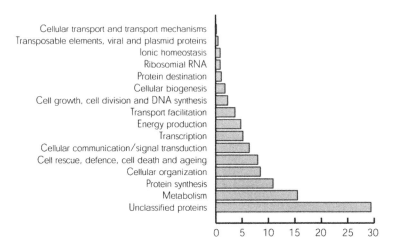

Figure 2.10 Functional classification of 1200 ESTs

The public databases report a total of 712 sequences of C. *arabica* and nine sequences of C. *canephora*, mainly microsatellites or ESTs. The remaining sequences can be classified as reported in Table 2.9.

Among the enzyme sequences, it is worth mentioning the complete sequence of N-methyltransferase, one of the enzymes involved in caffeine metabolism, which was provided by Kretschmar and Baumann (Genebank Accession Numbers: AF94411-AF94420). Also important are the studies on sugar metabolism (Marraccini *et al.*, 2001b), which, together with the aromatic compounds, are relevant for coffee quality.

The most studied gene encoded the 11S storage protein, the most abundant protein in the coffee bean (Acuña *et al.*, 1999; Marraccini *et al.*, 1999; Rogers *et al.*, 1999a). The authors identified the complete coding sequence as well as the promoter region by expressing the protein in transformed tobacco plants.

Notably, a number of nucleotide binding site–leucine rich repeat (NBS–LRR) sequences, which are believed to confer resistance to a wide spectrum of pests, have been identified (Noir *et al.*, 2001). This is the first step in the possible use of these genes in breeding programmes or transformations.

2.5.5 Introducing new genes

Classical breeding has been reviewed by a number of authors, the article by Van der Vossen (2001) being the most recent. Unfortunately, the coffee plant is perennial and breeding programs take 20–30 years. Such a long breeding time coupled with the peculiar genetic structure of C. *arabica* (tetraploidy, low heterozygosity and self-fertility) strongly suggests the use of an alternative route as genetic transformation.

Table 2.9 Categories of sequences of C. *arabica* and C. *canephora* other than microsatellite or EST categories

Category	No.
Enzymes	19
Structural proteins	5
Ribosomal DNA	4
Chloroplast	6

Successful transformation experiments have been reported by Leroy *et al.* (1997a, 1999, 2000) and Spiral *et al.* (1999). They introduced the *cry*IAc gene of *Bacillus thuringiensis* into *arabica* and *canephora* plants through a disarmed strain of *Agrobacterium tumefaciens*. The analysis of 60 transformed plants showed a correlation between expression levels of the *cry*IAc gene and the newly acquired resistance to leaf miner *Perileucoptera* spp. The same technique has been used by Ribas *et al.* (2001), who introduced two sequences for the resistance to the herbicide gluphosinate.

An elegant transformation experiment has been reported by Moisyadi *et al.* (1998, 1999). They transformed coffee plants with an antisense sequence of an enzyme apparently involved in the biosynthetic pathway of caffeine and they obtained plants producing little or no caffeine.

Even if the introduction of genes by transformation must be regarded as an important tool for improving coffee quality and production, possible drawbacks must be taken into account. There is little doubt that the introduction of a gene conferring resistance to one of the many coffee plant pests would enhance production and be a considerable help to farmers. Nevertheless, coffee is a perennial plant and a field of plants resistant to a specific insect, for instance, eventually would select for an insect insensitive to the toxin. This problem could be partially overcome by the introduction of more than one toxin for the same pest. In addition, an important consideration is the attitude of consumers who might be unwilling to consume a beverage derived from transformed plants.

2.5.6   Prospects

There is little doubt that in the near future more ESTs will be produced and more polymorphism assays will be developed. In fact, some public Brazilian funding agencies have launched a research programme for the production of 200 000 ESTs. Presumably, a number of 'important' genes will be identified, as those involved in flowering and ripening control, or those responsible for resistance to diseases and pests. These molecular tools should allow for the amelioration of the coffee plant through classical breeding. A good example is the introgression of resistance to the nematode *Meloidogyne exigua* reported by Bertrand *et al.* (2001).

Far more complex is the genetic approach to improving the quality of the coffee beverage. Taste and aroma are presumably under the control of a large number of genes and affected by environmental conditions. Nevertheless there is no possibility of having a good quality coffee without a good genetic base. To approach this problem we have to wait for the development of new tools. Genome-wide analysis and microarray analysis will allow the study of gene expression and gene–environment interactions and could become a diagnostic approach to assess quality.

2.5.7 Glossary

AFLP: Amplified fragment length polymorphism. This technique is based on the digestion of DNA by two restriction enzymes; adapters are added at both ends of the DNA fragments and some of the DNA fragments are polymerase chain reaction amplified (see after).

Alleles: Different variants of the same gene.

Allotetraploid: Organism with four sets of chromosomes, two sets inherited from one ancestral progenitor of a given species and the remaining two sets inherited from a different species.

Codominance: Two gene variants (alleles) are codominant when they are simultaneously expressed and visible.

Disomic inheritance: Normal way of inheriting chromosomes or genes.

EST: Expressed sequence tag. Partial sequence of an expressed gene.

Genome: The totality of all the chromosomes and genes of an organism.

Homoeologous chromosomes: Chromosomes carrying similar genes but originated from two ancestral plants of different species.

Homozygosity: Uniformity of alleles.

ITS: Internally transcribed spacer. Specific portion of DNA located between the genes coding for the ribosomal RNA.

Microarray: Hybridization technique to visualize the genome expression.

Microsatellite: DNA sequences repeated in tandem.

NBS–LRR: Nucleotide binding site−leucine rich repeat. Type of gene conferring resistance to diseases and pests.

PCR: Polymerase chain reaction. *In vitro* reaction allowing for the amplification of specific genes or DNA fragments.

Polyploid organism: Organism that contains more than two sets of chromosomes.

Primers: Short fragments of DNA allowing for the initiation of the PCR (see above).

RAPD: Random amplified polymorphic DNA. Enzymatic amplification of short DNA stretches lying between two inverted repeats.

RFLP: Restriction fragment length polymorphism. Variation of the recognition site of a restriction enzyme (see below).

Restriction enzyme: Enzyme able to cut DNA at or near specific sequences.

Southern blots: Chromatographic technique for isolating and identifying specific DNA fragments.

REFERENCES

Acuña R., Bassüner R., Beilinson V., Cortina H., Cadena G., Montes V. and Nielsen N.C. (1999) Coffee seeds contains 11S storage proteins. *Physiol. Plant.* 105, 122–131.

Aerts R.J. and Baumann T.W. (1994) Distribution and utilization of chlorogenic acid in *Coffea* seedlings. *J. Experiment. Botany* 45, 497–503.

Agwanda C.O., Lashermes P., Trouslot P., Combes M.C. and Charrier A. (1997) Identification of RAPD markers for resistance to coffee berry disease, *Colletotricum kahawae*, in arabica coffee. *Euphytica* 97, 241–248.

Alborn T., Turlings T.C.J., Jones T.H., Stenhagen G. and Loughrin J.H. (1997) An elicitor of plant volatiles from beet army worm oral secretion. *Science* 276, 945–949.

Alvim P. de T. (1973) Factors affecting flowering of coffee. *J. Plant. Crops* 1, 37–43.

Amorim H.V., Teixeira A.A., Moraes R.S., Reis A.J., Pimentel-Gomes F. and Malavolta E. (1973) Studies on the mineral nutrition of coffee. XXVII. Effect of N, P and K fertilization on the macro- and micronutrients of coffee fruits and on beverage quality. *Ann. Esc. Sup. Agric. LQ (Piracicaba)*, 20, 323–333 (in Portuguese).

Anthony F., Combes M.C., Astorga C., Bertrand B., Graziosi G. and Lashermes P. (2002) The origin of cultivated *Coffea arabica* L. varieties revealed by AFLP and SSR markers. *Theor. Appl. Genet.* 104, 894–900.

Anzueto F., Bertrand B., Sarah J.L., Eskes A.B. and Decazy B. (2001) Resistance to *Meloidogyne incognita* in Ethiopian *Coffea arabica*: detection and study of resistance transmission. *Euphytica* 118, 1–8.

Ashihara H. and Crozier A. (1999) Biosynthesis and metabolism of caffeine and related purine alkaloids in plants. *Adv Botanical Res.* 30, 117–205.

Bade-Wegner H., Bending I., Holscher W. and Wollmann R. (1997) Volatile compounds associated with the over-fermented flavour defects. *Proc. 17th ASIC Coll.*, pp. 176–182.

Bade-Wegner H., Holscher W. and Vitzhum O.G. (1993) Quantification of 2-methylisoborneol in roasted coffee by GC-MS. *Proc. 15th ASIC Coll.*, pp. 537–544.

Balzer H.H (2001) Acids in coffee. In R.J. Clarke and O.G.Vitzthum (eds), *Coffee: Recent Developments*. Oxford: Blackwell Science, pp. 18–32.

Barboza C.A. and Ramirez-Martinez J.R. (1991) Antocianinas en pulpa de café del cultivar Bourbon rojo. *Proc. 14th ASIC Coll.*, pp. 272–276.

Bardner R. (1985) Pest control. In M.N. Clifford and K.C. Willson (eds), *Coffee: Botany, Biochemistry and Production of Beans and Beverage*. London: Croom Helm, pp. 208–218.

Baumann T.W. and Gabriel H. (1984) Metabolism and excretion of caffeine during germination of *Coffea arabica* L. *Plant Cell Physiol.* 25, 1431–1436.

Baumann T.W. and Seitz R. (1992) Coffea. In R. Hänsel, K. Keller, H. Rimpler and G. Schneider (eds), *Hagers Handbuch der Pharmazeutischen Praxis*, Volume 4. Berlin: Springer, pp. 926–940.

Baumann T.W., Sondahl M.R., Mösli Waldhauser S.S. and Kretschmar J.A. (1998) Non-destructive analysis of natural variability in bean caffeine content of Laurina coffee. *Phytochem.* 49, 1569–1573.

Bellachew B. (1997) Arabica coffee breeding in Ethiopia: a review. *Proc. 17th ASIC Coll.*, pp. 406–414.

Beretta M.J.G., Harakawa R., Chagas C.M., Derrick K.S., Barthe G.A., Ceccardi T.L., Lee R.F., Paradela-F. O., Sugimori M. and Ribeiro I.A. (1996) First report of *Xylella fastidiosa* in coffee. *Plant Dis. (St Paul)* 80, 821–826.

Bergamin J. (1963) Pests and diseases in coffee. In C.A. Krug *et al.* (eds), *Cultivation and Fertilization of Coffee*. Saõ Paulo: Inst. Bras. Potassa, pp. 127–141 (in Portuguese).

Berthaud J. (1980) L'incompatibilité chez *Coffea canephora*: méthode de test et détérminisme génétique. *Café Cacao Thé* 24, 267–274.

Berthaud J. and Charrier A. (1988) Genetic resources of *Coffea*. In R.J. Clarke and R. Macrae (eds), *Coffee: Volume 4 – Agronomy*. London: Elsevier Applied Science, pp. 1–42.

Bertrand B., Aguilar G., Santacreo R., Anthony F., Etienne H., Eskes A.B. and Charrier A. (1997) Comportement d'hybrides F1 de *Coffea arabica* pour la vigueur, la production et la fertilité en Amérique Centrale. *Proc. 17th ASIC Coll.*, pp. 415–423.

Bertrand B., Anthony F. and Lashermes P. (2001) Breeding for resistance to *Meloidogyne exigua* of *Coffea arabica* by introgression of resistance genes of *Coffea canephora*. *Plant Pathol.* 50, 637–643.

Bertrand B., Guyot B., Anthony F., Selva J.C., Alpizar J.M., Etienne H. and Lashermes P. (2004) Selection can avoid accompanying the introgression of *Coffea canephora* gene of resistance with a drop in beverage quality, *20th ASIC Coll.* In press.

Bertrand B., Peña Duran M.X., Anzueto F., Cilas C., Etienne H., Anthony F. and Eskes A.B. (2000) Genetic study of *Coffea canephora* coffee tree resistance to *Meloidogyne incognita* nematodes in Guatemala and *Meloidogyne* sp. nematodes in El Salvador for selection of rootstock varieties in Central America. *Euphytica* 113, 79–86.

Bettencourt A.J. and Rodrigues C.J. (1988) Principles and practice of coffee breeding for resistance to rust and other diseases. In R.J. Clarke and R. Macrae (eds), *Coffee: Volume 4 – Agronomy*. London: Elsevier Applied Science, pp. 199–234.

Bhat S.S., Daivasikamani S. and Naidu R. (1995) Tips on effective management of black rot disease in coffee. *Indian Coffee* 59 (8), 3–5.

Bouharmount, J. (1959) Recherche sur les affinités chromosomiques dans le genre *Coffea*. *INEAC Séries Sci.* 77, Brussels.

Bradbury A.G.W. (2001) Chemistry I: Non-volatile compounds, 1A – carbohydrates. In R.J. Clarke and O.G. Vitzthum (eds), *Coffee – Recent Developments*. Oxford: Blackwell Science, pp. 1–17.

Bridson D.M. (1994) Additional notes on *Coffea* (Rubiaceae) from Tropical East Africa. *Kew Bulletin* 49 (2), 331–342.

Bridson D.M. and Verdcourt B. (1988) *Rubiaceae* (Part 2). In R.M. Polhill (ed.), *Flora of Tropical East Africa*. Rotterdam: Balkema, pp. 703–723.

Buchanan R.L., Tice G. and Marino D. (1981) Caffeine inhibition of ochratoxin A production. *J. Food Sci.* 47, 319–321.

Cannell M.G.R. (1985) Physiology of the coffee crop. In M.N. Clifford and K.C. Willson (eds), *Coffee: Botany, Biochemistry and Production of Beans and Beverage*. London: Croom Helm, pp. 108–124.

Capot J. (1972) L'amélioration du caféier robusta en Côte d'Ivoire: Les hybrides Arabusta. *Café Cacao Thé* 16, 3–18 and 114–126.

Carvalho A. (1946) Geographic distribution and botanical classification of genus *Coffea* with special reference to arabica species. *Bull. Superint. Serv. Café* 23, 3–33 (in Portuguese).

Carvalho A. (1988) Principles and practice of coffee plant breeding for productivity and quality factors: *Coffea arabica*. In R.J. Clarke and R. Macrae (eds), *Coffee: Volume 4 – Agronomy*. London: Elsevier Applied Science, pp. 129–165.

Carvalho A.C., Ferwerda F.P. *et al.* (1969). Coffee. In F.P. Ferwerda and F. Wit (eds), 'Outlines of perennial crop breeding in the tropics', Miscellaneous papers No. 4, Agricultural University, Wageningen, The Netherlands, pp. 189–241.

Charrier A. (1978) La structure génétiquedes caféiers spontanés de la région malgache (*Mascarocoffea*). Mémoires ORSTOM 87, Orstom Ed., Paris.

Charrier A. and Eskes A.B. (1997) Les Caféiers. In A. Charrier, M. Jacquot, S. Hamon and D. Nicolas (eds), 'L'Amélioration des plantes tropicales', pp. 171–196. Repères CIRAD and ORSTOM Montpellier (English edition, 2002).

Chevalier A. (1947) Les caféiers du globe. III. Systématique des caféiers et faux caféiers, maladies et insectes nuisibles. In *Encyclopédie biologique no. 28*. Paris: Paul Le Chevalier.

Clarke R.J. and Vitzthum O.G. (2001) *Coffee – Recent Developments*. Oxford: Blackwell Science.

Combes M.C., Andrzejewski S., Anthony F., Bertrand B., Rovelli P., Graziosi G. and Lashermes P. (2000) Characterisation of micro-satellite loci in *Coffea arabica* and related coffee species. *Mol. Ecol.* 9, 1178–1180.

Couturon E. (1980) Le maintien de la viabilité des graines de caféiers par le contrôle de leur teneur en eau et de la température de stockage. *Café Cacao Thé* 24, 27–31.

Couturon E., Lashermes P. and Charrier A. (1998) First intergeneric hybrids (*Psilanthus ebracteolatus* Hiern x *Coffea arabica* L.) in coffee trees. *Canadian J. Botany* 76, 542–546.

Cros J., Combes M.C., Chabrillange N., Duperray C., Monnot des Angles A. and Hamon S. (1995) Nuclear DNA content in the subgenus *Coffea* (Rubiacee): inter- and intra-specific variation in African species. *Canadian J. Botany* 73, 14–20.

Cros J., Lashermes P., Marmey F., Anthony F., Hamon S. and Charrier A. (1993) Molecular analysis of genetic diversity and phylogenetic relationship in *Coffea*. *Proc. 14th ASIC Coll.*, pp. 41–46.

Cros J., Trouslot P., Anthony F., Hamon S. and Charrier A. (1998) Phylogenetic analysis of chloroplast DNA variation in *Coffea* L. *Mol. Phylogenet. Evol.* 9, 109–117.

de Kochko A., Louarn J., Hamon P., Hamon S. and Noirot M. (2001) *Coffea* genome structure and relationship with evolution. *Proc. 19th ASIC Coll.*, CD-ROM.

Dedecca D.M. (1957) Anatomy and antogenetic development of *Coffea arabica* L. var. Typica Cramer. *Bragantia* 16, 315–366.

Dentan E. (1985) The microscopic structure of the coffee bean. In M.N. Clifford and K.C. Willson (eds), *Coffee: Botany, Biochemistry, and Production of Beans and Beverage.* London: Croom Helm, pp. 284–304.

Detzel A. and Wink M. (1993) Attraction, deterrence or intoxication of bees. *Chemoecology* 4, 8–18.

Dicke M. and Van Loon J.J.A. (2000) Multitrophic effects of herbivore-induced plant volatiles in an evolutionary context. *Entomol. Experimentalis Applicata* 97, 237–249.

Edwards P.J. (1992) Resistance and defence: the role of secondary plant substances. In P.G. Ayres (ed.), *Pest and Pathogens. Plant Responses to Foliar Attack.* Environmental Plant Biology Series, Abingdon: Bios Scientific Publishers, pp. 69–84.

Eskes A.B. (1989) Resistance. In A.C. Kushalappa and A.B. Eskes (eds), *Coffee Rust: Epidemiology, Resistance and Management.* Boca Raton, FL: CRC Press, pp. 171–291.

Etienne H., Anthony F., Dussert S., Fernandez D., Lashermes P. and Bertrand B. (2002) Biotechnological applications for the improvement of Coffee (*Coffea arabica* L.). *Vitro Cell. Dev. Biol. – Plant*, 38, 129–138.

Ferreira M.E. and Cruz M.C.P. (1988) Symposium on Micronutrients in Agriculture: I. Jaboticabal. Saõ Paulo, Brazil (in Portuguese).

Flood J. and Brayford D. (1997) Re-emergence of Fusarium wilt of coffee in Africa. *Proc. 17th ASIC Coll.*, pp. 621–628.

Fonseca H. and Gutierrez L.E. (1971) Study on content and composition of oil from some coffee varieties. *Ann. Esc. Sup. LQ, Piracicaba* 18, 313–322 (in Portuguese).

Fraenkel G.S. (1959) The *raison d'être* of secondary plant substances. *Science* 129, 1466–1470.

Frischknecht P.M., Eller B.M. and Baumann T.W. (1982) Purine alkaloid formation and CO_2 gas exchange in dependence of development and of environmental factors in leaves of *Coffea arabica*. *Planta* 156, 295–301.

Frischknecht P.M., Ulmer-Dufek J. and Baumann T.W. (1986) Purine alkaloid formation in buds and developing leaflets of *Coffea arabica*: expression of an optimal defence strategy? *Phytochem.* 25, 613–616.

Full G., Lonzarich V. and Suggi Liverani F. (1999) Differences in chemical composition of electronically sorted green coffee beans. *Proc. 17th ASIC Coll.*, pp. 35–42.

Gimenez-Martin G., Lopez-Saez J.F., Moreno P. and Gonzalez-Fernandez A. (1968) On the triggering of mitosis and the division cycle of polynucleate cells. *Chromosoma* 25, 282–296.

Golden K.D., John M.A. and Kean E.A. (1993) ß-Galactosidase from *Coffea arabica* and its role in fruit ripening. *Phytochemistry* 34, 355–360.

Grace S.C. and Logan B.A. (2000) Energy dissipation and radical scavenging by the plant phenylpropanoid pathway. *Phil. Trans. R. Soc. Lond. B* 355, 1499–1510.

Guerreiro Filho O. (1992) *Coffea racemosa* Lour – une revue. *Café Cacao Thé* 36, 171–186.

Guerreiro Filho O., Denoit P., Pefercen M., Decazy B., Eskes A.B. and Frutos R. (1998) Susceptibility to the coffee leaf miner (*Perileucoptera* spp.) to *Bacillus thuringiensis* d-endotoxins: a model for transgenic perennial crops resistant to endocarpic insects. *Curr. Microbiol.* 36, 175–179.

Harborne J.B. (2001) Twenty-five years of chemical ecology. *Natl Prod. Rep.* 18, 361–379.

Hartmann T. (1996) Diversity and variability of plant secondary metabolism: a mechanistic view. *Entomol. Experimentalis Applicata* 80, 177–188.

Herrera J.C., Combes M.C., Anthony F., Charrier A. and Lashermes P. (2002) Introgression into the allotetraploid coffee (*Coffea arabica* L.): segregation and recombination of the *C. canephora* genome in the tetraploid interspecific hybrid (*C. arabica* x *C. canephora*). *Theor. Appl. Genet.* 104, 661–668.

Hollingsworth R.G., Armstrong J.W. and Campbell E. (2002) Caffeine as a repellent for slugs and snails. *Nature* 417, 915–916.

Holscher W. (1996) Comparison of some aroma impact compounds in roasted coffee and coffee surrogates. In A.J. Taylor and D.S. Mottram (eds), *Flavour Science, Recent Developments*. Cambridge: The Royal Society of Chemistry, pp. 239–244.

Horman I. and Viani R. (1972) The nature and conformation of the caffeine-chlorogenate complex of coffee. *J. Food Sci.* 37, 925–927.

ICO (2002) *Annual Statistics on Coffee Production and International Trade*. London: International Coffee Organization.

Illy E. (1997) How science can help to improve coffee quality. *Proc. 17th ASIC Coll.*, pp. 29–33.

Illy A. and Viani R. (eds) (1995) The plant composition. In A. Illy and R. Viani (eds), *Espresso Coffee. The Chemistry of Quality*. London: Academic Press, pp. 23–38.

ITC (2002) *Coffee, an Exporter's Guide*. International Trade Centre. Product and marketing development. UNCTAD CNUCED, WTO OMC. Geneva.

Kammacher P. and Capot J. (1972) Sur les relations caryologiques entre *Coffea arabica* et *C. canephora*. *Cafè Cacao The* 16, 289–294.

Kappeler A.W. (1988) Der Purinalkaloid-Chlorogensäure-Komplex. Seine physikalisch-chemische Natur und seine Bedeutung in der Gewebekultur und in den Blättern von *Coffea arabica* L. Inaugural Dissertation. Institut für Pflanzenbiologie, Universität Zürich.

Keller H., Wanner H. and Baumann T.W. (1972) Kaffeinsynthese in Früchten und Gewebekulturen von *Coffea arabica*. *Planta* 108, 339–350.

Kende H. (1960) Untersuchungen über die Biosynthese und Bedeutung von Trigonellin in *Coffea arabica*. *Berichte Schweiz. Botanisch. Ges.* 70, 232–267.

Kessler A. and Baldwin T. (2001) Defensive function of herbivore-induced plant volatile emissions in nature. *Science* 291, 2141–2144.

Kihlman B.A. (1977) *Caffeine and Chromosomes*. Amsterdam: Elsevier.

Kretschmar J. and Baumann T. (1999) Caffeine in citrus flowers. *Phytochem.* 52, 19–23.

Krug C.A. and Carvalho A. (1951) The genetics of *Coffea*. *Adv. Genet.* 4, 127–158.

Krug C.A. and Mendes A.J.T. (1940) Cytological observation in *Coffea*. *J. Genet.* 39, 189–203.

Krug C.A. and de Poerck R.A. (1968) World Coffee Survey. FAO Agricultural Studies no. 76, Rome.

Ky C.L., Barre P., Lorieux M., Trouslot P., Akaffou S., Louarn J., Charrier A., Hamon S. and Noirot M. (2000) Interspecific genetic linkage map, segregation distortion and genetic conversion in coffee (*Coffea* sp.). *Theor. Appl. Genet.* 101, 669–676.

Ky C.L., Louarn J., Guyot B., Charrier A., Hamon S. and Noirot M. (1999) Relations between and inheritance of chlorogenic acid contents in an interspecific cross between *Coffea pseudozanguebariae* and *Coffea liberica* var. 'dewevrei'. *Theor. Appl. Genet.* 98, 628–637.

Lashermes P., Cros J., Marmey P. and Charrier A. (1993) Use of random amplified DNA markers to analyse genetic variability and relationship of *Coffea* species. *Genet. Resour. Crop Evol.* 40, 91–99.

Lashermes P., Couturon E., Moreau N., Pailard M. and Louarn J. (1996a) Inheritance and genetic mapping of self-incompatibility in *Coffea canephora*. *Theoret. Appl. Genet.* 93, 458–462.

Lashermes P., Cros J., Combes M.C., Trouslot P., Anthony F., Hamon S. and Charrier A. (1996b) Inheritance and restriction fragment length polymorphism of chloroplast DNA in the genus *Coffea* L. *Theor. Appl. Genet.* 93, 626–632.

Lashermes P., Trouslot P., Combes M.C. and Charrier A. (1996c) Genetic diversity for RAPD markers between cultivated and wild accessions of *Coffea arabica*. *Euphitica* 87, 59–64.

Lashermes P., Combes M.C., Trouslot P. and Charrier A. (1997) Phylogenetic relationships of coffee tree species (*Coffea* L.) as inferred from ITS sequences of nuclear ribosomal DNA. *Theoret. Appl. Genet.* 94, 947–955.

Lashermes P., Combes M.C., Robert J., Trouslot P, D'Hont A., Anthony F. and Charrier, A. (1999) Molecular characterization and origin of the *Coffea arabica* L. genome. *Mol. Gen. Genet.* 261, 259–266.

Lashermes P., Paczek V., Trouslot P., Combes M.C., Couturon F. and Charrier A. (2000a) Single-locus inheritance in the allotetraploid *Coffea arabica* L. and interspecific hybrid C. *arabica* x C. *canephora*. J. *Heredity* 91, 81–85.

Lashermes P., Andrzejewki S., Bertrand B., Combes M.C., Dussert S., Graziosi G., Trouslot P. and Anthony F. (2000b) Molecular analysis of introgressive breeding in coffee (*Coffea arabica* L.). *Theor. Appl. Genet.* 100, 139–146.

Lashermes P., Combes M.C., Prakash N.S., Trouslot P., Lorieux M. and Charrier A. (2001) Genetic linkage map of *Coffea canephora*: effect of segregation distortion and analysis of recombination rate in male and female meioses. *Genome* 44, 589–596.

Leroy J.F. (1980) Evolution et taxogénese chez les caféiers. Hypothese sur leur origine. *CR Acad. Sci. Paris* 291, 593–596.

Leroy T., Henry A.M., Royer M., Altosaar I., Frutos R., Duris D. and Philippe R. (2000) Genetically modified coffee plants expressing the *Bacillus thuringiensis* cry1Ac gene for resistance to leaf miner. *Plant Cell Reports* 19, 382–389.

Leroy T., Montagnon C., Cilas C., Yapo A., Charrier A. and Eskes A.B. (1997b) Reciprocal recurrent selection applied to *Coffea canephora* Pierre. III. Genetic gains and results of first cycle intergroup crosses. *Euphytica* 95, 347–354.

Leroy T., Philippe R., Royer M., Frutos R., Duris D., Dufour M., Jourdan I., Lacombe C. and Fenouillet C. (1999) Genetically modified coffee trees for resistance to coffee leaf miner. Analysis of gene expression, resistance to insects and agronomic value. *Proc. 18th ASIC Coll.*, pp. 332–338.

Leroy T., Royer M., Paillard M., Berthouly M., Spiral J., Tessereau S., Legavre T. and Altosaar I. (1997a) Introduction de gènes d'intérêt agronomique dans l'espèce *Coffea canephora* Pierre par transformation avec *Agrobacterium* sp. *Proc. 17th ASIC Coll.*, pp. 439–446.

Marraccini P., Allard C., André M-L., Courjault C., Gaborit C., LaCoste N., Meunier A., Michaux S., Petit V., Priyono P., Rogers J.W. and Deshayes A. (2001a) Update on coffee biochemical compounds, protein and gene expression during bean maturation and in other tissues. *Proc. 19th ASIC Coll.*, CD-ROM.

Marraccini P., Deshayes A., Pétiard V. and Rogers W.J. (1999) Molecular cloning of the complete 11S seed storage protein gene of *Coffea arabica* and promoter analysis in transgenic tobacco plants. *Plant Physiol. Biochem.* 37, 273–282.

Marraccini P., Rogers W.J., Allard C., Andre M.L., Caillet V., Lacoste N., Lausanne F., Michaux S. (2001b) Molecular and biochemical characterization of endo-beta-mannanases from germinating coffee (*Coffea arabica*) grains. *Planta* 213, 296–308.

Martin R., Lilley T.H., Falshaw P., Haslam E., Begley M.J. and Magnolato D. (1987) The caffeine-potassium chlorogenate molecular complex. *Phytochemistry* 26, 273–279.

Mayer F., Czerny M. and Grosch W. (1999) Influence of provenance and roast degree on the composition of potent odorantes in arabica coffees. *Eur. Food Res. Technol.* 209, 242–250.

McCready S.J., Osman F. and Yasui A. (2000) Repair of UV damage in the fission yeast Schizosaccharomyces pombe. *Mutat. Res.* 451, 197–210.

Mettulio R., Rovelli P., Antony F., Anzueto F., Lashermes P. and Graziosi G. (1999) Polymorphic microsatellites in *Coffea arabica*. *Proc. 18th ASIC Coll.*, pp. 344–347.

Minorsky P.V. (2002) The hot and the classic. Trigonelline: A diverse regulator in plants. *Plant Physiol.* 128, 7–8.

Mitchell H.W. (1988) Cultivation and harvesting of the arabica coffee tree. In R.J. Clarke and R. Macrae (eds), *Coffee: Volume 4 – Agronomy*. London: Elsevier Applied Science, pp. 43–90.

Moisyadi S., Neupane K.R. and Stiles J.I. (1998) Cloning and characterisation of a cDNA encoding xantosine-N^7-methyltransferase from coffee (*Coffea arabica*). *Acta Hort.* 461, 367–377.

Moisyadi S., Neupane K.R. and Stiles J.I. (1999) Cloning and characterisation of xantosine-N^7-methyltransferase, the first enzyme of the caffeine biosynthetic pathway. *Proc. 18th ASIC Coll.*, pp. 327–331.

Montagnon C., Leroy T. and Eskes A.B. (1998a) Amélioration variétale de *Coffea canephora*. I. Critères et méthodes de sélection. *Plantations, Recherche, Dév.* 5 (1), 18–33.

Montagnon C., Leroy T. and Eskes A.B. (1998b) Amélioration variétale de *Coffea canephora*. II. Les programmes de sélection et leurs résultats. *Plantations, Recherche, Dév.* 5 (2), 89–98.

Moreno G., Moreno E. and Cadena G. (1995) Bean characteristics and cup quality of the Colombia variety (*Coffea arabica*) as judged by international tasting panels. *Proc. 16th ASIC Coll.*, pp. 574–578.

Mösli Waldhauser S.S. and Baumann T.W. (1996) Compartmentation of caffeine and related purine alkaloids depends exclusively on the physical chemistry of their vacuolar complex formation with chlorogenic acids. *Phytochem.* 42, 985–996.

Mösli Waldhauser S.S., Kretschmar J.A. and Baumann T.W. (1997) N-methyltransferase activity in caffeine biosynthesis: biochemical characterisation and time course during leaf development of *Coffea arabica*. *Phytochem.* 44, 853–859.

Nathanson J.A. (1984) Caffeine and related methylxanthines: possible naturally occurring pesticides. *Science* 226, 184–187.

Nehlig A. (1999) Are we dependent upon coffee and caffeine? A review on human and animal data. *Neurosci Biobehav. Rev.* 23, 563–576.

Njoroge S.M., Morales A.F., Kari P.E. and Owuor J.B.O. (1990) Comparative evaluation of the flavour qualities of Ruiru II and SL28 cultivars of Kenya arabica coffee. *Kenya Coffee* 55 (643), 843–849.

Noir S., Combes M.C., Anthony F. and Lashermes P. (2001) Origin, diversity and evolution of NBS disease-resistance gene homologues in coffee trees (*Coffea* L.). *Mol. Gen. Genomics* 265, 654–662.

Orozco-Castillo C., Chalmes K.J., Waugh R. and Powell W. (1994) Detection of genetic diversity and selective gene introgression in coffee using RAPD markers. *Theor. Appl. Genet.* 87, 934–940.

Paillard M., Lashermes P. and Pétiard V. (1996) Construction of a molecular linkage map in coffee. *Theor. Appl. Genet.* 93, 41–47.

Pallavicini A., Del Terra L., De Nardi B., Rovelli P. and Graziosi G. (2001) A catalogue of genes expressed in *Coffea arabica* L. *Proc. 19th ASIC Coll.*, CD-ROM.

Pendergrast M. (1999) *Uncommon Grounds: the History of Coffee and How It Transformed Our World*. New York: Basic Books, Perseus Books Groups.

Pichersky E. and Gershenzon J. (2002) The formation and function of plant volatiles: perfumes for pollinator attraction and defense. *Curr. Opin. Plant Biol.* 5, 237–243.

Prakash S.N., Combes M.C., Naveen S.K., Graziosi G. and Lashermes P. (2001) Application of DNA marker technologies in characterizing genome diversity of selected coffee varieties and accessions from India. *Proc. 19th ASIC Coll.*, CD-ROM.

Raina S.N., Mukai Y. and Yamamoto M. (1998) *In situ* hybridization identifies the diploid progenitor species of *Coffea arabica* (*Rubiaceae*). *Theoret. Appl. Genet.* 97, 1204–1209.

Rhoades D.F. (1979) Evolution of plant chemical defense against herbivores. In *Herbivores*. New York: Academic Press, pp. 3–54.

Ribas A.F., Kobayashi A.K., Bespalhok Pilho J.C., Galvão R.M., Pereira L.F.P. and Vieira L.G.E. (2001) Trasformação genètica de cafè mediada por *Agrobacterium tumefaciens*. II Simposio de Pesquisa dos Cafes do Brasil, Victoria ES, pp. 420–428.

Rizvi S.J.H., Mukerji D. and Mathur S.N. (1981) Selective phyto-toxicity of 1,3,7-trimethylxanthine between *Phaseolus mungo* and some weeds. *Agric. Biol. Chem.* 45, 1255–1256.

Rogers W.J., Bézard G., Deshayes A., Pétiard V. and Marraccini P. (1999a) Biochemical and molecular characterization and expression of the 11S-type storage protein from *Coffea arabica* endosperm. *Plant Physiol. Biochem.* 37, 261–272.

Rogers W.J., Michaux S., Bastin M. and Bucheli P. (1999b) Changes to the content of sugars, sugar alcohols, myo-inositol, carboxylic acids and inorganic anions in developing grains from different varieties of robusta (*Coffea canephora*) and arabica (*C. arabica*) coffees. *Plant Sci.* 149, 115–123.

Rovelli P., Mettulio R., Anthony F., Anzueto F., Lashermes P. and Graziosi G. (2000) Microsatellites in *Coffea arabica* L. In T. Sera, C. R. Soccol, A. Pandey and S. Roussos (eds), *Coffee Biotechnology and Quality*. Dordrecht: Kluwer Academic, pp. 123–133.

Ruas P.M., Diniz L.E.C., Ruas C.F. and Sera T. (2000) Genetic polymorphism in species and hybrids of *Coffea* revealed by RAPD. In T. Sera, C. R. Soccol, A. Pandey and S. Roussos (eds), *Coffee Biotechnology and Quality*. Dordrecht: Kluwer Academic, pp. 187–195.

Schiestl F.P. and Ayasse M. (2001) Post-pollination emission of a repellent compound in a sexually deceptive orchid: a new mechanism for maximising reproductive success? *Oecologia* 126, 531–534.

Schiestl F.P., Ayasse M., Paulus H.F., Lofstedt C., Hansson B.S., Ibarra F. and Francke W. (1999) Orchid pollination by sexual swindle. *Nature* 399, 421–422.

Shimizu M.M. and Mazzafera P. (2000) A role for trigonelline during imbibition and germination of coffee seeds. *Plant Biol.* 2, 605–611.

Sondahl M.R. and Baumann T.S. (2001) Agronomy II: development and cell biology. In R.J. Clarke and O.G. Vitzthum (eds), *Coffee – Recent Developments*. Oxford: Blackwell Science, pp. 202–223.

Sondahl M.R. and Sharp W.R. (1979) Research in *Coffea* spp. and applications of tissue culture methods. In W.R. Sharp et al. (eds), *Plant Cell and Tissue Culture. Principles and Applications*. Columbus, OH: Ohio State University Press, pp. 527–584.

Sondheimer E. (1964) Chlorogenic acid and related depsides. *Botanical Rev.* 30, 667–712.

Sondheimer E., Covitz F. and Marquisée M.J. (1961) Association of naturally occurring compounds, the chlorogenic acid-caffeine complex. *Arch. Biochem. Biophys.* 93, 63–71.

Speer K. and Kolling-Speer I. (2001) Lipids. In R.J. Clarke and O.G. Vitzthum (eds), *Coffee – Recent Developments*. Oxford: Blackwell Science, pp. 33–49.

Spiral J., Leroy T., Paillard M. and Pétiard V. (1999) Transgenic coffee (*Coffea* species). In Y.P.S. Bajaj (eds), *Biotechnology in Agriculture and Forestry*. Berlin: Springer, pp. 55–76.

Srinivasan C.S., Prakash N.S., Padma Jyothi D., Sureshkumar V.B. and Subbalakshmi V. (2000) Coffee cultivation in India. In T. Sera, C.R. Soccol, A. Pandey and S. Roussos (eds), *Coffee Biotechnology and Quality*. Dordrecht: Kluwer Academic Publishers, pp. 17–26.

Stocker H. (1976) Epikutikuläre Blattwachse bei *Coffea*. Veränderungen während der Blattentwicklung; Blattwachse als chemosystematisches Merkmal. Inaugural Dissertation, Institüt für Pflanzenbiologie, Universität Zürich.

Taylor E., McGirr R., Bates J., Lee D., Rovelli P., Graziosi G. and Donini P. (2002) Modern technology for traceability and authenticity of coffee throughout food processing. *Plant, Animal and Microbe Genomes X Conference*, 12–16 January, San Diego, p. 174.

The Arabidopsis Genome Initiative (2000) Analysis of the genome sequence of the flowering plant *Arabidopsis thaliana*. *Nature* 408, 796–815.

Turlings T.C.J., Tumlinson J.H. and Lewis W.J. (1990) Exploitation of herbivore-induced plants odors by host-seeking parasitic wasps. *Science* 250, 1251–1253.

Urbaneja G., Ferrer J., Paez G., Arenas L. and Colina G. (1996) Acid hydrolysis and carbohydrates characterization of coffee pulp. *Renewable Energy* 9, 1041–1044.

Valio I.F.M. (1976) Germination of coffee seeds (*Coffea arabica* L. cv. Mundo Novo). *J. Exp. Bot.* 27, 983–991.

Van der Vossen H.A.M. (1980) Methods of preserving the viability of coffee seeds in storage. *Kenya Coffee* 45, 31–35.

Van der Vossen H.A.M. (1985) Coffee selection and breeding. In M.N. Clifford and K.C. Willson (eds), Coffee: Botany, Biochemistry and Production of Beans and Beverage. London: Croom Helm, pp. 48–96.

Van der Vossen H.A.M. (1997) Quality aspects in arabica coffee breeding programmes in Africa. *Proc. 17th ASIC Coll.*, pp. 430–438.

Van der Vossen H.A.M. (2001) Coffee breeding practices. In R.J Clarke and O.G. Vitzthum (eds), *Coffee – Recent Developments.* Oxford: Blackwell Science, pp. 184–201.

Van der Vossen H.A.M. and Browning G. (1978) Prospects of selecting genotypes of *Coffea arabica* which do not require tonic sprays of fungicide for increased leaf retention and yield. *J. Horticult. Sci.* 53, 225–233.

Van der Vossen H.A.M. and Walyaro D.J. (1981) The coffee breeding programme in Kenya: a review of progress made since 1971 and plan of action for the coming years. *Kenya Coffee* 46, 113–130.

Van der Vossen H.A.M., Soenaryo and Mawardi S. (2000) *Coffea* L. In H.A.M. van der Vossen and M. Wessel (eds), *Plant Resources of South-East Asia no.16. Stimulants.* Leiden: Backhuys Publishers, pp. 66–74.

Viani R. (1988) Physiologically active substances in coffee. In R.J. Clarke and R. Macrae (eds), *Coffee: Volume 3 – Physiology.* London: Elsevier Applied Science, pp. 1–31.

Vitzthum O.G., Weisemann C., Becker R. and Kohler H.S. (1990) Identification of an aroma key compound in robusta coffees. *Café Cacao Thé* 34, 27–33.

Willson K.C. (1985) Mineral nutrition and fertilizer needs. In M.N. Clifford and K.C. Willson (eds), *Coffee: Botany, Biochemistry and Production of Beans and Beverage.* London: Croom Helm, pp. 135–156.

Wrigley G. (1988) *Coffee* (Tropical Agriculture Series). Harlow: Longman Scientific and Technical.

Zezlina S., Soranzio M., Rovelli P., Krieger M.A., Sondhal M.R. and Graziosi G. (1999) Molecular characterisation of the cultivar Bourbon LC. *Proc. 18th ASIC Coll.*, CD-ROM.

CHAPTER 3

The raw bean

S. Bee, C.H.J. Brando, G. Brumen,
N. Carvalhaes, I. Kölling-Speer, K. Speer,
F. Suggi Liverani, A.A. Teixeira,
R. Teixeira, R.A. Thomaziello, R. Viani
and O.G. Vitzthum

3.1 INTRODUCTION

A.A. Teixeira

As seen in the previous chapter, the genotype, the macro- and microclimates, as well as agricultural practices, play a fundamental role in the quality of coffee. One of the main characteristics of quality coffee is the possession of good organoleptic properties. Ripe cherry coffee from the *Coffea arabica* species is the ideal raw material for obtaining fine quality coffee (Figure 3.1). In order to maintain the bean's inherent qualities, special care is required during processing.

3.2 HARVESTING

A.A. Teixeira, C.H.J. Brando, R.A. Thomaziello
and R. Teixeira

Preparation for harvesting should be made in advance in order to avoid last minute surprises and damage to the quality of the coffee. The cleaning

Figure 3.1 Coffee plantation (left) and cherry coffee on the tree (right)

and maintenance of the entire infrastructure and the checking of equipment to be used should be accomplished at least one month before the start of the harvest.

Harvesting should only start after a very careful examination of the level of maturation, when most of the cherries are ripe, with a minimum (5%) presence of unripe cherries. Unripe cherries lead to light-green beans, which, when dried at a temperature above 30 °C, become black-green; both types weigh less than a normal bean (Teixeira et al., 1991). The extended stay of the cherries on the tree or on the ground should be avoided in order not to increase the quantity of sour and black beans, which also weigh less than normal beans. Light-green, black-green, sour and black beans are the defects (see 3.7.2) that have the greatest negative effect on overall coffee quality, damaging type, appearance, roast, yield and beverage (Teixeira, 1978).

Harvesting can be accomplished in different ways: by stripping onto the ground (not recommended) or onto sheets, by selective hand picking or by mechanical means.

3.2.1   Stripping (Figure 3.2a)

This operation is accomplished by stripping all the cherries from the branches at the same time. The biggest problem is to evaluate the right time to perform this operation, therefore the need to collect samples of 100 cherries to determine the correct percentage at each stage of maturation. At the time of harvest, depending on the number and spacing between one flowering and the other, one can find on a tree immature, greenish, ripe, overripe and dry cherries.

In the preparation of natural coffee, cherries are dried whole, that is, the beans are dried while surrounded by the pulp (mucilage and husk). In this operation, the homogeneous maturation of the cherries is very important and, therefore, the percentage of green cherries collected should be as low as possible. In regions with a dry climate, stripping can begin when the majority of the cherries are in the overripe or partially dry stage. In this case, drying will be quite uniform and the risk of undesirable fermentation very low, in contrast to stripping with a high percentage of ripe cherries.

Harvesting by stripping should be done directly onto sheets placed under the tree beforehand, to avoid contamination and contact with cherries that have spontaneously fallen and which may have already deteriorated (Carvalho et al., 1972). For this reason, stripping onto the ground must be avoided. In general, the sheets are laid under the coffee trees, in the same direction as the rows of trees, and normally reach

(a)

(b)

Figure 3.2   Harvesting (a) by stripping and (b) by hand picking

beneath three or four plants. Once the stripping is complete, the cherries should be collected and screen tossed ('winnowed') before being bagged and taken to the processing infrastructure on the day of harvest. Mechanical winnowing can be used in the coffee plantation or at the processing installation.

Cherries that have fallen directly onto the ground before harvesting, called sweepings (*varrição* in Brazil), and those stripped directly onto the ground should always be processed separately in all the phases of preparation and should never be mixed with cherries stripped onto sheets.

3.2.2   Hand picking (Figure 3.2b)

In this operation only ripe cherries are picked, normally in baskets or bags. This type of harvesting is generally employed in regions with constant rain, where many flowerings occur throughout the year. Harvesting of the cherries should be performed as many times as necessary. The labour force should be sufficient to avoid the fall of cherries and their consequent deterioration. To avoid undesirable fermentation, the ripe cherries should be taken to the preparation infrastructure and processed on the day of harvest.

3.2.3   Mechanical harvesting (Figure 3.3)

This technique is mainly applied in Brazil and Hawaii. The operation can be done by different systems, all based on the vibration of the branches of the coffee tree:

■   self-propelled machines that remove the cherries from the tree, collect and winnow them and put the product into bags, hoppers, carts or trucks;

Figure 3.3 Mechanical harvesting

- mechanical stripping machines, pulled by tractors, that drop the cherries on the ground;
- portable stripping machines (*derriçadeiras* in Brazil) carried by the picker, using pneumatic or internal combustion power, that drop the cherries on the ground, whether covered (recommended) or not by cloths or plastic sheets.

All these systems are being constantly improved in order to increase selectivity and give priority to the harvest of ripe cherries. An increasingly common practice in mechanical harvesting is to conduct the stripping operation at two or three different times, concentrating each time on the part of the tree with the most ripe cherries.

Self-propelled stripping machines and those pulled by tractors can only be used on plantations that are flat or slightly undulated, from 10 to 20% of slope depending on the technology. Portable stripping machines are compatible with any topography and represent an important alternative to increase the efficiency of the harvest in mountainous areas.

Mechanical harvesting enables a substantial reduction in the operating costs that comprise one of the main components of the total cost of coffee production.

3.2.4 The myth of selective picking

It is widely believed that quality coffee can only be obtained if selective hand picking is used. This is certainly valid for small plantations, but becomes a long-standing myth on modern medium-sized to large estates. Quality coffee can be produced using a variety of harvesting systems, manual or mechanical. Selective coffee picking – the careful hand picking of only ripe coffee cherries – is indeed one way to ensure that quality on the tree is transferred to the cup. However, it is not the only way. In fact, selective picking is nothing more than an indicator that only sound, fresh, ripe coffee cherries are used as raw material to produce the finest beans from which a perfect cup is brewed. Sound, fresh, ripe cherries may be obtained from a variety of picking practices combined with processing techniques. Top-quality coffee can be produced regardless of the harvesting technique employed.

When unwanted products are picked, quality must be maintained by post-harvest separation techniques so that high-quality coffee may still be produced from the remaining ripe cherries. Coffee growers must therefore be prepared to cope with harvesting that brings in mixed cherries. Their ability to separate sound, fresh, ripe cherries from immature, over-ripe, semi-dry and dry cherries is becoming a crucial factor in the production of top-quality coffees at a reasonable cost. Modern post-harvest processing equipment now available can handle such mixed cherries efficiently while using less water and causing less pollution.

3.3 PROCESSING OF THE HARVEST

A.A. Teixeira, C.H.J. Brando, R.A. Thomaziello and R. Teixeira

Processing must start on the same day as the harvest to avoid undesirable fermentation and reduce the risk of mould contamination, starting with the pulp of the fruit, rich in nutrients and moisture (Figure 3.4). This is a critical point in the preparation of coffee and can often jeopardize an entire year of work caring for the plants.

In all three types of processing (natural, pulped natural or washed), whenever possible, it is important that the cherries pass through washer–separators. In this equipment, rocks and impurities are eliminated and the cherries are separated by density: on one side the lighter cherries (floaters – the dry and over-ripe cherries) and on the other side the heavier ones

Figure 3.4 Stereo microscope view of a coffee cherry: (left) entire; (centre) pulp (mucilage and husk) partially removed; (right) section

Figure 3.5   Cherries passing through washer–separators

(immature and ripe cherries), enabling the separation of cherries with different humidity levels, facilitating drying and making the lot of coffee more homogeneous (Figure 3.5).

3.3.1   The natural (dry) process (Figure 3.6)

Harvesting must begin when most of the cherries are at the over-ripe and dry stages, especially in regions characterized by a dry harvest season. The harvesting of a high percentage of ripe cherries can cause undesirable fermentation, especially if the drying process is not begun on the same day as the harvest (see also 3.11.11.1 for the risk of mould formation).

In the processing of natural coffee, the whole cherries (bean, mucilage and pulp) are dried on patios or racks under the sun or in mechanical dryers. By this process, if well conducted, it is possible to produce a good coffee with 'body' and a pleasant 'aroma' (see 3.8.8 for an explanation of these terms), greatly appreciated in the preparation of espresso coffees.

Figure 3.6   Patio drying of natural coffee: (left) just harvested; (right) almost dry

3.3.2   The washed (wet) process (Figure 3.7)

The so-called washed or wet process requires a raw material composed of only ripe cherries that have been selectively picked or are mechanically separated in the process itself. After passing through the washer–separators and before removal of the pulp, the separation of the green immature cherries from the ripe ones can be performed, using differences in pressure, in a separator of green cherries. This machine has a specially designed screen with long slotted holes forming a cylinder containing a rotor that forces the cherries against the stationary screen and towards the edge(s) of the cylinder. The soft, ripe cherries pass through the holes of the screen. The hard, unripe cherries, which cannot pass through the holes, go to the edge(s) of the cylinder where a counter-weight controls their outflow.

The pulping process consists in the removal of the pulp by a pulper, followed by the removal of the mucilage from the parchment, which can be accomplished either mechanically, by the use of chemical products, or by fermentation. The mucilage that adheres to the parchment, with thickness of 0.5–2.0 mm, is very slimy and composed of pectins and sugars.

Fermentation to remove it is carried out in tanks at ambient temperature in the presence of microorganisms for between 12 and 36 hours, depending on local temperature. The mass of coffee can be either immersed in water (wet fermentation) or not (dry fermentation). The latter type is increasingly employed since it reduces fermentation time. In wet fermentation, the water must just cover the coffee mass. This process, however, takes more time and, if longer than 72 hours, can increase the formation of fluorescent beans, which easily generate 'stinker' beans during storage (see 3.7.2.5 for a definition of the term 'stinker').

Fermentation ends when the parchment loses the slimy feel of the mucilage. This stage can be recognized by rubbing the beans in one's hand: when they no longer slip against each other and produce a characteristic noise, fermentation is concluded.

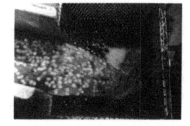

Figure 3.7   Steps of pulping process

Warm water rich in microorganisms is generally used in order to decrease fermentation time. The use of chemicals is less common and more risky, and is applied only in research laboratories. In the fermentation tanks, minerals and sugars are freed from the seeds (coffee bean with parchment plus mucilage) and can then become concentrated if the washing water is not changed regularly. The water can easily be contaminated by all types of microorganisms, leading to uncontrolled fermentation and a deterioration in cup quality: in this case a slight 'onion' flavour, which can reach a very bad 'stinker' taste, and can easily be detected in cup tasting.

Too high pH levels and the presence of ferric ions above 5 mg/l also lead to the production of off-flavours (Vincent, 1987). Volatile acidity, mostly owing to the formation of acetic acid, increases with fermentation, especially if the process lasts longer than 20 hours. After the conclusion of the pulping operation, the coffee must be very carefully washed so as to ensure the removal of any trace of mucilage. Incomplete washing can cause undesirable fermentation, with detrimental consequences to quality.

Removal of mucilage can be carried out mechanically by friction. In general the mucilage is not completely removed. There has been a strong trend in the adoption of mechanical removal of the mucilage after the advent of modern machines, which consume little water and energy and substitute the traditional *aquapulpas*. Modern mucilage remover machines avoid environmental pollution since they consume and contaminate less water than wet or dry fermentation.

3.3.3 The pulped natural process

A process called pulped natural (*cereja descascado* in Brazil), intermediate between the natural and washed processes, began to be used in Brazil in the early 1990s. In this process, the cherries are pulped and the beans in parchment dried while surrounded by the mucilage. Fermentation for the removal of the mucilage is not used in this process. As early as 1960, an experiment on good cup quality and increased germinative power demonstrated the validity of pulping cherries without submitting them to fermentation (Ferraz et al., 1960). The quality of pulped natural coffee, when well processed, has been shown to be excellent, with the advantage of producing coffee with greater body than coffee produced using the wet process.

In this case, harvesting can be conducted by stripping when the majority of the cherries are ripe. Processing must take place on the same day, starting with the washer–separators. The pulpers used for the preparation of the pulped natural coffees are equipped with separators of green (immature) cherries, to separate by pressure immature cherries from mature ones. In the preparation of pulped natural coffees, the pulp and remains of mucilage must be disposed of in a location far from the preparation area. After initial decomposition, it can be used on the farm as a fertilizer. The immature cherries must be dried separately and coffee in parchment with unremoved mucilage must be immediately put to dry.

3.3.4 Environmental impact

The processing of the harvest can produce liquid and solid residues that are aggressive to the environment in case they are not correctly disposed of or treated.

The natural process is the only one that does not attack the environment, since it does not require the use of water and does not produce solid and liquid residues rich in organic substances. Even when light and heavy cherries are separated by water, contamination is low (only suspended solids) or inexistent, because the contact between water and cherries is rapid and the whole cherry does not lose contaminating substances in the water. The natural process is, therefore, environmentally friendly by definition.

The pulped natural process produces solid residue (the pulp) and contaminates the water that enters into contact with the pulp and the parchment. Finally, the washed (wet) process is the most pollutant of all, since it requires, in addition to pulping, the removal of the mucilage, which is extremely rich in organic substances and corresponds to about half of the polluting load of the process. The washed process is particularly aggressive when the removal of the mucilage occurs by natural fermentation, which requires great volumes of water that consequently become contaminated.

The washed process is, therefore, attracting increasingly great attention among environmentalists whose pressure is provoking changes to protect the environment. The pulped natural process was conceived as a less polluting alternative since it does not require fermentation and was developed at a time when a greater concern for the environment already existed.

The tendency in wet processing is to use as little water as possible, transporting the products in a dry state and not in water channels, using modern machines that consume less water, substituting fermentation by mechanical removal of the mucilage and re-circulating used water after the removal of solids. The residual waters can be used for irrigation, infiltrated into the ground, or be treated before disposal in waterways.

The major recent changes to make coffee mills more environmentally responsible, which are still ongoing, can be grouped in three areas:

■ design or layout of the mills;
■ water-saving equipment;
■ recycling and safe disposal of wastewater.

Old coffee mills had many channels to convey coffee – and water that became contaminated in the process – from one machine to another. Modern wet mill design brings machines close together to enable gravity feeding without water (dry feeding) when slope is available. If slope is unavailable dry conveyors are used (e.g., bucket elevators, inclined conveyors, etc. that convey coffee without water) instead of pumps that can only move coffee that is mixed with water. New water-saving washer–separators, replacing water-intensive siphon tanks, pulpers and mechanical mucilage removers, consume much less water than conventional machines; this is particularly the case of mucilage removers used to replace traditional fermentation (the most water-intensive stage of the process and the most contaminating one). Finally, recycling of liquid and solid wastes produced by wet milling is becoming a required practice, as is the safe disposal of such water in a manner that does not harm the environment. Since wastewater treatment and disposal is cheaper and more convenient for smaller volumes with high pollution loads than for large volume with smaller pollution loads, the use of as little water as possible becomes a very critical issue.

Attempts to make the pulping process totally dry have been frustrated by a fall in the quality of the beans and have subsequently been abandoned.

3.4 DRYING

A.A. Teixeira, C.H.J. Brando, R.A. Thomaziello and R. Teixeira

The more homogeneous the mass of coffee, the better and more uniform will be the drying process. In green immature cherries, humidity reaches 70%, while in mature cherries it varies from 50 to 70%, in over-ripe cherries from 35 to 50% and in dry cherries from 16 to 30% (Tosello, 1946; Rigitano et al., 1963).

The drying of the coffee should be slow. The withdrawal of the bound water of the bean is difficult and, therefore, a drawn-out process. The slower the drying operation, including periods of rest, the more homogeneous will be the final product.

The drying can be conducted under the sun only, on patios or suspended tables; in mechanical dryers only; or initiated under the sun and concluded in mechanical dryers.

3.4.1 Patio drying of natural coffee

It is important that natural coffees, normally harvested by stripping, consisting of immature, greenish, mature, over-ripe and partially dry cherries, be dried separately. Therefore, the floaters (over-ripe and partially dry cherries) must be separated from the immature and mature cherries to form groups of similar humidity to achieve a uniform product, of good appearance and homogeneous dryness.

The patio area must be well calculated and large enough to avoid the coffee remaining in thick layers, especially at the beginning of the drying process, and so damaging its final quality. The drying patios must be constructed in areas without accumulations of cold air, with great exposure to sunlight, with a slope between 0.5 to 1.5% to facilitate water drainage and be coated with bricks, tiles, concrete or even asphalt (see Figure 3.6).

At the beginning of the drying process, the coffee must be spread in thin layers with a height of 2–3 cm, and be constantly turned, always observing the direction of the sun. If the sun is in front of the worker his shadow should be behind him or if the sun is behind the worker his shadow should be in front of him. The volume or kg/m^2 of coffee required to obtain such a thickness varies with the stage of maturation and the percentage of humidity: unripe, ripe, over-ripe and dry cherries (natural process) or coffee in parchment (wet process). An indicative correlation indicates that a layer of 5–6 cm of thickness takes 30–40 kg of fruits per m^2 of yard: 2–3 kg of cherry coffee (at 60–65% moisture) or 6–9 kg of parchment coffee per m^2 can be dried per day (Matiello et al., 2002).

The coffee must be constantly turned, 15–17 times per day, in both directions, to speed up the removal of the easily eliminated external water and to avoid the appearance of fermented or mouldy beans, easily identified during classification and rejected by the market.

Every evening during the initial days of drying the coffee should be heaped in thin rows, of 5–10 cm height, in the direction of the patio's slope. With the passing of the days, the rows can be thickened until the coffee reaches a stage of semi-dryness, or 20–30% of humidity. From now on, every evening, the coffee should be heaped and covered with cotton or waxed fabric that will facilitate the equalization and uniformity of drying.

3.4.2   Patio drying of pulped natural and washed coffees

Fully washed parchment coffee (parchment devoid of mucilage) is dried in drying patios or suspended tables (Figure 3.8). In the initial phase, the layer should be very thin, 2–3 cm, with constant turning. Only after skin drying (the elimination of the external water) can the coffee be dried in thicker layers. Once it is half dry, the coffee can be heaped and covered during the night.

It has been observed that the light of the sun, especially radiation between 400 and 480 nm, is important for the maintenance of coffee quality, improving aroma and acidity, and reducing mould and bacterial load, above all in the final phase of the intermediate stage of drying, when the beans have reached a moisture content below 40% and a dark translucid aspect (Northmore, 1969).

The care of pulped natural coffees must be greater than that of washed coffees due to the presence of the (sticky) mucilage. The layers should be 2–3 cm high and constantly turned until the coffee is skin-dry, that is, it has lost all external humidity. If this drying operation is well performed, it can be concluded in a single day. Later, the operation will be the same as the one for coffee in parchment without the mucilage, in patios or mechanical dryers.

Figure 3.8 Sun drying of pulped natural or washed coffee: (left) patio drying; (right) suspended table drying

3.4.3 Mechanical drying of coffee (Figure 3.9)

Some types of dryer can directly receive the coffee parchment without mucilage, while others require that the coffee be pre-dried in the sun. Drying can be accomplished solely by mechanical means in regions with adverse climate, but normally mechanical drying is used after partial drying in patios, when the coffee attains a moisture level between 20 and 30%. The dryers should be equipped with heat exchangers (indirect fire furnaces or steam or water radiators) or gas burner to avoid direct contact between combustion fumes and coffee and the smell of smoke in coffee. The use of coffee husks or parchment as fuel can reduce the cost of drying and the cutting down of trees. Firewood should always be dry, to enable a higher yield, and to avoid smoke and pollution of the environment. Mechanical dryers should be loaded with beans of the same moisture to improve uniformity of drying. All types and brands of dryers are designed to work with a full load and to avoid heat loss, increase in drying time and unnecessary energy and labour expenses. The temperature in the coffee mass should not exceed 40–45°C, to avoid death of the embryo, and unpleasant smells. When immature cherries are present, the temperature should not exceed 30°C to avoid their transformation into dark-green and black-green defects, of very low quality (Teixeira *et al.*, 1982). The 12% humidity level is reached after the coffee is removed from the dryer, which it must leave with a slightly higher (around 1%) humidity.

3.4.4 Comparison between natural and washed coffee

The biggest difference between the chemical composition of washed and natural coffee is in the soluble solids content, which is higher in the case of natural coffee. Pulped natural coffee presents an intermediate position.

Figure 3.9 Mechanical drying of coffee

The difference can be explained by osmotic phenomena. During cherry drying a gradient of concentration is established between the mucilage (mesocarp) and the bean (endosperm) with a migration to the bean of some sugars present in the mucilage. On the other hand, in the wet process, there occurs a loss of 0.7% of soluble solids during fermentation and 0.7% during washing, by the diffusion to the water with the mucilage of aqueous solids from the internal part of the bean (Wootton, 1971). Sugars consist of one-third of the total loss of soluble solids during fermentation with a reduction ranging between 0.34 and 0.51%. Studies show that soluble solids in arabica varieties cultivated in Campinas, Brazil, ranged from 26.7 to 30.5%; the highest values were measured with the Mokka variety. Dry-processed samples yielded 0.4–0.9% more solids than wet-processed ones, where leaching of water-soluble matter occurs during fermentation (Table 3.1) (Carvalho, 1988). The soluble solid content measured by the Navellier method (Navellier and Brunin, 1963; Wilbaux, 1967) ranged from 26.1 to 30.6%, with an average of 28.3% in robusta, while in arabica coffee it ranged from 23.8 to 27.3%, with an average of 25.6%. In Icatu, an interspecies hybrid between arabica and robusta coffee, this index reached levels of 26.8–28.6% (Moraes et al., 1973). The greater amount of soluble solids present in natural coffees is undoubtedly responsible for their increased body in comparison with washed ones (Table 3.1).

The mineral content is slightly reduced in wet-processed beans compared to dry-processed ones (Table 3.2).

The natural process does not require high volumes of water and is less polluting. However, in the wet process, about 60% of the cherry (pulp) volume is removed, significantly reducing the area and the time required for drying. In the wet process, where the pulp and mucilage are removed, the risk of undesirable fermentation is significantly reduced. A further advantage that must be considered is the possibility of obtaining a product with a more homogeneous appearance and drying in the cases of pulped natural and washed coffees. The body produced during the process of

Table 3.1 Soluble solids in washed and natural arabica coffees (% dry matter)

Variety	Washed	Natural	Difference
Caturra	26.71	27.12	+0.41
Mundo Novo	27.91	28.33	+0.42
Yellow Bourbon	27.81	27.85	+0.04
Maragogype	27.90	28.84	+0.94
Mokka	29.69	30.47	+0.83

Table 3.2 Mineral content of green coffee (% dry matter)

Type of coffee	Minerals	Potassium
Dry-processed robusta	4.14–4.39	1.84–2.00
Dry-processed arabica	4.11–4.27	1.77–1.88
Wet-processed arabica	3.58–3.95	1.63–1.70

drying in natural coffees is, however, basic to the sensorial quality of an espresso coffee, from which derives the recommendation for their use in this type of preparation. Pulped natural coffees present a body intermediate between the washed ones, with low body, and the natural ones, with good body. These coffees are also strongly appreciated in blends for espresso coffee.

3.5 FINAL PROCESSING FOR EXPORT AND ROASTING

C.H.J. Brando and A.A. Teixeira

Dry parchment coffee or dry cherry coffee must be stored at moisture levels below 12% in order to avoid the development of musty, earthy or fermented flavours. Dry coffee can be stored in bulk (silos) or in bags in a dry atmosphere, preferably at mild temperatures and in the dark, before final processing for export or roasting. Dry cherry and parchment coffee is stored in Brazil in large wooden cells, the 'tulhas', for up to several weeks to 'mature', i.e., to equilibrate the moisture before final processing (Figure 3.10).

Figure 3.10 *Tulhas* for maturation of coffee used in Brazil: (left) empty; (right) full

In areas where coffee is dry-processed, it is common for the farmer to deliver green (hulled) coffee to a mill. There it is submitted to rigorous processing which may include size grading, density separation and colour sorting. Small farmers may deliver semi-finished product in the form of dry cherries to cooperatives for further processing.

In areas where coffee is wet-processed, parchment coffee is delivered to a mill. Hulling and further processing of washed or pulped natural coffees should only take place shortly before coffee is exported.

3.5.1 Cleaning

Final dry coffee processing starts with the cleaning of the parchment or cherries, performed in two stages: pre-cleaning and destoning. Pre-cleaning is accomplished by the suction of light impurities and dust combined with the sieving of impurities larger or smaller than coffee. Destoning employs flotation of coffee to separate it from the heavier stones. A plate or rotary magnet may be used at either stage to separate iron particles.

3.5.2 Hulling and polishing

Hulling is the process by which the outer shell (husk) of parchment or dry cherry coffee is removed. Several systems are used depending on the product and on local practices.

Parchment coffee is often hulled by friction as it is forced to travel in the space between a screw or multi-facet rotor and its static case. These machines often remove the parchment husk as well as the 'silverskin' (the perisperm) in a process that is also called polishing or hulling-polishing. In either case, heat results from the friction and care must be taken to avoid overheating, which may have negative impact on quality.

Arabica cherry coffee is usually hulled in machines with rotating blades. These 'beaters' force coffee through the openings of a cylindrical case that encloses the rotor. Since little or no heat is generated in the process, the silverskin is not removed. Friction hullers for cherry coffee are used primarily for robusta coffee. They have a rotating cylinder with cleats and a static adjustable blade. In this case heat is developed and some polishing may take place.

Removal of silverskin (polishing) is commonly required for washed coffees but seldom for natural coffees. The cross-beater huller used for cherry coffee may be also used for parchment coffee if polishing is not required.

All types of hullers include an air column, the 'catador' that separates parchment or cherry husk from the hulled green coffee beans. A fan blows or aspirates husk and discharges it away from the machine, outside the building or into a husk cyclone and/or silo. Some hullers have a built-in larger catador, or may be coupled to one, that separates hulled coffee in different density fractions, e.g., the 'heavier', sound beans and the 'lighter', often defective beans. A built-in repassing device is often found in some types of hullers, especially those for cherry coffee.

The proportion between husk and green coffee is around 20% for parchment coffee and 50% for cherry coffee.

3.5.3 Grading

Green coffee may be separated by size not only for marketing purposes but also to enable better density and colour separation. Machines that grade coffee by size, i.e., coffee graders, can also be used to separate coffee beans by shape. Flat (regular) beans are the ones with one flat and one concave surface. Round beans from cherries holding just one seed, called 'peaberries', 'mocas' or 'caracoles', occur in much smaller proportions.

Coffee graders use sieves of different sizes and shapes to separate green beans according to size and shape. Though most countries rely on sieves with holes measured in multiples of $1/64$ s of one inch (e.g., beans size 18 are retained by a sieve $18/64$ in.), millimetres are also used. Very large beans are usually retained by sieve 19 (7.54 mm) and medium to large beans by sieve 16 (6.35 mm). Peaberries are separated by slotted (oblong) screens of several sizes (Figure 3.11).

3.5.4 Mechanical sorting

Densimetric separation aims at removing defects associated with less dense beans, such as malformed beans, insect-damaged beans, fermented

Figure 3.11 Cleaning and grading of coffee (courtesy of Sortex Ltd)

beans, some types of black beans, etc. Traditional catadors, that remove such defects by airflow, are being increasingly replaced by densimetric tables, also called gravity separators, using flotation for a more efficient separation.

Air flotation created by powerful fans located below the deck of a gravity separator causes the light, defective beans to float whereas the heavier sound beans lie at the bottom, in contact with the deck. Vibration causes these two fractions to leave the deck separately.

3.5.5 Electronic sorting
S. Bee and F. Suggi Liverani

Differences in bean colour are usually associated with defects (Illy *et al.*, 1982; Clarke, 1988). This allows the distinguishing of good beans from defective ones. Coffee sorting by hand is still practised in regions where labour rates remain low. However, when consumers began demanding increased quality, cost and complexity of this labour increased. Automated techniques were introduced since machines can maintain greater levels of consistency than hand sorting (Anon., 1987), thus providing a premium quality product at increased margins.

The widespread use of natural coffees in espresso blends makes electronic sorting a prerequisite for the production of high-quality blends.

The size, cost and complexity of electronic sorting machines varies, depending on the output and on the complexity of optical measurement and processing unit. Colour sorters (Vincent, 1987) generally consist of four principal subsystems (Figure 3.12): feeding system, optical system, ejection devices and processing unit.

3.5.5.1 Equipment subsystems

Feeding system: In the feeding system, bulk coffee is fed from a vibrating hopper onto a flat, or channelled, gravity chute. This method separates the product into a uniform 'curtain', or monolayer. The feeding systems commonly employed are: inclined gravity chute, a flat belt, inclined belt, contra-rotating rollers and a narrow grooved belt.

Optical system: The optical system measures the magnitude of light reflected from each single bean. Two or three cameras are used to view the product from different angles as it leaves the end of the chute, increasing the defect detection efficiency.

Ejection devices: The ejection process typically takes place while the product is in free fall. Accepted particles fall along their normal trajectory while rejected ones are deflected into a receptacle with a blast of

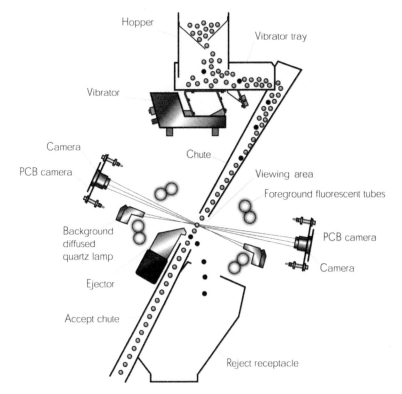

Figure 3.12 Schematic layout of an electronic sorting machine (courtesy of Sortex Ltd)

compressed air from a high-speed solenoid or piezoelectric valve, connected to a nozzle. Pneumatic ejector valves must have rapid action (on/off time of less than 0.25 msec, a typical duty cycle of 150–300 Hz, firing a pulse of air for 1–3 msec), reliability, a long lifetime (at least a billion cycles or more) and mechanical strength.

Processing unit: The processing unit manages the control of the machine and classifies particles as either 'acceptable' or 'rejected' on the basis of colour, or both colour and shape. The electronic processing systems in sorting machines have progressed from the simple analogue circuits of the early machines to the advanced digital microprocessor-based ones, as in the present generation of machines. Most of the setting up of the sorting parameters can be done by the machine itself including, in some cases, the ability of the machine to 'learn' the difference between a good and a bad product, this is usually represented as a two-dimensional

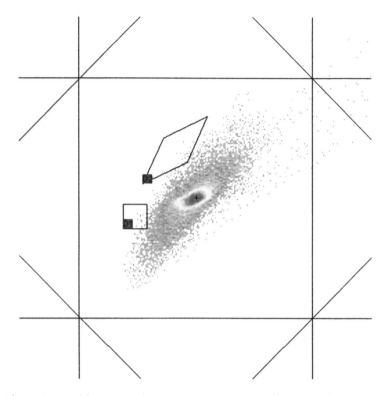

Figure 3.13 Bichromatic colour sorting map (courtesy of Sortex Ltd)

reflectivity 'colour map' (Figure 3.13), where the boundary curve is the contour, outlining the acceptable product. A sophisticated control system will track the average colour of the product so that, even though the average product colour may change with time, the machine will continue to remove only the predefined abnormal particles (Figure 3.14). Advanced sorting machines have the capability of providing information about the product, for example the number of rejects, or information about any drifts in colour in a certain batch of product.

3.5.5.2 Sorting techniques

Monochromatic sorting: Monochromatic sorting is based on the measurement of reflectance at a single isolated band of wavelengths. The removal of beans lighter or darker in hue than the average is a typical application of this system.

input   reject   accept

Figure 3.14   An example of coffee sorting (courtesy of Sortex Ltd)

Bichromatic sorting: Bichromatic sorting is only used when a simple monochromatic measurement is not adequate for effective optical sorting. Usually green and red filters are used for arabica beans and red or near infrared for robusta beans, allowing the elimination of unripe, waxy, chipped, insect-damaged or broken beans as well as of white and black beans (see 3.7) (Figure 3.15).

Trichromatic sorting: Trichromatic sorting enhances bichromatic sorting, adding the capability to sort beans by size or shape or for the detection of gross defects such as the presence of foreign material like glass, stones and insects. In this way, objects of the same colour but different shapes, or with holes or cracks, can be effectively removed.

UV fluorescence sorting: It has been found that certain non-visible defects, such as moulds or bacteria, fluoresce when irradiated with long-

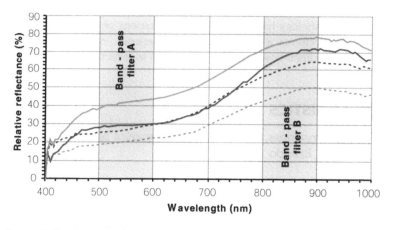

Figure 3.15   Spectral reflectance curves obtained from green arabica coffee. The solid lines represent acceptable beans; the dotted lines represent discoloured beans (courtesy of Sortex Ltd)

wave ultraviolet light (360 nm). This property may be used as a basis for sorting. This technique was originally developed for removing 'stinkers' from green arabica coffee beans, but has found applications in sorting other materials as well. However, the fluorescence effects can be best sorted on freshly harvested coffee. In the case of light beans after a long storage period following the harvest, this technique faces more difficulties.

Infrared sorting: Over the last decade, the wavelength range used by sorting machines has been extended from the visible into the infrared region. Here, both water absorption and other chemical effects play an important part in determining the reflectivity characteristics of beans.

Optical sorting with lasers: The use of lasers is a technique that is still in its relative infancy. A laser beam is used to illuminate the product and the magnitude of reflected light is affected by the amount of laser light that is either scattered from the surface, or diffused within an object. Since a laser produces narrow beams of coherent light at a single wavelength, there is no need to use optical band pass filters.

3.5.5.3 Future trends

Computer vision systems are increasingly being used in the food industry, pushing the development of new devices and software algorithms, offering many benefits over a conventional colour sorter, as the ability to simultaneously sort objects on the basis of multi-criteria. Furthermore, improved computer technology (Suggi Liverani, 1991) has enabled classification of green coffee lots automatically, using colour mapping together with specialized algorithm based on fuzzy logic (Suggi Liverani, 1995). This improved technology has overcome the limitations of man-made classification by operators who, for all their experience, may be subject to variations and imprecision in their perception in detecting defects.

3.6 LOGISTICS

N. Carvalhaes, R. Teixeira, A.A. Teixeira and C.H.J. Brando

3.6.1 Storage

Correct storage is the maintenance of the product for a certain period while preserving its original characteristics. Storage can serve the interests of a nation's economy by compensating for cyclical and other variations of agricultural production.

Raw coffee, when stored under appropriate conditions of ambient moisture, temperature and humidity, is quite stable and deterioration is slow (see 3.11.7 for the changes in amino acid content during storage).

The warehouse must be constructed in a well-ventilated area not subject to accumulations of cold air. A temperature close to 20 °C, and a relative humidity of the air never exceeding 60%, are optimal. There should be as little natural light as possible and artificial lighting must be controlled, preferably located in passageways and corridors. The warehouse must have good ventilation, without draughts, especially from the south, with well-located doors. To avoid development of a woody taste and colour bleaching of the beans (already observed at moisture levels above 12%) the coffee must not contain more than $11\pm0.5\%$ humidity when prepared by the dry process and $12\pm0.5\%$, by the wet process, because excess humidity in the beans facilitates attacks of mould and bacteria (see also 3.11.11.1). The increase in the humidity level of the beans is on average 0.18% for each percentage unit of increase in the relative humidity of the air. The correct waterproofing of the floor prevents the deterioration of the lowest bags in the piles. The coffee bags can be placed on wooden pallets, or be protected by plastic sheets, to prevent direct contact with the floor and should never be stacked against walls to prevent any type of humidity. Coffee warehouses should not contain other agricultural goods or chemicals that can transmit foreign flavours.

General warehouses with flow for commerce and exportation are substituting the traditional 60.5 kg bags by so-called 'big bags' (with capacity equivalent to 20 bags), which are made of crossed polypropylene and are odourless and long-lasting. These serve to store coffee, for transport to mills, to stuff containers in bulk and for delivery to those industries that are structured for receiving the product in silos. The factors that stimulate use of the big bag are: economy of space, ease in internal transport with forklifts, greater speed and improved stock control.

Some of the changes that take place in the early stages of storage, about 30–40 days of warehousing, may improve quality with the disappearance of the slightly green taste of fresh recently harvested coffee.

During prolonged storage of hulled coffee, colour changes in the bean from green to white are common. This discoloration presents a serious problem since, in addition to depreciating the product in terms of appearance; such beans can present an inferior cup quality, with corresponding losses for producers and exporters. It has been verified that beans proceeding from washed coffee are more affected than those coming from the dried cherry (pod). The first report published on this subject (Bacchi, 1962) indicated that bruising of the beans, such as normally caused by mechanical hulling, is the indirect cause of the whitening of the coffee. Among the extrinsic factors studied, the relative humidity of the air was the most important. The higher this factor is, especially at levels above 80%, the faster and the more intense is the discoloration of the beans.

Another experiment carried out to verify the behaviour of stored beans in different types of packing and storage time disclosed the occurrence, after the sixth month of storage, of an irreversible whitening and an increase in the volume of beans stored in burlap, cotton and paper bags, while beans stored in cans and plastic bags maintained their green colour and a normal volume. The whitening of the beans of stored coffees was more influenced by type of packing than by storage time. This bean discoloration was accompanied by physical (humidity level, level of dry matter, absolute density and bean weight) and chemical (soluble nitrogen content) changes and depreciation in cup quality. The alterations in the physical and chemical properties are attributed to enzymatic degradation processes by the polyphenoloxidases and proteases present in the coffee bean, and are possibly stimulated by atmospheric oxygen and variations in the humidity of the air, according to the results obtained in paper, cotton and burlap packing (Mello et al., 1980).

Several factors affect whitening, such as temperature and relative humidity of the air, ambient light and bean humidity. A study of these variables indicated that, under constant conditions, only at 10 °C of temperature and 52–67% of air relative humidity did the stored coffee beans not lose their original colour after 192 days. In some conditions, even maintaining the bean humidity below 13%, discoloration occurred during the period. The interaction between temperature and relative humidity at higher levels makes the beans start to lose colour within a few days from the beginning of storage (Vilela et al., 2000).

The beans stored in cells of a silo or in bags in warehouses are presented as a porous mass, constituted by the beans and the interstitial, also called inter-granulate, space. The oxygen present in the inter-granulate space is used in the beans' respiratory process. After harvesting, the beans continue to live and, like all living organisms, to breathe. The breathing process is accompanied by the decay of the product's nutritional substances. The rate of decay of the stored beans is accelerated as temperature increases.

Beans stored in dry conditions, with humidity between 11 and 13%, maintain a discreet breathing process. However, if the humidity level is increased, the breathing and, consequently, the deterioration are considerably accelerated (Puzzi, 1973).

In Bacchi's type graphs (Figure 3.16) one can observe that the hygroscopicity curves are of the sigmoid type, whose final portion, corresponding to high relative humidity, presents a sharp rise.

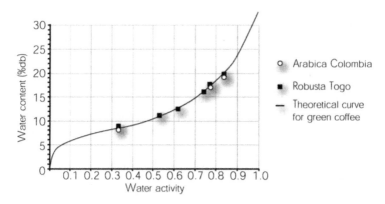

Figure 3.16 Isotherm of moisture content

Temperature, storage time and, most important, moisture are the critical parameters to control for the prevention of spoilage. About 1% of the moisture is present as bound water, a further 4% is weakly bound and the remainder is present as free water. Water availability is measured in terms of activity (a_w), i.e., the ratio between the partial pressure of water in the bean (P) and the partial pressure of pure water (P_0) at the same temperature ($a_w=P/P_0$), and ranges between 0 and 1. Figure 3.17 shows

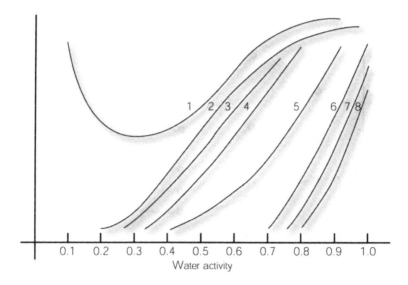

Figure 3.17 Relationship between water activity and main causes of degradation

the dependence of lipid oxidation (1), non-enzymatic browning (2), hydrolytic reactions (3), ascorbic acid loss (4), enzymatic activity (5) and the growth of moulds (6), yeasts (7) and bacteria (8) on a_w (Labuza *et al.*, 1972). The most serious damage to green coffee derives from mould contamination, the most common moulds belonging to the *Penicillium* and *Aspergillus* species. The principal fungal species found in the storage of green coffee are: *Penicillium spp.*, *Aspergillus ochraceus*, *A. niger*, *A. carbonarius*, *A. fumigatus*, *A. tamarii*, *A. flavus*, *Wallemia sebi* and *Eurotium spp.*

Humidity, temperature and storage time are dependent variables. The shorter the storage time, the better tolerated are extreme conditions of humidity and temperature. Such interdependence can be well described by the isochronous storage diagram (Figure 3.18). For instance, for a storage time of 100 days, the relative humidity must be below 75% and temperature below 30 °C.

Relatively minor variations in storage conditions may lead to marked differences: for example, at 35 °C, coffee changes colour after just one month even when it is stored under the best possible conditions, whereas at 30 °C it is still unaltered after four months. After one year, quality is generally reduced, and the beans change colour, becoming white–yellow

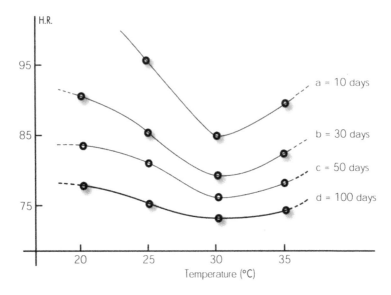

Figure 3.18 Isochronous storage diagram

and fluorescent, their aroma becomes woody and stale while sourness increases. Such decay is matched by a chemical degradation of the free amino acids and sugars, and by oxidation of the lipids; it has also been linked to attack by *Streptococci* and oxidation (Multon *et al.*, 1973).

Laboratory tests have shown that coffee placed under inert gas maintains colour and cup quality for years.

Apart from keeping humidity as low as possible in the storage room, another way to prevent the microbial contamination that damages coffee could be irradiation with gamma rays before the beginning of storage; this method, however, has not been implemented, as it may lead to rejection by consumers. Tests carried out showed that doses ranging between 7.5 and 10 kGy can be used to sterilize coffee in parchment and effectively reduce moulds, and at the same time delay, even eliminate re-contamination, without producing structural or organoleptic alterations (Lopez Garay *et al.*, 1987).

Coffees of different qualities must be kept apart in the warehouse; otherwise defective coffees may rapidly contaminate the good ones. For example, after three and a half months good coffees stored near stinker beans pick up the defect. 'Rio' coffee (see 3.7.2.6) can also contaminate sound coffees.

Green coffee at 11–12% moisture is stored in jute or sisal bags (Figure 3.19). These fibres are treated with paraffin and emulsifying agents to soften and facilitate spinning and weaving into fabrics. The batching oil content of the bag is normally between 2 and 4%, and, if unclean oil is used, it can impart a taint to the coffee beans, with loss of cup quality (Grob *et al.*, 1991a, 1991b). The replacement of mineral oil by neutral vegetable oils is now common, and the risk of contamination has disappeared.

The density of green beans is about $0.7 \, gl^{-1}$, and a lot of one hundred 60.5 kg bags requires $8.3 \, m^3$ of warehouse space (Rigitano *et al.*, 1963).

The market now demands current crop coffees and no longer accepts old crop coffees for blending. There exists, however, a small market

Figure 3.19 Storage

segment that purchases such old crop coffees, where price, not quality, is the most relevant factor. Depending on climatic conditions and ware-housing, the washed coffees and the pulped natural coffees must be commercialized at most 6 months after processing, because the loss in quality is greater than that of natural coffees.

3.6.2 Transport

In recent years, with the delivery of merchandise to consuming countries at the moment of consumption (just-in-time), the operations of transport in the ports and container terminals have become very efficient and shorter in duration. The logistical and transportation conditions vary in the diverse producing origins, from highly developed to rudimentary. Most of the producing countries have their coffee transported in 20 feet containers, usually filled with 60.5 kg bags, sometimes in bulk, rarely in big bags of 1200 kg, while shipment of loose bags in the hold of ships is disappearing due to the high risk of damage to the load.

For shipment, coffee is normally transferred to new bags, with a capacity of 60.5 kg, as used in Brazil (in most countries the bag is 60 kg, while in Colombia it weighs 70 kg). Arabica coffee is still traded according to the number of defects (see 3.7) and to cup quality. Robusta coffee is rarely cup tested by the trade, but the exported quality must contain few defects and foreign materials.

Up to 320 bags can be shipped in 20 feet containers. After arrival at destination, they are stored at roasting installations for a period of 1 to 3 months. Coffee can be transported in bulk in 20 feet containers that can hold the equivalent of 360 bags of 60.5 kg. The use of this system is only possible when the industry at the receiving end has adequate installations. High-quality (speciality or gourmet) coffee should always be transported in 60.5 kg bags in containers, for enhanced quality control.

Containers must be well ventilated and equipped with an appropriate ventilation system, drainage and moisture-collecting system, free from smells or strange odours.

The critical parameters that need to be taken into consideration for transport are once again the relative air humidity, the humidity of the coffee at the time of loading, the changes in the environmental conditions from the port of loading to the port of discharge and the duration of the journey. Controls over the loading and unloading operations are very important. Below follow the most important conditions in the transportation.

■ During container shipment, relative humidity should not exceed 70%; in such conditions the coffee travels safely and the risk of damage is reduced; in case of high relative humidity, the container must be loaded with a smaller load (maximum of 285 bags) and/or stowed in well-ventilated areas.

■ The shorter the duration of the journey, the lower the risk of damage.

■ The more extreme the climatic conditions at loading and unloading (for example: usual tropical conditions when loading and European cold season when unloading), the shorter the duration of the journey should be (Jouanjan, 1980).

Transport in bulk offers several advantages, such as:

■ Enabling the loading of the container with 10–15% more coffee (up to 360 bags).

■ The moisture in the mass of coffee is more constant.

■ There is a saving of US$1 to US$1.50 of the cost of each bag, depending on the origin of the shipment.

■ Loading and unloading are simplified.

■ General costs (insurance, freight, port costs and others) are reduced.

When a container is exposed to the sun in the daytime, water evaporates and then condenses at night in cold spots, giving rise to pockets of moist coffee, which can go mouldy and contaminate the remaining portion of the lot during unloading. This problem can be avoided by completely lining the container with special paper that protects the coffee mass in bulk. The same problem can occur with 60.5 kg bags, but in this case the outer surface of the contaminated bags would present blackish or whitish spots that can easily be recognized and separated. It is important to observe the quality of the container. Before loading, the coffee must be checked, weighed and a phytosanitary certificate issued. This procedure is very important because phytosanitary controls in consuming countries are becoming stricter and more demanding. Some producing countries, such as Brazil, are adopting the use of baits in the interior of the container to control insects.

Once unloaded, the coffee is again weighed, the figures compared with those indicated on the certificate of loading and a phytosanitary certificate is issued. Any resulting difference should be within the limits of the humidity of coffee respiration. Moreover, samples are taken to verify if the lot complies with local import regulations.

Figure 3.20 gives the flow-chart of coffee processing.

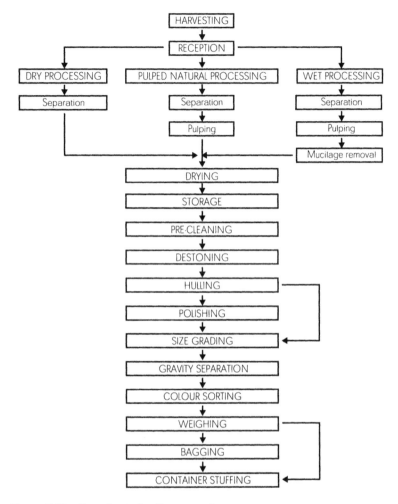

Figure 3.20   Flow-chart of coffee processing

3.7 DEFECTS

R. Teixeira and A.A. Teixeira

Espresso coffee is rich in lipids, which are excellent aroma carriers, making espresso an optimal way to increase the aroma of the cup. Unfortunately, taints coming from defects already present in the green bean can also be enhanced, so that defects may constitute a major problem for the quality of an espresso brew, since all aromas, both good and poor, are obvious before drinking, by smelling the foam. Indeed, it takes about 50 coffee beans to make a cup of espresso coffee, therefore, a low percentage of defects of the order of 1–2%, common in fair quality lots on the market, implies the presence of at least one defective bean among those used to prepare the cup. Furthermore, volatile substances contributing to the defect are often perceived at extremely low levels, masking the pleasant aroma of the sound beans (Table 3.3). For this reason, a brew, still acceptable in spite of defects when prepared by another technique, can become undrinkable with the espresso system.

The dimension of the problem has led the coffee scientific community to devote much effort to the characterization of defects from a chemical, physical and morphological point of view, to the definition of their causes and to the development of suitable techniques that can prevent the appearance of defects. This section will provide a detailed characterization of defects and then discuss means of eliminating them. In addition to the defects that lead to loss of organoleptic quality of an espresso, the physical and morphological defects that hamper good roasting and proper cup preparation will also be discussed.

3.7.1 Characterization of defects

Coffee beans are considered normal when they produce a drink capable of satisfying the consumer. Common market usage is that coffee beans without defects make a sound lot. A defect is anything that diverges from a normal bean inside the lot and that can be produced in the field or during the harvest, processing, transport and storage. In commercial use, it is defined as the number of defective beans and foreign matter present in specific green coffee samples, normally in 300–500 g, or 1000 beans. Most producing countries have their own criteria for the classification of coffee, based on the sum of specific defects, each of which is evaluated in

Table 3.3 Threshold level in water of the most odorous chemicals present in defective beans

Odorant	Odour threshold in water ($\mu g l^{-1}$)
2,4,6-trichloroanisole	0.001
2-methylisoborneol	0.0025
geosmin	0.005

Source: R. Viani, 2003

accordance with established commercial criteria. In an effort to cover all coffee types in a general norm, the International Organization for Standardization (ISO) has produced in the past Standard 10470-1993 – Green Coffee – Defect Reference Chart (ISO, 1993), which met with little utilization. This standard is currently in the final stage of a revision process that aims at a simple and more precise enforcement. Along with a reorganized chart of defects, now also shown in photographs, the main new concepts introduced will be:

- ■ Defects are grouped in five classes, namely
  - ■ non-coffee defects (foreign matter) (see 3.7.1.1 and Table 3.4);
  - ■ defects of non-bean origin (e.g., husks/hulls) (see 3.7.1.2 and Table 3.5);
  - ■ irregularly formed beans (e.g., ears/shells/broken beans/nipped beans) (see 3.7.1.3 and Table 3.6);
  - ■ beans with an irregular visual appearance, with risk of influencing cup taste) (see 3.7.1.4 and Table 3.7);

Table 3.4   Defects associated with foreign matter

Name	Characteristics/ definition	Causes	Brew flavour/ roasting	Origin
Stone*	Stone of any diameter found in a green coffee lot	Inadequate separation/cleaning	Effect mainly economic	H/P
Stick*	Twig of any diameter found in a green coffee lot	Inadequate separation/cleaning	Non-specific downgrading	H/P
Clod*	Granulated lump of soil particles	Inadequate separation/cleaning	Effect mainly economic	H/P

*Found in WPA/WPR/DPA/DPR.
Abbreviations: WPA, wet-processed arabica; WPR, wet-processed robusta; DPA, dry-processed arabica; DPR, Dry-processed robusta.
H, Harvest-damaged beans: inadequate crop management (picking cherries before or after maturation, cherries from the ground, etc.)
P, Process-damaged beans: defects due to imperfect processing operations (pulping, washing, drying, cleaning, hulling, etc.).

Table 3.5 Defects associated with fruit parts

Name	Characteristics/ definition	Causes	Brew flavour/ roasting	Origin
Bean in parchment*	Coffee bean entirely or partially enclosed in its parchment (endocarp)	Faulty hulling or husking of the dry parchment	Non-specific downgrading	P
Piece of parchment*	Fragment of dried endocarp (parchment)	Inadequate separation of the parchment after hulling or husking	Non-specific downgrading	P
Dried cherry[†] (pod)	Dried cherry of the coffee tree, comprising its external envelopes and one or more beans	Incorrect dehusking, allowing whole dried cherries through, not removed subsequently	Foul flavour	P
Husk fragment[†]	Fragment of the dried external envelope (pericarp)	Inadequate separation after dehusking	Foul flavour	P

*Found in WPA/WPR/DPA/DPR.
[†]Found in DPA/DPR.
P, Process-damaged beans: defects due to imperfect processing operations (pulping, washing, drying, cleaning, hulling, etc.).

■ off-tastes (defects identifiable on cup testing only) (see 3.7.1.5 and Table 3.8).
■ Quantifying defects by weight, to allow for exact calculations to be applied to any contract of purchase of green coffee that may be negotiated between provider and client.

Defects can modify cup quality leading to unpleasant flavours, the loss of product due to the presence of foreign matter or the burning of fragments or small beans in the roaster. Some defects, such as mouldiness, which can produce toxins, can also affect consumers' health. Many defects may appear equally in wet-processed arabica (WPA), wet-processed robusta (WPR), dry-processed arabica (DPA) and dry-processed robusta (DPR). Other

Table 3.6 Defects associated with regularity/integrity of bean shape

Name		Characteristics/definition	Causes	Brew flavour/ roasting	Origin
Shell or ear*; shell core*		Part of bean originated from the elephant bean: shell or ear – external part of the elephant bean presenting a cavity; shell core – internal part of the elephant bean	Splitting of the elephant bean (growth defect) generally through handling (dehulling or dehusking), producing the separation of the inner and outer parts	Uneven roast; slight bitterness; less acidity	F/P
Bean fragment*		Fragment of a coffee bean of volume less than half a bean	Formed mainly during faulty dehulling, dehusking or pulping operations. Over-drying making beans easily breakable during handling	Uneven roast; bitterness; less acidity	P
Broken bean*		Fragment of a coffee bean of volume equal to or greater than half a bean	Formed mainly during faulty dehulling, dehusking or pulping operations. Over-drying making beans easily breakable during handling	Uneven roast; bitterness; less acidity	P
Insect-damaged bean*		Coffee bean damaged internally or externally by insect attack	Attack on cherries by *Hypothenemus haempei* (coffee berry borer)	Increased bitterness; less acidity	F

Defect		Description	Cause	Effect	
Insect-infested bean*		Coffee bean containing one or more live or dead insects at any stage of development	Bean attacked by storage pests, generally *Araecerus fasciculatus* insect	Loss of flavour; slight bitterness	S
Pulper-nipped† pulper-cut bean†		Wet-processed coffee bean cut or bruised during pulping, often with brown or blackish marks	Faulty adjustment of pulping machine or feeding with under-ripe cherries or malformed beans	Non-specific downgrading to slightly putrid or stinker	P
Crushed bean*		Crushed beans often partly split and faded with centre-cut largely open	Treading on the beans during drying. Hulling or polishing of soft under-dried beans	Non-specific downgrading to rancid, slightly fermented	P

*Found in WPA/WPR/DPA/DPR.
†Found in WPA/WPR.
Abbreviations: WPA, wet-processed arabica; WPR, wet-processed robusta; DPA, dry-processed arabica; DPR, Dry-processed robusta.
F, Field-damaged bean: defects originating in the field, the coffee tree (genetic problems), the environment (climate, soil, water and nutrient stress), attacks by pests and diseases.
P, Process-damaged beans: defects due to imperfect processing operations (pulping, washing, drying, cleaning, hulling, etc.).
S, Storage damaged beans: defects due to deficient storage (faulty storage practices and storage pests).

Table 3.7 Defects associated with beans irregular in visual appearance (colour and surface texture)

Name		Characteristics/definition	Causes	Brew flavour/roasting	Origin
Black bean*		Coffee bean whose interior is partly (partly black bean) or totally black (endosperm).	Bean from over-ripe cherry, fallen on the ground. Due to *Colletotrichum coffeeanum* attack, other fungi species and pests. Carbohydrate deficiency bean due to poor cultural practices. Mature cherries subjected to over-fermentation by moulds/yeasts and subsequent drying	Slow to roast; beans tend to be yellowish; loss of acidity; harsh; ashy flavour	F/P
Black-green bean*		Unripe coffee bean often with a wrinkled surface, with dark-green (dark-green bean) to black-green silverskin colour	High temperatures affecting immature beans, causing chemical transformation (like fermentation) with the silverskin becoming dark or black-green	Slow roast; rotten fish flavour	H/P
Light-green immature bean*		Unripe coffee bean often with a wrinkled surface. The bean has a greenish or metallic-green silverskin colour. Cell walls and internal structure are not fully developed	Beans from cherries picked before ripening	Slow roast; bitterness; less acidity; astringent; metallic; sometimes fermented	H
Sour (ardido) bean*		Coffee bean deteriorated by excess fermentation, with a range of colours: light to dark brown-reddish, dark brown or yellowish green internally (endosperm). In some cases it has a waxy appearance or a brown silverskin colour	Bean from immature, mature or overripe cherries that have been in adverse conditions, becoming fermented by bacteria or xerophilic moulds, with embryo death. More frequent in fruits fallen to the ground, can also appear in fruits on the tree, and when there is an excessive time between harvesting and drying or pulping	Sour; fermented; acetic; fruity; sulphurous; vinegar flavour	F/H/P

	Description	Cause	Effect	Class
	in weight	physiological problems, nutritional lack, drought or heavily stressed tree	slight loss of acidity	F
Spongy bean*	Coffee bean of consistency like cork that may be verified by pressure of fingernail. Whitish in colour	Moisture absorption during storage or transportation leading to deterioration by enzymatic activity	Roasts rapidly; lack of acidity; woody	S
Whitish bean*	Coffee bean with whitish colour with tissue of normal density	Discoloration of surface due to bacteria of the genus Coccus, during storage or transportation. Associated with old crop. Faulty drying of the bean	Woody; stale; bitterness taste	P/S
Mouldy bean*	Coffee bean with mould growth and evidence of attack by mould visible to the naked eye	Temperature and humidity conditions favourable to mould growth	Musty flavour	P/S
Frost-damaged bean*	Coffee bean with spotted silverskin like a 'spotted quail egg'	Frost damaged coffee bean	Rancid flavour	F

*Found in WPA/WWPR/DPA/DPR.

Abbreviations: WPA, wet-processed arabica; WPR, wet-processed robusta; DPA, dry-processed arabica; DPR, Dry-processed robusta.

F, Field damaged bean: defects originating in the field, the coffee tree (genetic problems), the environment (climate, soil, water and nutrient stress), attacks by pests and diseases.

H, Harvest-damaged beans: inadequate crop management (picking cherries before or after maturation, cherries from the ground, etc.).

P, Process-damaged beans: defects due to imperfect processing operations (pulping, washing, drying, cleaning, hulling, etc.).

S, Storage damaged beans: defects due to deficient storage (faulty storage practices and storage pests).

Table 3.8 Defects associated with beans irregular in cup taste after proper roasting and brewing (off-taste coffee)

Name	Characteristics/definition	Causes	Brew flavour/ roasting	Origin
Foul/ dirty bean*	The bean presents a normal appearance. Unpleasant foul flavour is detected in the cup like earthy, woody, musty or jute-bag	Unfavourable conditions of temperature, humidity and time in processing, storage or transportation. Use of bad quality jute-bag	Musty; earthy, woody, jute-bag flavours	P/S
Stinker bean*	The bean presents usually a normal appearance. A very unpleasant flavour is detected in the cup like stinker, over-fermented or rotten. Stinker smell when cut or scratched	Delay in pulping; too long period of fermentation; wild fermentation in beans trapped in the pulping machines or tanks. Abrasion of beans during pulping, losing the superficial protective layer, leaving them susceptible to microorganism attack. Contamination with recycled polluted water	Stinker; over-fermented, rotten flavour	F/P/S
Rioy bean*	The bean presents a normal appearance. Medicinal smell when cut or scratched.	Overripe cherry left on branch, possibly contaminated by microorganisms. The drying patio soil is heavily contaminated by microorganisms and/or TCA (trichloroanisole)	Medicinal, phenolic, flavour of iodine	F/P/S

*Found in WPA/WPR/DPA/DPR.
Abbreviations: WPA, wet-processed arabica; WPR, wet-processed robusta; DPA, dry-processed arabica; DPR, dry-processed robusta.
F, Field-damaged bean: defects originating in the field, the coffee tree (genetic problems), the environment (climate, soil, water and nutrient stress), attacks by pests and diseases.
H, Harvest-damaged beans: inadequate crop management (picking cherries before or after maturation, cherries from the ground, etc.).
P, Process-damaged beans: defects due to imperfect processing operations (pulping, washing, drying, cleaning, hulling, etc.).
S, Storage-damaged beans: defects due to deficient storage (faulty storage practices and storage pests).

defects are characteristic of a certain type of processing. The different types of defects can be divided into categories, which are discussed below.

3.7.1.1   Defects associated with foreign matter

These include defects such as stones, sticks, clods (agglomerations of earth), metal and foreign matter. They must be removed at an appropriate stage, for example during the cleaning of green coffee, by sieving, classifying or by removal of metals.

3.7.1.2   Defects associated with coffee fruit parts

These include defects such as beans in parchment, pieces of parchment, husk fragments and dried cherries (pods). In general, these are removed by sieving or by air classifying, leading to a loss of physical volume.

3.7.1.3   Defects associated with regularity/integrity of bean shape

These include defects like malformed beans (shell/ear and shell core/body of the elephant), bean fragments, broken beans, insect-damaged beans, pulper-nipped/pulper-cut beans and crushed beans. They are in general removed by densimetric sorting and in some cases by optical sorting.

3.7.1.4   Defects associated with beans irregular in visual appearance (colour and surface texture)

These include defects like black beans, dark- and black-green immature beans, light-green immature beans, sour beans etc. Some of these can be removed by hand or by optical sorting techniques.

3.7.1.5   Defects associated with beans irregular in cup taste after proper roasting and brewing (off-taste coffee)

Defects of sensory concern, to be identified after sample roasting and cup testing, also entailing a further risk of contamination of other beans. These include defects like 'stinker' beans, foul/dirty beans and 'rio' beans. Off-tastes are hard to remove by sorting. They can be identified after cupping a sample of roast and ground coffee, following proper roasting and brewing.

3.7.2 Definition of defects

Many defects have been studied from the physical, chemical, micro-biological and flavour quality points of view.

3.7.2.1 Light-green immature beans

The presence of immature beans, characterized by a light-metallic-green silverskin colour, is due to the harvest of unripe cherries. The flavour of the defect, described as an increased bitterness in both arabica and robusta beans (ISO, 1993), is probably better characterized as metallic and astringent. The evolution of the phenolic constituents during ripening (Clifford et al., 1987; Guyot et al., 1988a, 1988b) may give some indication on the nature of this defect, which has been explained as being due to an excess of dichlorogenic acids with respect to monochlorogenic acids, particularly important before the last five weeks of ripening (Clifford and Ohiokpehai, 1983). This may not be the full explanation since recent studies have shown that 5-caffeoylquinic acid, the major component of the family present in coffee, has an astringent taste of its own (Naish et al., 1993) and so the question remains open. The microscopic examination of immature beans confirms their incomplete ripeness (Dentan and Illy, 1985).

The green grassy flavour perceived in some beans has been shown to be due to incompletely mature beans, which have already reached the size of ripe beans (Dentan, 1991). These beans appear to be normal, but are still unripe, even though they are riper than immature beans.

A concentration of light-green immature beans greater than 15% causes noticeable harm to the brew (Teixeira et al., 1969).

Unripe beans ferment very easily when exposed to temperatures around 40 °C, producing dark-green immature beans or, worse, black-green immature beans (Teixeira et al., 1982). It has been observed that light-green immature beans are sometimes also fermented, which can be perceived when the beans are scratched or during cupping.

Light-green immature beans are characterized by:

- light-metallic-green silverskin colour;
- metallic and astringent taste, reminiscent of the taste of some amino acids, perceivable when the proportion of unripe beans exceeds 1% in espresso;
- vivid silverskin, adhering well to the bean surface;
- cell-walls much thinner than in ripe beans, with a lower cellulose content;
- absence of saccharose, a constituent of sound ripe beans;

- presence of arabinose, absent in ripe beans;
- complete absence of serotonin, formed from tryptophane during ripening;
- greater than average content of chlorogenic acids (up to 17% in unripe beans as opposed to 7% in ripe ones);
- reduced lipid content;
- reduced level of oleic acid and increased level of linoleic acid in total lipids;
- oleic/stearic acids ratio less than 1;
- rather marked minimum at 672 nm in the reflectance spectrum attributable to the presence of chlorophyll.

3.7.2.2 Black-green/dark-green immature beans

Black-green and dark-green immature beans are formed when unripe beans (light-green immature beans) are dried at or exposed to an excessive temperature, above 40 °C (Teixeira *et al.*, 1982). They have black to dark-green silverskin colour and are characterized by astringency and a taste reminiscent of rotten fish, attributed to high levels of cis-4-

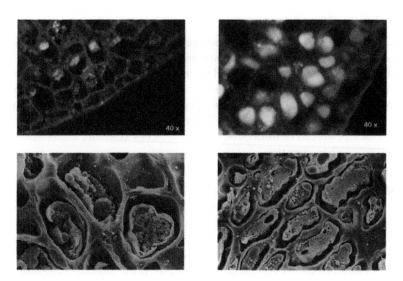

Figure 3.21 Arabica beans
Above: Optical microscope images: (left) immature – many cells are still empty, the cell-walls are thin, the lipids (red) surround proteins (yellow); (right) mature – the cells are full; the lipids (red) have migrated towards the walls
Below: SEM images: (left) immature; (right) mature

heptenal (Full *et al.*, 1999), which makes this a very serious defect. This compound is a product of the auto-oxidation of linoleic acid (Grosch, 1998). An experiment was made of removing the silverskin from beans with these defects and they were classified as sour beans (Teixeira *et al.*, 1971).

Black-green immature beans can be distinguished under the microscope (Dentan, 1989) by:

■ dark to black-green silverskin colour and wrinkled appearance;
■ no microbial attack is apparent, and the content of the cell is congealed into an undistinguished mass;
■ total degradation of proteins in the cells of the surface layers, smaller in the central area;
■ absence of lipids in the peripheral cells, owing to lipase activity;
■ very low reflectance; the minimum present in unripe beans is almost absent;
■ density well below that of a healthy bean, which facilitates densimetric separation.

3.7.2.3 Black beans

Coffee beans whose interior (endosperm) is partially (partially black beans) or totally black. These beans are more frequently found in over-ripe cherries on the tree or fallen on the ground, but can also be encountered at other phases of maturation (Carvalho *et al.*, 1972) or during processing. More than 10% of black beans cause noticeable harm to the brew (Gomes *et al.*, 1967). They have a harsh and ashy flavour.

Black beans can be distinguished by microscopy (Dentan, 1989) by the following characteristics:

■ black beans, where the interior of the bean is also more or less completely black depending on the severity of the attack, are beans having undergone a yeast fermentation starting at the epidermis; the surface of the bean is covered with minute holes surrounded by mineral micro crystals, left after enzymatic degradation of cellulose. The more serious the damage, the blacker is the interior of the bean. When the fermentation has progressed long enough, the black beans are just a light empty shell, which can be separated densimetrically from normal beans;
■ particularly in Africa, black beans, characterized by more or less homogeneously black surface, are due to fungal attack by

Colletotrichum kahawae (*coffeanum*); in this case the beans are often infected by *Aspergilli*, *Penicillia* or other moulds;
■ easy peeling off of the silverskin.

3.7.2.4   Sour beans

Sour (*ardido*) beans are deteriorated by excess fermentation, with a sour taste. They have a range of colours: light to dark reddish-brown, dark brown or yellowish-green internally (endosperm) and sometimes have a waxy appearance, typical of dead beans. This defect has different causes. A study was made removing the silverskin from black-green immature beans and they were classified as sour beans (Teixeira *et al.*, 1971). Further studies indicated that sour beans constitute one phase of coffee deterioration, which ends by reaching a black colour (Gomes *et al.*, 1967; Carvalho *et al.*, 1972). Microscopic examination indicates that they are ripe beans probably killed by overheating during processing, or by infection with xerophilic moulds, such as *Aspergilli* and *Eurotia* (Dentan, 1989, 1991).

They are characterized by:

■ yellow-green to dark reddish-brown colour, sometimes with waxy appearance;
■ sour and fermented smell and taste.

3.7.2.5   Stinker beans

Stinkers are over-fermented beans, usually with normal appearance but a rotten smell and flavour. Morphological analyses show that some stinkers have lost the embryo. Microscopic examination often reveals multiple contaminations with bacteria and moulds (Dentan, 1991). In one specific case only one bacterium, *Bacillus brevis*, was identified. The defect has been associated with the presence of excessive amounts of dimethylsulphide, >0.60 mg/kg instead of <0.30 mg/kg found in healthy arabica coffee, and of dimethyldisulphide, >0.40 instead of 0.15 (Guyot *et al.*, 1991).

Short chain aliphatic esters and acids, formed during lactic fermentation, have also been associated with this defect, and a count of lactic acid bacteria above $10^6/g$ is often associated with a fermented, rotten flavour.

A useful indicator of the presence of stinker beans is the development of red colour by litmus paper suspended above a sample for an hour due to the presence of acetic acid. This reaction is specific to stinker beans and does not occur with other defects or with healthy beans, the faster the reaction, the more serious the defect.

Under examination using ultraviolet light, stinker beans show a white or white-blue fluorescence, probably associated with larger quantities of free caffeic acid compared with healthy beans. Nevertheless, other defects and old crop beans are also fluorescent, making it difficult to sort stinker beans solely by this method.

Stinker beans can contaminate a large batch, even when present at very low levels, turning healthy beans into stinkers at a lower level of intensity (Gibson and Butty, 1975). Beans that have become stinkers only by contact are not fluorescent. A further indicator is headspace gaschromatography, which shows peaks that do not appear in healthy beans.

This defect has been attributed to alcoholic fermentation by *Saccaromyces cerevisae*.

Possible causes of stinkers are the following:

- too long fermentation;
- wild fermentation of beans trapped in the interstices of pulping machines, fermentation tanks, etc.;
- wild fermentation with recycled fermentation water, contamination with polluted water;
- abrasion during pulping, with loss of the superficial protective layer, which renders the bean easily susceptible to attack by microorganisms;
- contamination of healthy beans by stinkers;
- delay in pulping;
- over-heating during processing: at 55 °C, stinkers develop in 4 hours, whereas at 30 °C 4 days are necessary; the most critical processing steps are the storage of cherries before processing, fermentation, and the white stage of drying;
- exposure to over-heating in the sun of the heaps of coffee covered by sheets during drying;
- presence of *Ceratitis capitata* (as larva) in the cherries.

3.7.2.6   Rioy beans

These beans have a flavour described as medicinal and iodine-like. Rioy beans were first detected in Brazilian coffee grown in the Rio de Janeiro area and subsequently in other areas and countries. The substance responsible for this important defect is 2,4,6-trichloroanisole (TCA), which in coffee has an odour threshold of 8/ng/l and a flavour threshold of 1–2/ng/l. By direct olfaction it has been described as dusty, musty, earthy, woody, corky, cereal, iodine-like, phenolic and an oral/retronasal perception of bitterness, burned, rubbery, phenolic, acrid, pungent, earthy, stale and medicinal (Spadone *et al.*, 1990). Rio-tainted beans are heavily infested with moulds (*Aspergilli, Fusaria, Penicillia, Rhizopus*), and bacteria (*Lactobacilli, Streptococci*), accompanied by a degradation of the cell structure (Dentan, 1987). The levels of TCA found in Rioy coffee are usually well above the threshold. 2,4,6-trichlorophenol (TCP) appears to be the direct precursor of TCA in coffee, probably by microbial degradation. The origin of TCP itself is still unclear: some authors maintain that it derives from the chlorophenols used as fungicides, others that it is biosynthesized by a yet unknown microorganism. Of the two hypotheses, the latter is more likely, for only one TCP isomer has been found in Rioy beans, whereas synthetic products should contain all the possible isomers (Spadone *et al.*, 1990). Furthermore, moulds with chloroperoxidase activity are known. *Aspergillus fumigatus*, found in Rioy beans but absent in sound beans, could be the source of the defect; yet, studies carried out on wine, whose corky smell is also due to the presence of TCA, have shown that *Penicillia* can synthesize TCA in the presence of chlorine and methionine. Rioy beans are characterized (Amorim *et al.*, 1977, Dentan, 1987) as follows (Table 3.9):

- external appearance similar to that of a normal bean;
- full or excessive ripeness; the defect is never present in unripe beans;
- contamination with moulds (*Aspergilli, Fusaria, Penicillia, Rhyzopus*) and bacteria (*Lactobacilli, Streptococci*) degrading the cellular structure;
- adjacent cells no longer attached to one another, and both the volume and the thickness of the cell-walls well below those of healthy beans;
- density lower than that of normal beans;
- different microbial populations present inside (*Fusaria, Pseudomonas* and yeasts) and outside (*Aspergillus versicolor, Wallemia sebi*);
- low polyphenoloxidase activity;

Table 3.9 Morphological differences between sound and rioy beans

Sample	Bean weight (mg)	Density	Cell-wall thickness (μm)
Strictly soft	73.03	1.085	6.2
Soft	57.69	1.014	5.6
Rioy	58.21	0.967	5.0
Strong rio flavour	49.24	0.761	4.3

■ lower than average content of hydrolysable phenols;
■ lower than average content of hydrolysable proteins;
■ absence of high and medium molecular weight (150 000 and 64 000 Daltons) proteins, usually present in sound coffees;
■ level of low molecular weight proteins higher than in sound coffees;
■ presence of some low molecular weight (9000 Daltons) proteins that usually do not appear in sound coffees;
■ UV-VIS reflectance spectrum identical to that of a normal bean.

Figure 3.22(a) shows that the cell-walls have been opened by moulds forming channels (the white spots are moulds).

The origin of the defect could be due to several factors:

■ cherries that are contaminated by moulds and/or specific bacteria while still on the plant;
■ the drying stage after harvesting is too slow;
■ the drying patio soil is heavily contaminated by microorganisms and/ or TCA.

3.7.2.7 Whitish beans

The surface discoloration of whitish beans is due to fermentation by *Streptococcus* bacteria that in the most severe cases may also reach the cells under the epidermis (Dentan, 1991). The attack can occur if storage

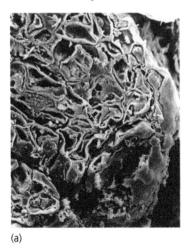
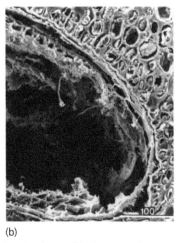

(a)   (b)

Figure 3.22   (a) Degraded cell structure of a rioy bean; (b) electron microscope section of a mould-invaded coffee bean

is too long or in conditions of excessive humidity. White beans are not very aromatic and give slightly bitter and woody cups. Under ultraviolet radiation they show a blue fluorescence probably attributable to chlorogenic acids and caffeic acid, which makes them difficult to sort out from stinkers because of the fluorescent backgrounds they produce.

3.7.2.8 Mouldy beans

A mould/yeast level above 10^5/g is always associated with mustiness in flavour. Geosmin, identified in a heavily rioy and musty tasting sample of Portorican coffee, is probably the substance responsible for mouldiness in beans (Spadone et al., 1990).

These beans have a greyish colour, and under microscopic examination the epidermis, silverskin and central cut appear covered with moulds, mostly Aspergilli (A. tamarii, A. niger, A. ochraceus and A. Flavus, essentially). Pseudomonas bacteria living in symbiosis are also visible. Although the cell-walls are intact, only the peripheral lipids remain inside (Dentan, 1991). In Figure 3.22b hyphes formed by Aspergillus moulds are clearly visible in the central split. The presence of moulds or yeasts (more than 10^5 units per gram) always produces a putrid smell due to the presence of geosmin. The cause of contamination can be either inadequate drying or storage in overly humid and poorly ventilated conditions.

3.7.2.9 Earthy beans

The presence of 2-methylisoborneol, a secondary metabolite of Actinomycetes, Cyanobacteria and moulds, has been associated with the earthy flavour of robusta coffee, with a threshold level in water of 1–100 ng/l, and a flavour described as earthy, musty, robusta-like.

The levels present in robusta coffee are at least three times as high as in arabica (Vitzthum et al., 1990). These data indicate that robusta taste results, at least partially, from contamination by microorganisms rather than from specific aroma components.

Defective beans may have colours varying from clear brown to almost black; the silverskin, but neither the epidermis nor the central cut, is always infected by various microorganisms, bacteria, yeasts and moulds, Fusarium, Geotrichum, Eurotium, are often the major populations present, as well as different Aspergilli (A. flavus, A. fumigatus, A. niger). Although the cell walls are intact, only the peripheral lipids remain inside (Dentan, 1991).

A patent has been registered for a method of eliminating 2-methylisoborneol from robusta. By this method, coffee is treated with saturated vapour at 138 °C, at a pressure of 3.8–3.9 bar, for 75–90 minutes (Vitzthum *et al.*, 1990).

3.7.2.10   Peasy beans

This defect, encountered only in Central African arabica coffees, has been identified as 2-isopropyl-3-methoxypyrazine with a perception threshold of about 300 ppb in air and 0.1 ppb in water (Becker *et al.*, 1987). The defect is due to a contamination of the cherry by a bacterium of the *entherobacteriaceae*, probably transmitted by *Antestiopsis orbitalis ghesquierei* feeding on the cherry (Bouyjou *et al.*, 1993). In the most seriously defective beans, the concentration of 2-isopropyl-methoxypyrazine can reach 2500 ppb as against 70–90 ppb in non-defective beans (a ratio of 35:1).

Peasy beans have a smell strongly reminiscent of fresh green peas. Morphologically, peasy beans appear normal, though they show some fluorescence.

3.8   CLASSIFICATION: PHYSICAL AND SENSORIAL ANALYSIS

R. Teixeira, A.A. Teixeira and C.H.J. Brando

The classification of coffee, which includes physical and sensorial analysis, is a very important phase in the commercialization process. Price quotations, as well as national regulations governing importation into consuming countries, are established on the basis of such classifications.

In order to discuss the quality and production of fine coffees, there is a fundamental need for technicians, producers and purchasers to be aware of the factors that enter into the evaluation of the final item, and to be able to judge its merits and demerits. Unfortunately, the existence of a variety of classification systems means that each country adopts a different classification, requiring equivalency norms for use at the international level, such as ISO 10470-1993 – Green Coffee – Defect Reference Chart (ISO, 1993) and ISO 4149-1980 – Green Coffee – Olfactory and Visual Examination and Determination of Foreign Matter and Defects (ISO, 1980), which are currently being revised.

In Brazil, it is customary to classify by type or defects as well as by cup quality. In Colombia, coffee is classified by the characteristics of the plantation, with regard to altitude, by bean size and by region of origin. Countries of Central America (El Salvador, Honduras, Mexico, Nicaragua, Guatemala, Costa Rica and Panama) have altitude as the main criterion, but also take into account appearance, bean size and cup quality. In Africa, the classification varies according to the area of production. Angola, Cameroon and the Ivory Coast have adopted systems similar to that of Brazil, which uses defects to establish types. In Kenya and Tanzania, coffee is classified by ranking in classes, in accordance with bean size and cup quality, in some of which defects are permissible. Indonesia classifies its coffees in accordance with the species, method of preparation, origin and number of defects (Jobin, 1982).

Overall, classification is based on the evaluation of all of the parameters related to the coffee bean and each one of them must be examined, separately, by physical methods and by sensory analysis (Table 3.10).

3.8.1 Classification by species and varieties

Coffee is classified by species and variety of origin. Only two easily distinguished species are important in economic terms: arabica and robusta. Arabica fetches a higher price in the commercial market, is more delicate, demands greater care in its culture, and possesses superior organoleptic characteristics in comparison with robusta, which is more resistant, has a lower cost of production, but results in an inferior cup quality.

3.8.2 Classification by screen (size and shape)

Different producing countries use different classifications of bean size. In Brazil, beans are classified according to size and shape. With regard to shape, the coffees receive the denomination of *chatos* or flats (long shape), and peaberries (round shape), also called *mocas* or *caracoles*. The size or

Table 3.10 Parameters used in commercial classification

Parameters	Description
Species and varieties	Arabica /robusta
Classification by screen size	Size and shape of the beans
Classification by type (number of defects)	Defective beans and foreign matter
Density	Specific weight of the beans
Appearance and dryness	Bean uniformity
Colour	Coloration of the beans
Processing	Natural (dry), pulped natural and washed (wet)
Roast	Roast regularity, smell, etc.
Cup quality	Characteristic aroma and flavours

screen of the beans is measured by the dimension of the holes in the screen that holds them back, designated by numbers that, divided by 64, express the size of the holes in fractions of an inch (see 3.5.3 above). Screens with round holes are used for *chatos* (varying from 19 to 10) and ones with elongated holes are used for peaberries (13 to 8). Some countries use screens with holes measured in millimetres.

Separation by screens is important, since it allows the selection of beans by size, separating them into groups adequate for uniform roasting. In Colombia, coffee is classified as 'Supremo', 'Excelso' and 'Pasilla' (the latter not exportable); in Kenya the designations used are AA, AB, PB, C, E, TT, T, UG, which indicate size and shape (Jobin, 1982).

3.8.3   Classification by type (Figure 3.23)

Most producing or importing countries have their own classification system for defects. Commercial classifications are established in accordance with the number of visual defects in a certain sample. At the international level, the 'New York Coffee and Sugar Exchange' introduced the concept of black bean equivalent. This system uses the black bean as a standard of measurement for all defects (for example: 1/2 means 2 defects are equivalent to 1 black bean). Other bodies have introduced similar scales for counting green coffee defects (Table 3.11).

The International Standardisation Organisation has established several norms regarding olfactory and visual examinations, determination of defects and foreign matter (ISO, 1980), determination of the proportion of beans damaged by insects (ISO, 1985), and the defect reference table (ISO, 1993), that deal with the majority of defects. Currently, several norms are being revised, including the defect reference table now in the draft stage, where each defect is defined and characterized according to its degree of influence (minimum, medium or maximum) in relation to the

Figure 3.23   Classification of coffee: (left) size determination; (right) olfactory examination

Table 3.11 Equivalency ratings of green coffee defects in different countries (ISO, 1993)

Defect	USA Brazils	Centrals	Colombian	UK Liffe	France 1965	Brazil	Tanzania 1992	Indonesia 1982	Ethiopia 1973
Dried cherry (pod)	1	1	1	1	1	1	1	1	2
Black bean	1	1	1	1	1	1	1	1	2
Semi-black bean	–	1/2–1/5	1/2–1/5	–	1/2	–	1/2	1/2	1/2
Sour bean	1	1	1	1/2	1	1/2	1/8	–	–
Insect-damaged	1/5–1/10	–	–	1/2–1/5	1/10	1/2–1/5	1	1/5–1/10	1/2
Immature bean	1/5	–	–	1/5	1/5	1/5	1/2	1/5	1/5
Floater (white light)	1/5	1/5	1/5	1/5	1/3	–	–	–	1/5
Bean in parchment	–	–	–	–	1/2	1/2	1/5	1/2	–
Broken bean									
More than half	1/5	1/5	1/5	1/5	–	1/5	1/5	1/5	1/10
Less than half	1/5	1/5	1/5	1/5	1/5	1/5	–	1/5	1/5
Shell	1/3	1/3	1/5	1/5	1/5	1/3	1/5	–	–
Husk fragments									
Large	–	–	–	1/2	1	1	1	1	3
Medium	1/2	1/2	1/3	–	–	1/2	1/2	1/2	1
Small	–	–	–	1/5	–	1/3	–	1/5	1
Parchment fragments									
Large	–	–	–	1/2	–	–	–	1/2	1
Medium	1/2	1/2	1/3	–	–	–	–	1/3	1
Small	–	–	–	–	1/3	–	–	1/10	–
Twigs									
Large	2–3	2–3	2–3	5	2	5	5	5	10
Medium	1	1	1	2	1	2	2	2	5
Small	1/2–1/3	1/3	1/3	1/2	1/3	1	1	1	3
Stones									
Large	2–3	2–3	2–3	5	–	5	5	5	10
Medium	1	1	1	2	–	2	2	2	5
Small	1/2–1/3	1/3	1/3	1/2	–	1	1	1	3

Figures in columns give defect values (i.e. 1/2 means: 2 defects equivalent to 1 black bean as the reference).

loss of mass and to organoleptic properties (sensorial concern). A formula is being introduced that intends to measure the total impact on quality, to be calculated by multiplying the percentage of defects in a sample by the coefficient of sensorial concern, according to the values found in the table.

Brazilian type classification, widely used in the international trade, provides for seven types with decreasing values from 2 to 8 resulting from the analysis of a 300 g sample of milled coffee, according to the norms established in the 'Brazilian Official Classification Table' (Table 3.12). Each type has a greater or lesser number of defects consisting of beans that have been altered in the field, harvesting, processing or storage (e.g., physiological or genetic origin and presence of foreign matter).

3.8.4 Appearance and humidity

Appearance and humidity are fundamental factors in the evaluation of quality, since they serve to predict a good or bad roast and other indications of the quality of the final product. Appearance is good when the majority of the beans are perfectly formed, uniform in size, colour and humidity, and bad when the majority of the beans are not uniform and defective beans are found. In coffee that has been correctly dried, the humidity level should be $11\pm0.5\%$ for natural coffees, washed and pulped natural coffees. Incorrect drying can be identified by spotted or humid beans.

3.8.5 Colour

Colour is always associated with quality. A bluish-green colour is very desirable in washed coffee, being considered a sign of high quality and freshness, while a yellowish colour is a sign of old coffee and low quality.

Table 3.12 Official Brazilian classification of green coffee (*Classificação Oficial Brasileira – COB*)

Grade	Number of defects in 300 g of coffee
COB 2	4
COB 3	12
COB 4	26
COB 5	46
COB 6	86
COB 7	160
COB 8	360
Unsuitable for export	>380

Factors that contribute to colour variations are: degree of dryness, time of exposure to light, processing method, storage conditions, bruising, polishing, etc. The polishing of the bean improves its appearance but hides defects, due to the removal of the silverskin (Teixeira *et al.*, 1985). The ISO recommends the following colour classification: blue, greenish, whitish, yellowish and brownish. The Brazilian classification adopted for export purposes is: green, greenish, pale, yellow and old (Teixeira *et al.*, 1970).

3.8.6 Processing

Coffee can be classified according to processing method as natural (dry-processed), pulped natural, and washed (wet-processed) coffee. The processing system can be recognized by the bean colour and by the appearance of the silverskin. Washed coffees possess a characteristic shiny, translucent and green-bluish colour. Natural coffees have a semi-opaque colour and a yellowish or even brown skin. Pulped natural coffee has an intermediate aspect.

3.8.7 Roast

The analysis of the roast is an important step, and an important gauge of quality. Defects that are not observed in the raw beans come out in the roast. The immature and sour beans become yellowish while black beans appear to be burnt. Broken beans, shells and shell core beans, due to their reduced volume in relation to perfectly formed beans, become darker. In Brazil, the roast is classified in accordance with appearance and, for natural coffees, it can be measured as: fine, good, regular and bad according to the degree of uniformity. The roast of washed *and* fermented coffee is considered to be characteristic when the majority of the beans present a clear and distinct silverskin in the ventral ridge of the bean (Teixeira and Ferraz, 1963). The roasting of natural coffee presents a brown silverskin. Pulped natural coffees and those with mechanically removed mucilage present a roast intermediate between those of natural and washed coffees.

3.8.8 Cup quality (Figure 3.24)

Even today, greater importance is attached to physical rather than organoleptic classification. A possible reason is the fact that when these criteria were introduced well-selected washed arabicas prevailed in the

Figure 3.24 Cup tasting

market and such a classification corresponded closely to the quality of the coffee in the cup. Another reason is that classification systems were introduced many years ago when knowledge about the subject of cup quality was limited. The obsolescence of classification systems can sometimes lead to paradoxical situations, where good coffees are classified as low grade, due to bean size, while mediocre coffees are well evaluated and expensive, even if they are poor from an organoleptic point of view.

According to the International Trade Centre (ITC, 2002) the cup or liquor can be characterized by the following terms:

■ 'Acidy – A desirable flavour that is sharp and pleasing but not biting. The term 'acid' as used by the coffee trade refers to coffee that is smooth and rich, and has verve, snap and life as against heavy, old and mellow taste notes.'
■ 'Aroma – Usually, pleasant-smelling substances with the characteristic odour of coffee.'
■ 'Body – A taste sensation or mouth feeling ... of a drink corresponding to a certain consistency or an apparent viscosity ... Sought after in most if not all coffees.'

Organoleptic characteristics vary according to the producing country and this parameter must be considered as specific to each commercial origin. In general, all washed arabica coffees from Eastern Africa (such as Kenya, Tanzania and Ethiopia) as well as from Central and South America (such as Guatemala, Costa Rica and Colombia), are marked by a degree of acidity and an intense aroma, although each origin has different characteristics. Natural dry-processed arabica coffees, especially those from Brazil, are less acid and have a less marked aroma but are endowed with a richer body ideal for espresso coffee. Finally, robusta coffees are characterized by a wooden and earthy flavour.

Classification systems that take organoleptic characteristics into account are used in some producing countries; unfortunately, this type of classification has only a small influence on the price. Brazilian coffees are classified as to cup quality according to the following ranking (from best to worst): strictly soft, soft, softish, hard (or hardish), rioy and rio. Kenyan coffees are graded by size and density into seven grades, divided into 10 classes by liquoring (fine, fair, fair to good, etc.). Colombia and Central American countries take altitude into consideration (Jobin, 1982). In addition to this lack of uniformity in classification systems, each producing country weighs organoleptic defects in a different way, which may lead to discrepancies in sensorial evaluation between producing and consuming countries.

3.9 BLENDING

G. Brumen

The blending of coffee is as old as coffee itself. Although the techniques vary, blending is used to optimize aroma, body and flavour: the goal is to make a coffee that is higher in cup quality than any of the ingredients individually, and, extremely important, maintain consistency in the final roasted product.

Each batch has it own personality in terms of taste, smell, body, chemical resilience to the hydrolytic action of water, etc., and blending can complete it and round it up or level it off.

Most espresso blends are based on high quality Brazil arabicas, some washed, some dry-processed. They often involve some African coffees for winey acidity or flowery fruitiness, or a high-grown Central American for a clean acidity. Some roasters add a little robusta to increase body.

Dry-processed coffees are responsible for the attractive 'crema' on the cup, among other mechanical factors in the extraction process (see 8.1.1). Wet-processed Central Americans add positive aromatic qualities. Robustas are used in cheaper blends to increase body and produce more foam.

Besides subjective quality (see Chapter 1), blending also assists in maintaining objective quality, because the more complex a blend, the easier it is to maintain constant quality when some ingredients change.

With the exception of a few countries that pay considerable attention to quality, the majority of producer countries often add up small batches produced by different growers to form larger ones of a size required by roasters. Although care is taken so that only batches of equivalent quality are blended, the result of this deplorable practice is often a quality downgrading to a level below that of the best fractions.

Coffee history records a number of popular blends that are published and available for public consumption. Other 'proprietary' blends tend to be closely guarded, with the information staying within a company structure. Proprietary or signature blend leads consumers to equate a particular coffee profile with a particular brand image. Blending requires the expert skill of knowing each ingredient coffee, having in mind a clear cup profile as the goal, and knowing how to achieve it.

Blending may be done before or after roasting. Blending before roasting is traditionally used by retail and institutional roasters. In this method coffees with similar characteristics are combined and roasted to the same development. Generally, professional in-house 'cuppers' evaluate the results of the blend, adjusting components if necessary to satisfy taste requirements and standards.

■ *Advantage*: Consistency of product.
■ *Disadvantage*: Inability to optimize the character of each coffee.

Blending after roasting is the method traditionally used by many speciality coffee roasters. The flavour profile development requires that each individual coffee used in the blend be roasted separately to optimize flavour. In other words, each coffee will have a different time and temperature setting. Consequently, the final roast development will be different for each coffee used in the blend. After roasting, each component of the blend is individually tasted (cupped), as is the final blend composition.

■ *Advantage*: Ability to optimize the character of each coffee.
■ *Disadvantage*: Inconsistency of product.

3.10 DECAFFEINATION

O.G. Vitzthum

A major impact compound, as well as the pleasant aroma and flavour components in roast coffee, is caffeine; it is the physiologically most active substance in coffee.

Caffeine stimulates the central nervous system, and is responsible for the vigilant effect of the coffee beverage. However, some people can suffer from insomnia and restlessness after drinking coffee in the evening. People who habitually drink coffee may develop a certain tolerance to the drink, whereas non-coffee drinkers experience a stimulating effect even from a single cup of coffee (for a discussion of the physiological effects of caffeine see 10.3.2).

It was the German writer J.W. von Goethe who, in 1820, gave some coffee beans to the chemist Runge (1820) requesting him to isolate the pharmacologically active compound therein. He found the alkaloid caffeine, a colourless, slightly bitter-tasting substance. The name was derived from the botanical name *Coffea* for coffee.

The idea of reducing or eliminating this physiologically active ingredient in coffee originated at the end of the nineteenth century. It was after the frustrating experiments of his chemist Meyer, who tried to extract the caffeine out of green coffee beans with caffeine-dissolving solvents, that Bremen coffee merchant L. Roselius had the successful idea of using raw coffee beans pre-swollen with water for decaffeination with solvents (Roselius, 1937). He knew that by using a steam treatment the volume of green coffee could be increased by about 100%. Consequently the solvent was able to penetrate more easily into the swollen beans and dissolve the caffeine. Water-immiscible solvents are necessary to avoid the extraction of the flavour precursor water-soluble components. The hydrophilic caffeine fortunately dissolves well both in water and in various solvents.

The first commercially decaffeinated coffee was produced in Bremen as Kaffee HAG at the beginning of the twentieth century. Due to the patent protection for this invention, many attempts were made in the early 1920s and 1930s to apply other solvents and processes in search of a legal circumvention of this technique. Until 1970 two solvents were successfully in use for the decaffeination of raw coffee beans or aqueous extracts thereof: methylene chloride or dichloromethane (DCM) and ethylacetate (EA). A new approach for the decaffeination of coffee was initiated by Zosel (1971) when, in 1970, he suggested the use of supercritical carbon dioxide. This activity spurned the development of other processes such as water extraction and the use of adsorbents or fats and oils.

Nowadays conventional processes such as extraction with DCM and EA are in use as well as various technologies based on decaffeination with carbon dioxide or water and certain extraction aids.

3.10.1 Processing

A comprehensive description of decaffeination principles with technical details has recently been published by W. Heilmann (2001). Here, only the various decaffeination techniques will be described.

3.10.1.1 Conventional decaffeination

The procedures for decaffeination by the solvents dichloromethane CH_2Cl_2 (DCM) and ethylacetate $CH_3COOC_2H_5$ (EA) are similar, although more coffee solids are lost during EA decaffeination. There are direct and indirect solvent-based decaffeination processes.

In the **direct solvent decaffeination** process (Figure 3.25), after cleaning, green beans are swollen by addition of water or steam into the extraction vessel. Thereafter solvent is added and decaffeination started by recycling of the solvent under elevated temperature conditions (70–100 °C). Processing continues for 8–12 hours, whereby the solvent is continually replaced by fresh re-distilled material. In a separate vessel the caffeine-rich solvent is distilled off and led back to the decaffeination vessel. After a corresponding refining process, the crude caffeine is pure enough for use in caffeinated beverages. Subsequent to completion of the decaffeination process, the residual solvents are eliminated by steaming. The beans are dried and are then ready for roasting. Decaffeinated beans are usually lighter in colour than the original green raw coffee beans.

In the **indirect solvent decaffeination** process (Figure 3.26), beans do not come into direct contact with DCM (Berry and Walter, 1943). A saturated aqueous extract of green beans, containing all water-soluble components including caffeine, is decaffeinated by the solvent in a liquid–liquid extraction process in a separate vessel. The caffeine-free

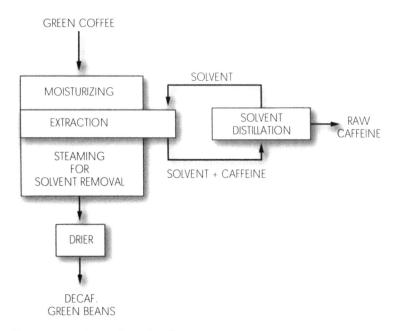

Figure 3.25 Direct solvent decaffeination

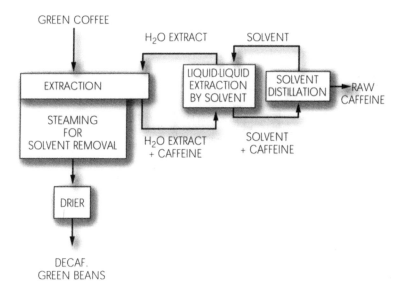

Figure 3.26   Indirect solvent decaffeination

extract after decaffeination is returned to the green beans and the process repeated using a new batch of green beans. Any traces of solvents present in the returned caffeine-free extract must be removed by subsequent steaming. After drying, the beans are ready for roasting.

3.10.1.2   Modern decaffeination

A revolution in decaffeination evolved with the invention of caffeine removal by **supercritical carbon dioxide** (Figure 3.27) (Martin, 1982; Lack and Seidlitz, 1993). This process uses the specific solvation power of gaseous carbon dioxide under elevated pressure conditions for caffeine. CO_2 is in a supercritical physico-chemical state above its critical temperature of $31\,^{\circ}C$ and, although still a gas, it has nearly the density of a liquid if pressures of about 200 atm are applied.

In comparison with conventional solvents, the solubility of caffeine in supercritical CO_2 is low; but multiple recycling of the gas through the green beans eventually leads to complete caffeine extraction. Caffeine is removed in every cycle from the gas stream by an effective adsorber such as active carbon or an ion-exchange resin. The advantage of the process lies in the fact that CO_2 is highly selective in extracting the caffeine out of the beans so that components that contribute to the aroma-yielding fraction in coffee are not solubilized.

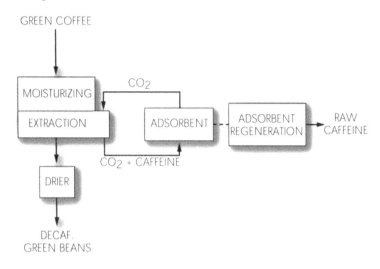

GREEN COFFEE

MOISTURIZING

EXTRACTION

CO_2

ADSORBENT

ADSORBENT REGENERATION

RAW CAFFEINE

CO_2 + CAFFEINE

DRIER

DECAF. GREEN BEANS

Figure 3.27   Supercritical CO_2 decaffeination

The **indirect adsorbent decaffeination** process (Figure 3.28) (Fischer and Kummer, 1979; Blanc and Margolis, 1981) is similar to indirect solvent decaffeination, except that the solvent is replaced by an extract that is saturated with the water-soluble compounds of the coffee material. Such a heated aqueous extract extracts caffeine from the fresh beans; thereafter the caffeine is selectively removed from the solution in a separate vessel by a 'caffeine-selective' adsorbent such as coated active carbon or non-ionic microporous resins. These serve to more or less selectively extract all the caffeine but not the aroma precursor containing water-soluble substances. This extract, which now contains no caffeine, is recycled back to the extraction vessel and caffeine is extracted again. The treated active carbon may be pH-adjusted or coated by carbohydrates to block the hydrophilic sites on the adsorbent. After use the adsorbents must be regenerated accordingly.

3.10.1.3   *Organoleptic comparison of the different techniques*

Although slight differences in taste are noticeable between the same blend, decaffeinated by MC, supercritical CO_2 or water, if the decaffeination process and the roasting are carried out correctly the difference in the cup with the corresponding non-decaffeinated blend is minimal.

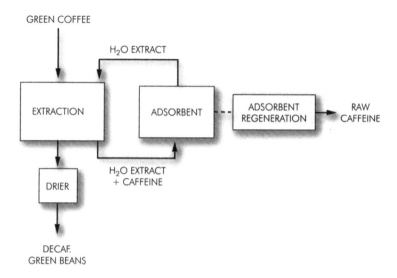

Figure 3.28 Indirect adsorbent decaffeination

If the process is not carried out under optimal conditions, typical defects appear at cupping:

■ MC-decaffeinated coffee has the 'cooked' flavour, associated with decaffeinated coffee.
■ CO_2-decaffeinated coffee is flat.
■ Water-decaffeinated coffee is thin, due to the loss and/or poor re-incorporation of the soluble solids co-extracted with caffeine (this defect is often masked by increasing roasting level).
■ EA-decaffeinated coffee has always the typical 'cooked' flavour associated with (poorly) decaffeinated coffee.

3.10.2 Perspectives for caffeine elimination

Whereas no major developments are expected for decaffeination processing in the near future, the application of molecular biology and biotechnology to coffee may yield new naturally caffeine-free coffee plants within 5–10 years. The deciphering of sequences of coffee genes is progressing and biochemical pathways for the formation of caffeine in the plant are being elucidated (Stiles *et al.*, 2000).

The possibility of cultivating coffee plants without caffeine by recombinant DNA technology is approaching. Ogawa *et al.* (2001)

recently reported on the successful cloning of a corresponding methyl-transferase, which is an important regulator enzyme in the biochemical pathway of caffeine formation in the plant.

3.10.3 Regulatory aspects

3.10.3.1 Caffeine residue

In European countries the maximum allowed caffeine content in coffee beans is 0.1%, in soluble coffees, 0.3%. In the USA at least 97% decaffeination (equivalent to 99.7% caffeine-free coffee) is common in decaffeinated coffees.

3.10.3.2 Solvent residue

No regulations exist for indirect decaffeination via coated activated carbon or fats and oils. No restrictions are necessary for ubiquitous liquid or supercritical carbon dioxide.

The following regulations regarding solvent residues in roasted coffee beans exist for dichloromethane:

■ 2–5 ppm in Europe, 10 ppm in South America. Practical observations under the provision of good manufacturing practices (GMP) yielded residues of <1 ppm in European DCM decaffeinated coffees.
■ GMP have been requested specifically for EA; in Italy the residues for EA are restricted to 15 ppm in roasted coffee.

3.10.4 Decaffeination capacity

The approximate (best estimate) decaffeination capacities from 1998 are given in Table 3.13 (opposite). Consumption of decaffeinated coffees is highest in Germany, Switzerland and the USA (12–18%).

3.11 RAW BEAN COMPOSITION

I. Kölling-Speer and K. Speer

The beans of *Coffea arabica* (arabica coffee) and *Coffea canephora* (robusta coffee) vary in mass from 100 mg to over 200 mg, with some evidence of variation associated with the geographical origin. Both species show quantitative and qualitative differences in their chemical compositions. Arabica contains more lipids and trigonelline, while robusta contains more caffeine and chlorogenic acids. Minor constituents specific to one

Table 3.13 Estimated decaffeination capacity by country

Country	Solvent used	Capacity (tonnes)
Germany	DCM; supercritical CO_2; liquid CO_2; EA	160 000
France	DCM; water	45 000
Italy	DCM; EA	16 000
Portugal	DCM	5 000
Spain	EA; water	18 000
Switzerland	DCM; EA; water	8 000
Total Europe		**252 000**
Canada	DCM; water	12 000
USA	DCM; EA; water; supercritical CO_2	120 000
South America	DCM; EA; water; fats and oils	50 000
Asia	DCM; EA	6 000
World total		**440 000**

species have also been identified. The composition of raw coffee beans is indicated in Table 3.14.

3.11.1 Water content

The water content of green beans, from dry or wet processing, influences the water activity and stability during storage and varies in Europe from 9 to 13%. Dry-processed robusta coffees (Vietnam) sometimes show higher values. For

Table 3.14 Composition of raw coffee beans (% dry matter)

Class	Arabica	Robusta
Caffeine	1.2	2.2
Lipids	16	10
Chlorogenic acids	6.5	10
Aliphatic acids	1.7	1.6
Total aminoacids	10.3	
• free	0.5	0.8
Trigonelline	1.0	0.7
Glycosides	0.2	Traces
Minerals	4.2	4.4
• potassium	1.7	1.8
Carbohydrates (by difference)	58.9	60.8

storage and transportation the moisture level proposed by the European Coffee Federation (ITC, 2002, p. 254) is 12.5%. Below 10% the germinating power decreases and, furthermore, the beans show crack formation. At water contents >12.5% there is a significant risk of microbiological spoilage. An international reference method for moisture analysis (ISO, 1983) should be used to standardize the rapid methods now available.

3.11.2 Ash and minerals

Coffee contains about 4% of mineral constituents (arabicas: 3.6%–4.5%; robustas 3.6%–4.8%), with potassium amounting to 40% of the total. The level is slightly higher in dry-processed robusta and arabica than in wet-processed arabica (see Table 3.2). The main minerals of raw coffee beans are given in Table 3.15; fertilizing may change the contents (Amorim et al., 1973).

Coffee has the highest content of rubidium, which has been analysed in food (arabica: 25.5–182 mg/kg dry matter; robusta: 6.6–95.2 mg/kg dry matter). Quijano Rico and Spettel (1975) determined higher content of copper, strontium und barium in robustas than in arabicas.

Table 3.15 Mineral content of raw coffee (dry base)

Mineral	Percent
Potassium	1.63–2.00
Calcium	0.07–0.035
Magnesium	0.16–0.31
Phosphate	0.13–0.22
Sulphate	0.13

3.11.3 Carbohydrates

The total amount of carbohydrates represents about 50% dry base of green coffee. The composition is complex with a range of different poly-, oligo- and monosaccharides (Table 3.16).

Lüllmann and Silwar (1989) analysed the low molecular weight carbohydrate profile of 20 green coffee samples from 13 different producer countries. Sucrose content varied from 6.25% to 8.45% for the arabica samples, and from 0.9% to 4.85% for the robusta samples. Besides sucrose,

Table 3.16 Carbohydrates in
raw coffee (% dry base)

Constituent	Arabica	Robusta
Monosaccharides	0.2–0.5	
Sucrose	6–9	3–7
Polysaccharides	43.0–45.0	46.9–48.3
Arabinose	3.4–4.0	3.8–4.1
Mannose	21.3–22.5	21.7–22.4
Glucose	6.7–7.8	7.8–8.7
Galactose	10.4–11.9	12.4–14.0
Rhamnose	0.3	
Xylose	0–0.2	

small amounts of the monosaccharides fructose, glucose, mannose, arabinose and rhamnose have been identified, traces of reducing sugars being higher in robustas.

There was no evidence in green coffees of other simple oligosaccharides such as raffinose or stachyose. Maltose was determined only in one robusta sample.

It would be expected that sucrose content would increase with degree of ripening and this is apparent with defective coffee beans, where for both immature-black and immature-green Brazilian robusta beans, sucrose levels were found to be one-third and one-fifth of the level of normal beans, respectively (Mazzafera, 1999). This was also observed for Vietnamese robustas, where black beans sorted from a sample contained 0.9% sucrose; normal beans from the same batch contained 4.0% sucrose.

Glucose content has been negatively correlated with aroma level and positively with cup sweetness, while fructose content has been negatively correlated with sweetness (Illy and Viani, 1995). Franz and Maier (1993, 1994) identified inositol tri-, tetra-, penta- and hexaphosphate (phytic acid). Total inositol phosphate content in robusta tended to be higher than in arabica, ranging from 0.34 to 0.40% and 0.28 to 0.32%, respectively. Polyalcohols have also been found in green coffee. Noyes and Chu (1993) found low levels of mannitol in blends of Brazilian robustas and arabicas (average content 0.027%).

The chemical structure of green coffee polysaccharides has been extensively studied by Bradbury and Halliday (1987, 1990) and Fischer *et al.* (2000). The soluble fraction is composed of sucrose and of polymers of galactose, arabinose and mannose (galactomannans and arabinogalactans). The insoluble cell constituents include 'holocelluloses', which contain, in addition to some cellulose, the very hard β-1,4-mannan, and a small amount of 'hemicellulose', mainly arabinogalactan. Other polysaccharides common to the plant kingdom such as starch or pectin are present, if at all, at only low levels in mature coffee beans. No lignine, characterized as polymers of xylose, is present in the seed.

3.11.4 Glycosides

A series of diterpene glycosides, the atractylglycosides, was identified in green and roasted coffee in the 1970s by Spiteller's group (Ludwig *et al.*, 1975; Obermann and Spiteller, 1976). The contents in coffee of the principal components of this compound class, termed KAI, KAII and KAIII were determined by Maier and Wewetzer (1978) and Aeschbach *et al.* (1982). Bradbury and Balzer (1999) have shown that these glycosides in green coffee actually contain two carboxyl groups at C-4, and consequently, the compounds were identified as carboxyatractylglycosides. The carboxyl group is labile, thus explaining why earlier studies only observed the atractyl forms. Arabica beans contain significantly more of the glycosides than robusta beans.

3.11.5 Carboxylic acids

Data on acids in green coffee (Table 3.17) came from Kampmann and Maier (1982), Scholze and Maier (1983, 1984), Engelhardt and Maier (1984, 1985) and Wöhrmann *et al.* (1997). The total content is similar in arabica and robusta coffees (average 1.7%) according to Van der Stegen and Van Duijn (1987).

Quinic acid, the aliphatic acid moiety in the chlorogenic acids, is also present in the free state. Its content is presumably affected by factors such as processing, fermentation and age. Free quinic acid content can increase to 1.5% in old beans. Besides quinic acid, the major acids in green coffee are malic and citric acids and phosphoric acid. Minor coffee acids were identified by Bähre (1997) and Bähre and Maier (1999).

The acidity of brewed coffees is an important organoleptic character and is associated with the best high grown arabicas, such as Kenyans.

3.11.6 Chlorogenic acids

The chlorogenic acids are a group of phenolic acids esterified to quinic acid (Clifford, 1999; Parliment, 2000). This class of acidic compounds

Table 3.17 Contents (g/kg) of quinic, malic, citric and phosphoric acids of different raw coffees

Species	Origin	Quinic acid	Malic acid	Citric acid	Phosphoric acid
Arabica	Santos	5.6	6.1	13.8	1.1
	Burundi	5.7	5.1	13.0	1.1
	Kenya	4.7	6.6	11.6	1.4
	Colombia	5.5			
	Mocca		4.6	10.5	1.5
Robusta	Burundi	3.5	3.8	10.0	1.4
	Angola		2.8	9.2	2.2
	Togo	3.1	2.5	6.7	1.6
	Guinea	3.9			
Arabusta	Ivory Coast		4.5	9.1	1.7
Excelsa	Ivory Coast		5.2	10.9	1.5
Stenophylla	Ivory Coast		2.9	10.0	2.2
Liberica	Unknown		3.9	10.6	1.8

Source: Kampmann and Maier, 1982; Scholze and Maier, 1983, 1984

accounts for up to 10% of the weight of green coffee. Mono- and di-caffeoylquinic acids have been identified in coffee with substitution at the 3-, 4- and 5-position of quinic acid. The phenolic acid fraction of coffee is composed of caffeoyl-, p-coumaroyl- and feruloyl-acids (Figure 3.29, see page 154).

In coffee these components are essentially mono- and diesters. The monoesters decrease during the unripe to semi-ripe stage then increase to the ripe stage. This is followed by a decrease in the slightly overripe stage, more important with wet than with dry processing (Castel de Menezes and Clifford, 1988).

3.11.7   Amino acids, peptides and proteins

Free amino acids are present in raw coffee beans at levels about 1% (Trautwein, 1987). Post-harvest treatments influence the contents of individual free amino acids. This has been estimated for eight samples of arabica by Arnold (1995): The total free amino acids do not show clear changes after drying at $40\,^{\circ}$C, but the individual contents change for some acids, especially glutamic acid, which shows an increase of about 50%, and aspartic acid, which mainly decreases; the hydrophobic acids (valine,

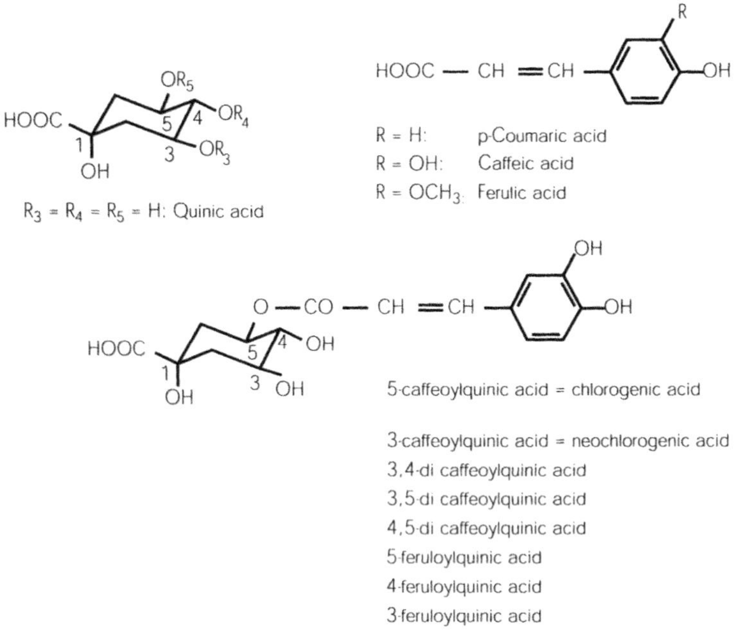
HOOC
OR5
OR4
OR3
OH
$R_3 = R_4 = R_5 = H$: Quinic acid
HOOC — CH ═ CH
R
OH
R = H: p-Coumaric acid
R = OH: Caffeic acid
R = OCH_3: Ferulic acid
O — CO — CH ═ CH
HOOC
OH
OH
OH
OH
OH
5-caffeoylquinic acid = chlorogenic acid
3-caffeoylquinic acid = neochlorogenic acid
3,4-di caffeoylquinic acid
3,5-di caffeoylquinic acid
4,5-di caffeoylquinic acid
5-feruloylquinic acid
4-feruloylquinic acid
3-feruloylquinic acid

Figure 3.29 Phenolic acids in raw coffee

phenylalanine, leucine, isoleucine) generally increase. Lipke (1999) isolated and characterized several coffee peptides.

During storage of the green beans, particularly at elevated temperatures, there are changes due to proteolysis (e.g. increases in alanine, isoleucine and tyrosine) and to losses of free amino acids by non-enzymatic browning reactions (Pokorny *et al.*, 1975).

The proteins consist of a water-soluble (albumin) and a water-insoluble fraction, present in approximately equal amounts. The majority of the proteins have molecular weights above 150 000 Daltons. Crude protein content, calculated from total nitrogen content, must be corrected for caffeine and ideally also for trigonelline nitrogen. If such corrections are made the protein content in green coffee is close to 10% with little quantitative and qualitative differences between species and there is no significant effect that can be attributed to green bean processing. Mazzafera (1999) found a higher protein content in mature beans than in the immature ones.

Several enzymes have been characterized in green coffee: α-galactosidase, malate dehydrogenase, acid phosphatase, peroxidase and more extensively polyphenol oxidase/tyrosinase/catechol oxidase as indicators of green coffee bean quality (Clifford, 1985). Polyphenol oxidase is responsible for the discoloration of defective beans, which is caused by catalysing the oxidation of the chlorogenic acids.

3.11.8 Non-protein nitrogen

3.11.8.1 Trigonelline

Raw coffee contains trigonelline (arabica: 0.6–1.3%, robusta: 0.3–0.9%). Dewaxing, decaffeination procedure and steaming (Stennert and Maier, 1994) do not lead to appreciable changes in the trigonelline content (Figure 3.30).

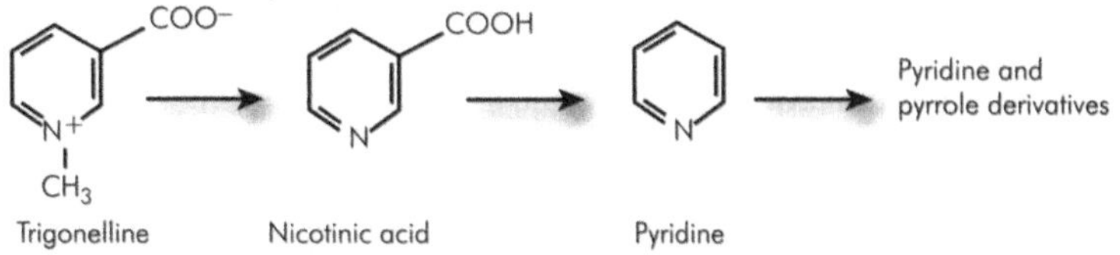

Trigonelline   Nicotinic acid   Pyridine

Figure 3.30 Trigonelline, a precursor of many aroma compounds

3.11.8.2 Caffeine and trace purine alkaloids (Figure 3.31)

Caffeine is the major purine in green coffee, where it is probably associated with the chlorogenic acids by a π-electron complex (Horman and Viani, 1972). The average content of caffeine is 1.2–1.3% in arabica and 2.2–2.4% in robusta, showing a marked inter-specific difference as well as high intra-specific variability. Caffeine is absent or present only in traces in some wild non-commercial species.

In addition to caffeine, coffee also contains the three dimethyl-xanthines – paraxanthine, theobromine and theophylline – and other trace purines, present in larger quantities in robusta – particularly in immature beans – than in arabica. Kappeler and Baumann (1986) found that unripe beans have higher contents of theophylline, theobromine, liberine and theacrine than ripe ones. Weidner and Maier (1999) also identified paraxanthine and theacrine. Prodolliet et al. (1998) tried to determine the geographic origin of coffee using the isotopic ratios (C, N, H) in samples of caffeine extracted from green arabicas and robustas from 16 countries. Both univariate and multivariate analysis allowed neither the determination of the species nor of the country of origin.

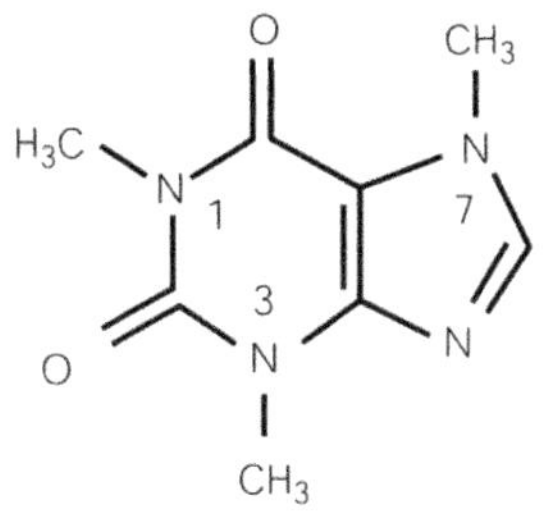

Theophylline: 1,3 dimethylxanthine

Theobromine: 3,7-dimethylxanthine

Paraxanthine: 1,3 dimethylxanthine

Caffeine
(1,3,7-trimethylxanthine)

Figure 3.31   Purine alkaloids in raw coffee

3.11.9   Lipids

The lipid content of green arabica coffee beans averages some 15%, whilst robusta coffees contain much less, around 10%. Most of the lipids – the coffee oil – are located in the endosperm; only a small amount, the coffee wax, is on the outer layer of the bean. The yield of crude lipids depends not only on bean composition, but also on extraction conditions, particularly particle size and surface area, choice of solvent and duration of extraction.

Coffee oil is composed mainly of triglycerides with fatty acids in proportions similar to those found in common edible vegetable oils. The relatively large unsaponifiable fraction is rich in diterpenes of the kaurane family, mainly cafestol, kahweol and 16-O-methylcafestol, which have received increasing attention in recent years because of their physiological effects (see 10.3). Furthermore, 16-O-methylcafestol can be used as a reliable indicator for robusta content in coffee blends. Among the sterols, that are also part of the unsaponifiable matter, various desmethyl-, methyl- and dimethylsterols have been identified. The composition of the lipid fraction of green coffee is given in Table 3.18.

3.11.9.1   Fatty acids

For the most part, the fatty acids are present as glycerol esters in the triglycerides, some 20% are esterified with diterpenes, and a small proportion as sterol esters. Folstar et al. (1975) and Speer et al. (1993) investigated the fatty acids of the triglycerides and the diterpene esters in detail. The fatty acids in sterol esters were determined by Picard et al. (1984).

Table 3.18 Composition of the lipid fraction of raw coffee (percentages of total lipids, average)

Component	Percent
Triglycerides	75.2
Esters of diterpene alcohols and fatty acids	18.5
Diterpene alcohols	0.4
Esters of sterols and fatty acids	3.2
Sterols	2.2
Tocopherols	0.04–0.06
Phosphatides	0.1–0.5
Tryptamine derivatives	0.6–1.0

Source: Maier, 1981

The presence of free fatty acids (FFA) in coffee has been described by various authors (Kaufmann and Hamsagar, 1962b; Calzolari and Cerma, 1963; Carisano and Gariboldi, 1964; Wajda and Walczyk, 1978). Speer *et al.* (1993) developed a method for the direct determination of free fatty acids: Nine different free fatty acids were detected, uniformly distributed in both robusta and arabica coffees. In both species the main fatty acids are $C_{18:2}$ and C_{16}. It was also possible to detect large proportions of C_{18}, $C_{18:1}$, C_{20} and C_{22}, whereas there were no more than traces of C_{14}, $C_{18:3}$ and C_{24}. While the proportion of stearic acid is noticeably smaller than that of oleic acid in robustas, the percentages of these two acids in arabica coffees are almost equal. The ratio stearic acid/oleic acid may give a first indication of robusta content in coffee blends.

3.11.9.2 Diterpenes

The main diterpenes in coffee are pentacyclic diterpene alcohols based on the kaurane skeleton. Working groups under Bengis and Anderson (1932), Chakravorty *et al.* (1943a, 1943b), Wettstein *et al.* (1945), Haworth and Johnstone (1957) and Finnegan and Djerassi (1960) elucidated the structure of two of the coffee diterpenes, namely kahweol and cafestol. In 1989, 16-O-methylcafestol was isolated from robusta coffee beans (Speer and Mischnick, 1989; Speer and Mischnick-Lübbecke, 1989). 16-O-methylkahweol, another O-methyl diterpene, has recently also been found in robusta coffee beans (Speer *et al.*, 2000). The structural formulae of these diterpenes are shown in Figure 3.32.

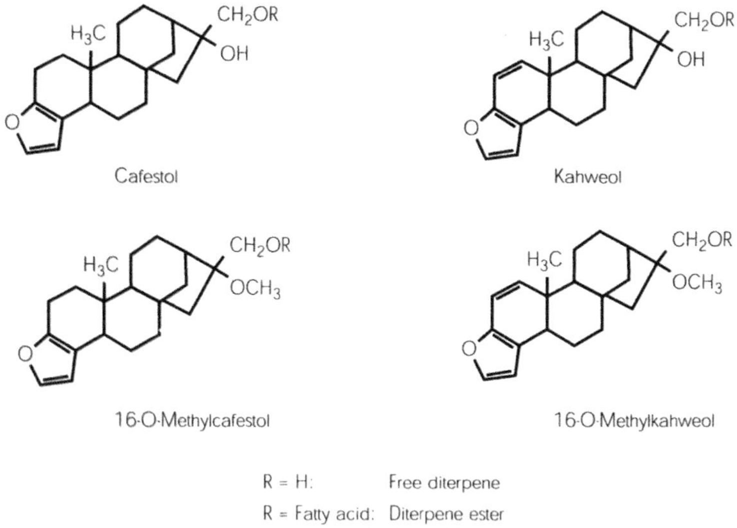

Cafestol

Kahweol

16-O-Methylcafestol

16-O-Methylkahweol

R = H: Free diterpene
R = Fatty acid: Diterpene ester

Figure 3.32   Structural formulae of raw coffee diterpenes

Arabica coffee contains cafestol and kahweol. Robusta coffee contains cafestol, small amounts of kahweol and, additionally, 16-O-methylcafestol (16-OMC) and traces of 16-O-methylkahweol, which were found only in robusta coffee beans (Speer and Mischnick-Lübbecke, 1989; Speer and Montag, 1989; Speer et al., 1991b; Kölling-Speer et al., 2001). Absence of 16-OMC in arabica coffee beans has been confirmed by Frega et al. (1994), White (1995) and Trouche et al. (1997). Because of its stability during the roasting process, 16-OMC can be used to detect the presence of down to less than 2% robusta in arabica blends (Speer et al., 1991b).

The diterpenes cafestol, kahweol and 16-OMC are mainly esterified with various fatty acids. In their free form, they occur only as minor components in coffee oil (Speer et al., 1991a; Kölling-Speer et al., 1999). Both free cafestol and kahweol are present in arabica coffees, cafestol being the major component. In robusta coffee, free cafestol content is slightly higher than 16-OMC content, and free kahweol is either present only in traces or absent. Comparison of free diterpenes with total diterpenes after saponification shows that in arabicas proportions ranged from 0.7 to 2.5%; in robustas the proportions of free diterpenes were slightly higher with 1.1 to 3.5%; experiments to verify if the difference is due to the processing technique or to the species are ongoing.

Fatty acid esters have been reported (Kaufmann and Hamsagar, 1962a; Folstar *et al.*, 1975; Folstar, 1985; Pettitt, 1987; Speer, 1991, 1995; Kurzrock and Speer, 1997a, 1997b): Cafestol, 16-OMC and kahweol esters with fatty acids such as C_{14}, C_{16}, C_{18}, $C_{18:1}$, $C_{18:2}$, $C_{18:3}$, C_{20}, C_{22}, C_{24} were identified, as well as esters with the fatty acid $C_{20:1}$ and some odd-numbered fatty acids such as C_{17}, C_{19}, C_{21} and C_{23} (Kurzrock and Speer, 2001a, 2001b). The individual diterpene esters were irregularly present in the coffee oil. The odd-numbered fatty acid esters were minor components, whereas the diterpenes esterified with palmitic, linoleic, oleic, stearic, arachidic, and behenic acid, were present in larger amounts. Table 3.19 indicates the distribution of the six main esters (summing up to nearly 98% of the respective diterpenes) for arabica coffee.

The total content of these six cafestol esters in sum ranged between 9.4 and 21.2 g/kg dry weight, corresponding to 5.2 to 11.8 g/kg cafestol in different arabica coffees, and between 2.2 and 7.6 g/kg dry weight in robusta, corresponding to 1.2 to 4.2 g/kg cafestol, notably less than in arabica.

Table 3.19 Distribution (%) of the diterpene esters in raw arabica coffee

	Cafestol + kahweol[1] ($n^2 = 1$)	Cafestol + kahweol[2] ($n = 1$)	Cafestol[3] ($n = 10$)	Kahweol[4] ($n = 10$)
C_{16}	42.5	51.4	40–49	46–50
C_{18}	17.5	9.1	9–11	8–11
$C_{18:1}$	11.0	7.4	9–15	8–12
$C_{18:2}$	20.5	26.4	24–30	25–29
C_{20}	6.0	4.6	3–6	3–6
C_{22}	2.5	1.1	0.6–1.2	0.7–1.3

Kahweol esters calculated as cafestol esters.
n = Number of samples analysed.
Sources: [1]Kaufmann and Hams, 1962a; [2]Folstar, 1985; [3]Kurzrock and Speer, 1997a; [4]Kurzrock and Speer, 2001a,b

3.11.9.3 Sterols

Coffee contains a number of sterols that are common in other seed oils as well (Figure 3.33, see page 160). In addition to 4-desmethylsterols, various 4-methyl- and 4,4-dimethylsterols have been identified. Coffee sterols have been found both in the free (around 40%) and in the esterified form (around 60%) (Nagasampagi *et al.*, 1971; Itoh *et al.*, 1973a, 1973b; Tiscornia *et al.*, 1979; Duplatre *et al.*, 1984; Picard *et al.*, 1984; Mariani

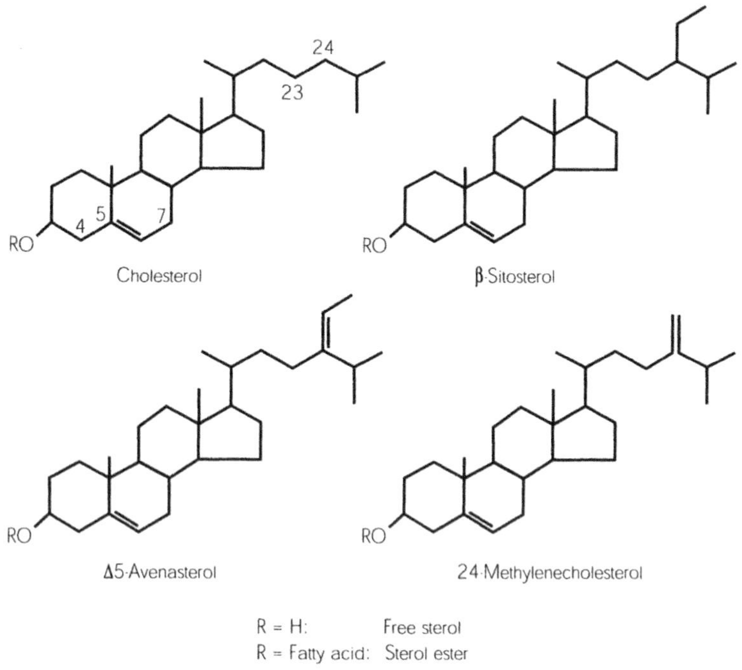

Cholesterol

β-Sitosterol

Δ5-Avenasterol

24-Methylenecholesterol

R = H: Free sterol
R = Fatty acid: Sterol ester

Figure 3.33   Sterols of raw coffee beans

and Fedeli, 1991; Frega *et al.*, 1994). The desmethylsterols represent 90% of the total sterol fraction, which ranges from 1.5% to 2.4% of the lipids (Picard *et al.*, 1984). Nagasampagi *et al.* (1971) found higher proportions (5.4%).

Table 3.20 presents the distribution of the main desmethylsterols in different robusta and arabica coffee samples. The main sterol is β-sitosterol, with about 50%. 24-methylenecholesterol and Δ5-avenasterol, occurring in larger amounts in robusta than in arabica coffee beans, are suitable for coffee blend studies (Duplatre *et al.*, 1984; Frega *et al.*, 1994). However, because of their varying natural contents, they can be only used for determining proportions of robusta in a mixture with arabica from 20% onward.

3.11.9.4   Tocopherols (Figure 3.34)

The presence of tocopherols in coffee oil was described for the first time by Folstar *et al.* (1977). They found concentrations of α-tocopherol of

Table 3.20 Mean distribution (%) of desmethylsterols in raw arabica and robusta coffees (30 samples)

Sterols	Arabica	Robusta
Cholesterol	0.3	0.2
Campesterol	15.8	16.9
Stigmasterol	21.9	23.1
β-Sitosterol	51.6	45.4
Δ5-Avenasterol	2.7	9.1
Campestanol	0.4	0.2
24-Methylenecholesterol	0.2	1.9
Sitostanol	2.0	0.8
Δ7-Stigmastenol	2.2	0.2
Δ7-Avenasterol	1.5	0.4
Δ7-Campesterol	0.6	0.2
Δ5.23-Stigmastadienol	0.3	0.5
Δ5.24-Stigmastadienol	0.1	Traces
Clerosterol	0.5	0.7

Source: Mariani and Fedeli, 1991

89 to 188 mg/kg oil, for β- + γ-tocopherol 252–53 mg/kg oil. The predominance of α-tocopherol is a prominent feature of coffee beans, in contrast to other vegetables and fruits (Aoyama *et al.* 1988; Ogawa *et al.* 1989).

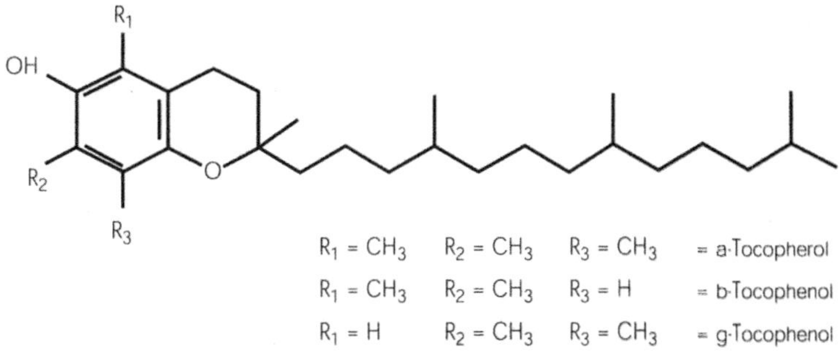

$R_1 = CH_3$ $R_2 = CH_3$ $R_3 = CH_3$ = a-Tocopherol
$R_1 = CH_3$ $R_2 = CH_3$ $R_3 = H$ = b-Tocophenol
$R_1 = H$ $R_2 = CH_3$ $R_3 = CH_3$ = g-Tocophenol

Figure 3.34 Tocopherols of raw coffee beans

3.11.9.5 Coffee wax

A thin wax layer, corresponding to 2–3% of the total lipids, covers the surface of green coffee beans. The wax content is generally defined as the material obtained from the intact beans by extraction with chlorinated solvents such as dichloromethane. The first investigations about wax composition of green arabica coffee beans were done by Wurziger and co-workers (Dickhaut, 1966; Harms, 1968). They isolated and identified three carbonic acid 5-hydroxytryptamides (C-5-HT) (Figure 3.35). Arachidic acid ($n = 18$), behenic acid ($n = 20$) and lignoceric acid ($n = 22$) are combined with the primary amino group of 5-hydroxytryptamine. In addition, Folstar et al. (1980) reported the presence of stearic acid 5-hydroxytryptamide ($n = 16$), later ω-hydroxyarachidic acid 5-hydroxytryptamide ($n = 18$), and ω-hydroxybehenic acid 5-hydroxytryptamide ($n = 20$). Arachidic acid and behenic acid 5-hydroxytryptamide are dominant, but the other amides are only minor constituents.

The content ranges between 500 and 2370 mg/kg in green arabica, and from 565 to 1120 mg/kg in green robusta (Maier, 1981). Polishing, dewaxing and decaffeination of the beans leads to a substantial reduction in C-5-HT (Harms and Wurziger, 1968; Van der Stegen and Noomen, 1977; Hunziker and Miserez, 1979; Folstar et al., 1980). The C-5-HT have antioxidant properties (Lehmann et al., 1968; Bertholet and Hirsbrunner, 1984).

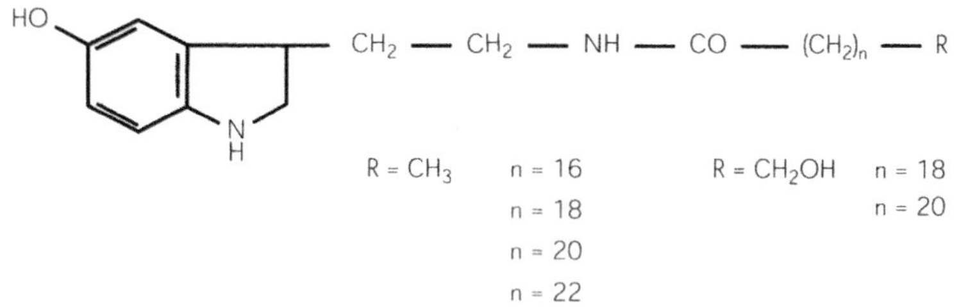

Figure 3.35 Structural formulae of carbonic acid 5-hydroxytryptamides (C-5-HT)

3.11.10 Volatile constituents

Some 180 volatile substances have been identified in green coffee beans (Boosfeld and Vitzthum, 1995; Holscher and Steinhart, 1995; Cantergiani et al., 1999). Arabicas are similar to robustas, but arabicas are distinguished by a large content of terpenes and fewer aromatic compounds. The range is affected by green bean processing, and could be used for evaluating green bean quality. The odour of defective green beans has been linked with specific components (see 3.7.2).

3.11.11   Contaminants

3.11.11.1   Mycotoxins

R. Viani

The occasional occurrence of several mycotoxins that may contaminate green coffee beans has been known since the early 1970s: they are sterigmatocystine, aflatoxin B and ochratoxin A (OTA) (Levi *et al.*, 1974).

Sterigmatocystine was only found in extremely contaminated beans, clearly unfit for consumption, and is of no practical relevance in coffee.

The aflatoxins, aflatoxin B in particular, have sometimes been detected in green coffee beans, and – in extremely small amounts, apparently of no practical concern for public health – also in the beverage (Nakajima *et al.*, 1997).

OTA contamination of coffee has been the subject of intensive research during the past 15 years, after it was realized, thanks to improved analytical techniques, that not all the OTA present in green beans was destroyed during roasting (Tsubouchi *et al.*, 1988). Much information is available through the three workshops, organized in Nairobi (Anon., 1997), Helsinki (Van der Stegen and Blanc, 1999) and Trieste (Van der Stegen *et al.*, 2001) by the *Association Scientifique Internationale du Café* (ASIC). A comprehensive review has been published by Bucheli and Taniwaki (2002).

The mould species, which are known to produce (OTA) in coffee, have been identified as *Aspergillus ochraceus*, *A. carbonarius* and, occasionally, *A. niger* (Frank, 2001; Ismayadi and Zaenudin, 2001; Joosten *et al.*, 2001; Paneer *et al.*, 2001; Pitt *et al.*, 2001). The specific time and place of contamination have not yet been clearly identified. Both natural and wet-processed coffees appear to be at risk of attack by the mould, although wet-processed coffees are rarely found contaminated by OTA.

The risk of OTA contamination can clearly be minimized by following good agricultural and good manufacturing practices at the harvest, post-harvest and storage-transportation stages (Teixeira *et al.*, 2001; Viani, 2002):

- **Storage of fresh cherries** – ripe and overripe cherries, stored over days piled up or, worse, in plastic bags, often in the sun, are at increased risk of OTA contamination (Bucheli *et al.*, 2000); therefore, avoiding storage of fresh cherries before processing is important not only for coffee organoleptic quality (see 3.2), but also to reduce the risk of OTA contamination (Figure 3.36).
- **Drying** – after drying begins, the length of time spent at a water activity above 0.80 at any time until roasting determines the risk of

mould growth and OTA production, which may explain the increased risk for sun drying with respect to mechanical drying, and of cherry sun drying with respect to the shorter parchment sun drying. Organoleptic quality considerations require a regular long drying process (see 3.3), while at the same time reduction of the risk of OTA contamination asks for quick drying. The drying process therefore needs to be optimized to adjust to both requirements (Ismayadi *et al.*, 2001; McGaw *et al.*, 2001).

- **Husks** – contamination appears to start within the husk, and, in the 'appropriate' moisture conditions, it can reach the bean in 3–4 days; therefore, freeing green coffee beans from all husk material during cleaning and grading is important to reduce the OTA load (Pittet *et al.*, 1996; Blanc *et al.*, 1998; Bucheli *et al.*, 2000).

- **Re-wetting** – storage in the tropics of green robusta coffee for up to one year at 70% relative humidity and at temperatures up to 30 °C has not produced any increase in the OTA load (Bucheli *et al.*, 1998). Transportation trials from the producer countries have shown that risk of condensation and wetting of coffee occur mainly during transport overland to the harbour for shipping, and on arrival at destination in the winter season; this may occur even with properly dried coffee if too large temperature variations cause evaporation followed by water condensation on the upper layer of coffee (Blanc *et al.*, 2001). OTA contamination can also occur during the decaffeination process (see 3.9), when decaffeinated wet beans are not properly and rapidly dried (unpublished data). Avoiding re-wetting at any stage before roasting is therefore very important (Figure 3.37).

Figure 3.36   Storage of fresh cherries must be short!

Figure 3.37 Drying coffee – cherry or parchment – must be protected from re-wetting

3.11.11.2   *Polycyclic aromatic hydrocarbons (PAH)*
I. Kölling-Speer and K. Speer

Different PAH, e.g. fluoranthene, pyrene, benz[a]anthracene, chrysene, triphenylene, benzo[b]fluoranthene, benzo[j]fluoranthene, benzo[k]fluoranthene, benzo[a]fluoranthene, benzo[e]pyrene, benzo[a]pyrene, perylene, indeno[1,2,3-cd]pyrene, dibenz[ah]anthracene, dibenz[ac]anthracene and benzo[ghi]perylene were identified and quantified in green and roasted coffees. The individual PAH can be formed both by the fuel and by the coffee bean under sub-optimal roasting conditions. On comparing the PAH contents of the unroasted coffee with those of the roasted product it is seen that it is not so much the roasting process after its optimization but the PAH content of the raw material which determines the content in the roasted product (Klein *et al.*, 1993). The contamination of the green coffee may be due to exhaust gases of cars or to transportation of the coffee beans in contaminated jute and sisal sacks. Their fibres are treated before spinning with batching oil. Before bean contamination was discovered the batching oil commonly consisted of an uncontrolled mineral oil fraction, which could even be discarded motor oil (Grob *et al.*, 1991a, 1991b; Moret *et al.*, 1997). The contamination was identified by analysing the PAH profile (Speer *et al.*, 1990; Moret *et al.*, 1997). Strict rules have been suscribed and are now adhered by bag manufacturers, who generally mark by a tag sacks prepared according to GMP.

3.11.11.3 Pesticide residues
I. Kölling-Speer and K. Speer

Due to the protective layers covering the beans in the fruit (skin, pulp and parch), the level of pesticides present in green coffee beans is very low. Cetinkaya (1988) could not detect either organochlorine or organophosphorus pesticide residues in 50 green coffee samples imported from 11 different countries.

REFERENCES AND FURTHER READING

Aeschbach R., Kusy A. and Maier H.G. (1982) Diterpenoide in Kaffee. *Z. Lebensm. Unters. Forsch.* 175, 337–341.
Amorim H.V., Cruz A.R., St Angelo A.G. *et al.* (1977) Biochemical, physical and organoleptic changes during raw coffee quality deterioration. *Proc. 8th ASIC Coll.*, pp. 183–186.
Amorim H.V., Teixeira A.A., Moraes R.S., Reis A.J., Pimentel-Gomes F. and Malavolta E. (1973) Studies on the mineral nutrition of coffee. XXVII. Effect of N, P and K fertilization on the macro and micro nutrients of coffee fruits and on the beverage quality. *An. Esc. Sup. Agr. LQ (Piracicaba)* 20, 323–333.
Anon. (1987) Electronic sorting reduces labour costs. *Food Technol. NZ*, p. 47.
Anon. (1997) Special workshop on the enhancement of coffee quality by reduction of mould growth. *Proc. 17th ASIC Coll.*, pp. 367–369, and following articles.

Aoyama M., Maruyama T., Kanematsu H., Niiya I., Tsukamoto M., Tokairin S. and Matsumoto T. (1988) Studies on the improvement of antioxidant effect of tocopherols. XVII. Synergistic effect of extracted components from coffee beans. *Yukagaku* 37, 606–612.

Arnold U. (1995) Nachweis von Aminosäureveränderungen und Bestimmung freier Aminosäuren in Rohkaffee, Beiträge zur Charakterisierung von Rohkaffeepeptiden. Thesis, University of Dresden.

Bacchi O. (1962) O branqueamento dos grãos de café. *Bragantia (Campinas)* 21 (28), 467–484.

Bähre F. (1997) Neue nichtflüchtige Säuren im Kaffee. Thesis, Technical University of Braunschweig.

Bähre F. and Maier H.G. (1999) New non-volatile acids in coffee. *Dtsche Lebensm.-Rundsch.* 95, 399–402.

Becker R., Döhla B., Nitz S. and Vitzthum O.G. (1987) Identification of the 'peasy' off-flavour note in central African coffees. *Proc. 12th ASIC Coll.*, pp. 203–215.

Bee S.C. (2001) Optical sorting for the coffee industry. *Proc. 19th ASIC Coll.*, CD-ROM.

Bengis R.O. and Anderson R.J. (1932) The chemistry of the coffee bean. I. Concerning the unsaponifiable matter of the coffee bean oil. Extraction and properties of kahweol. *J. Biol. Chem.* 97, 99–113.

Berry N.E. and Walter R.H. (1943) Process of decaffeinating coffee. *US Patent* 2,309,092.

Bertholet R. and Hirsbrunner P. (1984) Preparation of 5-hydroxytryptamine from coffee wax. *European Patent* 1984–104 696.

Blanc M. and Margolis G. (1981) Caffeine extraction. *European Patent* 0049 357.

Blanc M., Pittet A., Muñoz-Box R. and Viani R. (1998) Behavior of ochratoxin A during green coffee roasting and soluble coffee manufacture. *J. Agric. Food Chem.* 46, 673–675.

Blanc M., Vuataz G. and Hickmann L. (2001) Green coffee transport trials. *Proc. 19th ASIC Coll.*, CD-ROM.

Boosfeld J. and Vitzthum O.G. (1995) Unsaturated aldehydes identification from green coffee. *J. Food Sci.* 60, 1092–1096.

Bouyjou R., Fourny G. and Perreaux D. (1993) Le goût de pomme de terre du café arabica au Burundi. *Proc. 15th ASIC Coll.*, pp. 357–369.

Bradbury A.G.W. and Balzer H.H. (1999) Carboxyatractyligenin and atractyligenin glycosides in coffee. *Proc. 18th ASIC Coll.*, pp. 71–77.

Bradbury A.G.W. and Halliday D.J. (1987) Polysaccharides in green coffee beans. *Proc. 12th ASIC Coll.*, pp. 265–269.

Bradbury A.G.W. and Halliday D.J. (1990) Chemical structures of green coffee bean polysaccharides. *J. Agric. Food Chem.* 38, 389–392.

Bucheli P. and Taniwaki M.H. (2002) Research on the origin, and the impact of postharvest handling and manufacturing on the presence of ochratoxin A in coffee. *Food Addit. Contam.* 19 (7), 655–665.

Bucheli P., Kanchanomai C., Meyer I. and Pittet A. (2000) Development of ochratoxin A during Robusta (*Coffea canephora*) coffee cherry drying. *J. Agric. Food Chem.* 48, 1358–1362.

Bucheli P., Meyer I., Pittet A., Vuataz G. and Viani R. (1998) Industrial storage of green robusta coffee under tropical conditions and its impact on raw material quality and ochratoxin A content. *J. Agric. Food Chem.* 46, 4507–4511.

Calzolari C. and Cerma E. (1963) Sulle sostanze grasse del caffè. *Riv. Ital. Sostanze Grasse* 40, 176–180.

Cantergiani E., Brevard H., Amado R., Krebs Y., Feria-Morales A. and Yeretzian C. (1999) Characterisation of mouldy/earthy defect in green Mexican coffee. *Proc. 18th ASIC Coll.*, pp. 43–49.

Carisano A. and Gariboldi L. (1964) Gaschromatographic examination of the fatty acids of coffee oil. *J. Sci. Food Agric.* 15, 619–622.

Carvalho A. (1988) Principles and practice of coffee plant breeding for productivity and quality factors: *Coffea arabica*. In R.J. Clarke and R. Macrae (eds), *Coffee Volume 4 – Agronomy*. London: Elsevier Applied Science, pp. 129–165.

Carvalho A., Garruti R.S., Teixeira A.A., Pupo L.M. and Mônaco L.C. (1972) Ocorrência dos principais defeitos do café em várias fases de maturação dos frutos. *Bragantia (Campinas)* 29 (20), 207–220.

Castel de Menezes H. and Clifford M.N. (1988) The influence of stage of maturity and processing method on the relation between the different isomers of caffeoilquinic acid in green coffee beans. *Proc. 12th ASIC Coll.*, pp. 377–381.

Cetinkaya M. (1988) Organophosphor- und Organochlorpestizidrückstände in Rohkaffee. *Dtsch. Lebensm. Rundsch.* 84, 189–190.

Chakravorty P.N., Levin R.H., Wesner M.M. and Reed. G. (1943b) Cafesterol III. *J. Am. Chem. Soc.* 65, 1325–1328.

Chakravorty P.N., Wesner M.M. and Levin R.H. (1943a) Cafesterol II. *J. Am. Chem. Soc.* 65, 929–932.

Clarke R.J. (1988) International standardization. In R.J. Clarke and R. Macrae (eds), *Coffee: Volume 6 – Commercial and Technological Aspects*. London: Elsevier Applied Science, pp. 112–121.

Clarke R.J. and Vitzthum O.G. (eds) (2001) *Coffee – Recent Developments*. Oxford: Blackwell Science.

Clifford M.N. (1985) Chemical and physical aspects of green coffee and coffee products. In M.N. Clifford and K.C. Willson (eds), *Coffee: Botany, Biochemistry and Production of Beans and Beverage*. Westport, CT: AVI, pp. 314–315.

Clifford M.N. (1999) Chlorogenic acids and other cinnamates. Nature, occurrence and dietary burden. *J. Sci. Food Agric.* 79, 362–372.

Clifford M.N. and Ohiokpehai O. (1983) Coffee astringency. *Analyt. Proc.* 20, 83–86.

Clifford M.N., Kazi T. and Crawford S. (1987) The content and washout kinetics of chlorogenic acids in normal and abnormal green coffee beans. *Proc. 12th ASIC Coll.*, pp. 221–228.

Dentan E. (1987) Examen microscopique de grains de café riotés. *Proc. 12th ASIC Coll.*, pp. 335–352.

Dentan E. (1989) Etude microscopique de quelques types de café défectueux: grains noirs, blanchâtres, cireux et 'ardidos'. *Proc. 13th ASIC Coll.*, pp. 283–301.

Dentan E. (1991) Etude microscopique de quelques types de café défectueux. II: grains à goût d'herbe, de terre, de moisi; grains puants, endommagés par les insectes. *Proc. 14th ASIC Coll.*, pp. 293–311.

Dentan E and Illy A. (1985) Etude microscopique de grains de café matures, immatures et immatures fermentés arabica Santos. *Proc. 11th ASIC Coll.*, pp. 341–368.

Dickhaut G. (1966) Über phenolische Substanzen in Kaffee und deren analytische Auswertbarkeit zur Kaffeewachsbestimmung. Thesis, University of Hamburg.

Duplatre A., Tisse C. and Estienne J. (1984) Contribution à l'identification des espèces arabica et robusta par étude de la fraction stérolique. *Ann. Fals. Exp. Chim.* 828, 259–270.

Engelhardt U.H. and Maier H.G. (1984) Nichtflüchtige Säuren im Kaffee. Thesis, Technical University of Braunschweig.

Engelhardt U.H. and Maier H.G. (1985) Säuren des Kaffees. XII. Anteil einzelner Säuren an der titrierbaren Gesamtsäure. *Z. Lebensm. Unters. Forsch.* 181, 206–209.

Ferraz M.B. and Veiga A.A. (1960) Melhor bebida e maior poder germinativo do café. *Boletim de Superintendência dos Serviços do Café* 05–18, 398–399.

Finnegan R.A. and Djerassi C. (1960) Terpenoids XLV. Further studies on the structure and absolute configuration of cafestol. *J. Am. Chem. Soc.* 82, 4342–4344.

Fischer A. and Kummer P. (1979) Verfahren zum Entcoffeinieren von Rohkaffee. *European Patent* 008 398.

Fischer M., Reimann S., Trovato V. and Redghwell R.J. (2000) Structural aspects of polysaccharides from arabica coffee. *Proc 18th ASIC Coll.*, pp. 91–94.

Flament I. (2002) *Coffee Flavour Chemistry.* New York: John Wiley.

Folstar P. (1985) Lipids. In R.J. Clarke and R. Macrae (eds), *Coffee: Volume 1 – Chemistry.* London: Elsevier Applied Science, pp. 203–222.

Folstar P., Pilnik W., de Heus J. G. and Van der Plas H. C. (1975) The composition of fatty acids in coffee oil and wax. *Lebensm. Technol.* 8, 286–288.

Folstar P., Schols H.A., Van der Plas H.C., Pilnik W., Landherr C.A. and Van Vildhuisen A. (1980) New tryptamine derivatives isolated from wax of green coffee beans. *J. Agric. Food Chem.* 28, 872–874.

Folstar P., Van der Plas H.C., Pilnik W. and de Heus J.G. (1977) Tocopherols in the unsaponifiable matter of coffee bean oil. *J. Agric. Food Chem.* 25, 283–285.

Frank J.M. (2001) On the activity of fungi in coffee in relation to ochratoxin A production. *Proc. 19th ASIC Coll.*, CD-ROM.

Franz H. and Maier H.G. (1993) Inositol phosphates in coffee and related beverages. I. Identification and methods of determination. *Dtsche Lebensm.-Rundsch.* 89, 276–282.

Franz H. and Maier H.G. (1994) Inositol phosphates in coffee and related beverages. II. Coffee beans. *Dtsche Lebensm.-Rundsch.* 90, 345–349.

Frega N., Bocci F. and Lercker G. (1994) High resolution gas chromatographic method for determination of Robusta coffee in commercial blends. *J. High Resolution Chromatogr.* 17, 303–307.

Full G., Lonzarich V. and Suggi L.F. (1999) Differences in chemical composition of electronically sorted green coffee beans. *Proc. 18th ASIC Coll.*, pp. 35–42.

Gibson A. and Butty M. (1975) Overfermented coffee beans ('stinkers') a method for their detection and elimination. *Proc. 7th ASIC Coll.*, pp. 141–152.

Gomes F.P., Cruz V.F., Castilho A., Teixeira A.A. and Pereira L.S.P. (1967) A influência de grãos pretos em ligas com cafés de bebida mole. *Anais da E.S.A. 'Luiz de Queiroz' (Piracicaba)* 24, 71–81.

Grob K., Biedermann M., Artho A. and Egli J. (1991a) Food contamination by hydrocarbons from packaging materials determined by coupled LC–GC. *Z. Lebensm. Unters. Forsch.* 193, 213–219.

Grob K., Lanfranchi M., Egli J. and Artho A. (1991b) Determination of food contamination by mineral oil from jute sacks using coupled LC-GC. *J. Assoc. Off. Anal. Chem.* 74, 506–512.

Grosch W. (1998) Welche Verbindungen bevorzugt der Geruchssinn bei erhitzten Lebensmitteln? *Lebensmittelchemie* 52, 143–146.

Guyot B., Petnga E. and Vincent J.C. (1988a) Analyse qualitative d'un café *Coffea canephora* var. robusta. I. Evolution des caractères physiques et organoleptiques. *Café Cacao Thé* 32, 127–140.

Guyot B., Petnga E., Lotodé R. and Vincent J.C. (1988b) Analyse qualitative d'un café *Coffea canephora* var. Robusta en fonction de la maturité. II. Application de l'analyse statistique multidimensionnelle. *Café Cacao Thé* 32, 229–242.

Guyot B., Cochard B. and Vincent J.C. (1991) Détermination quantitative du diméthylsulfure et du diméthyldisulfure dans l'arôme de café. *Café Cacao Thé* 35, 49–56.

Harms U. (1968) Beiträge zum Vorkommen und zur Bestimmung von Carbonsäure-5-hydroxy-tryptamiden in Kaffeebohnen. Thesis, University of Hamburg.

Harms U. and Wurziger J. (1968) Carboxylic acid 5-hydroxytryptamides in coffee beans. *Z. Lebensm. Unters. Forsch.* 138, 75–80.

Haworth R.D. and Johnstone R.A.W. (1957) Cafestol Part. II. *J. Chem. Soc.* *(Lond.)*, pp. 1492–1496.

Heilmann W. (2001) Decaffeination of coffee. In R.J. Clarke and O.G. Vitzthum (eds), *Coffee – Recent Developments*. Oxford: Blackwell Science, pp. 108–124.

Holscher W. and Steinhart H. (1995) Aroma compounds in green coffee. In G. Charalambous (ed.), *Food Flavours: Generation, Analysis and Process Influence. 37 A*. Amsterdam: Elsevier Science, pp. 785–803.

Horman I. and Viani R. (1972) The nature and conformation of the caffeine-chlorogenate complex of coffee. *J. Food Sci.* 37, 925–927.

Hunziker H.R. and Miserez A. (1979) Bestimmung der 5-Hydroxytryptamide in Kaffee mittels Hochdrück-Flüssigkeitschromatographie. *Mitt. Geb. Lebensum. Unters. Hyg.* 70, 142–152.

Illy A. and Viani R. (eds) (1995) *Espresso Coffee. The Chemistry of Quality*. London: Academic Press, p. 29.

Illy E., Brumen G., Mastropasqua L. and Maughan W. (1982) Study on the characteristics and the industrial sorting of defective beans in green coffee lots. *Proc. 10th ASIC Coll.*, pp. 99–128.

Ismayadi C. and Zaenudin. (2001) Toxigenic mould species infestation in coffee beans taken from different levels of production and trading in Lampung – Indonesia (2001). *Proc. 19th ASIC Coll.*, CD-ROM.

Ismayadi C., Zaenudin and Priyono S. (2001) Mould species infestation during sun drying of sound and split coffee cherries. *Proc. 19th ASIC Coll.*, CD-ROM.

ISO (1980) Green Coffee – Olfactory and Visual Examination and Determination of Foreign Matter and Defects. ISO 4149-1980. Geneva: International Organization for Standardization.

ISO (1983) Green Coffee – Determination of Loss in Mass at 105 °C. ISO 6673-1983. Geneva: International Organization for Standardization.

ISO (1985) Green Coffee – Determination of Proportion of Insect-damaged Beans. ISO 6667-1985. Geneva: International Organization for Standardization.

ISO (1993) Green Coffee – Defect Reference Chart. ISO 10470-1993. Geneva: International Organization for Standardization.

ITC (2002) *Coffee: an Exporter's Guide*. Geneva: International Trade Centre, UNCTAD-WTO.

Itoh T., Tamura T. and Matsumoto T. (1973a) Sterol composition of 19 vegetable oils. *J. Am. Oil Chem. Soc.* 50, 122–125.

Itoh T., Tamura T. and Matsumoto T. (1973b) Methylsterol compositions of 19 vegetable oils. *J. Am. Oil Chem. Soc.* 50, 300–303.

Jobin P. (1982) *Les Cafés produits dans le monde*. Le Havre: Jobin.

Joosten H.M.L.J., Goetz J., Pittet A., Schellenberg M. and Bucheli P. (2001) Production of ochratoxin A by *Aspergillus carbonarius* on coffee cherries. *Int. J. Food Microb.* 65, 39–44.

172   Espresso Coffee

Jouanjan F. (1980) Transport maritime du café vert et conténeurisation. *Proc. 9th ASIC Coll.*, pp. 177–188.

Kampmann B. and Maier H.G. (1982) Säuren des Kaffees. I. Chinasäure. *Z. Lebensm. Unters. Forsch.* 175, 333–336.

Kappeler A.W. and Baumann T.W. (1986) Purine alkaloid pattern in coffee beans. *Proc. 11th ASIC Coll.*, pp. 273–279.

Kaufmann H.P. and Hamsagar R.S. (1962a) Zur Kenntnis der Lipoide der Kaffeebohne. I. Über Fettsäure-Ester des Cafestols. *Fette Seifen Anstrichmittel* 64, 206–213.

Kaufmann H.P. and Hamsagar R.S. (1962b) Zur Kenntnis der Lipoide der Kaffeebohne. II. Die Veränderung der Lipoide bei der Kaffee-Röstung. *Fette Seifen Anstrichmittel* 64, 734–738.

Klein H., Speer K. and Schmidt E.H.F. (1993) Polycyclic aromatic hydrocarbons (PAH) in raw and roasted coffee. *Bundesgesundheitsblatt* 36, 98–100.

Kölling-Speer I. and Speer K. (1997) Diterpenes in coffee leaves. *Proc. 17th ASIC Coll.*, pp. 150–154.

Kölling-Speer I., Kurzrock T. and Speer K. (2001) Contents of diterpenes in green coffees. *Proc. 19th ASIC Coll.*, CD-ROM.

Kölling-Speer I., Strohschneider S. and Speer K. (1999) Determination of free diterpenes in green and roasted coffees. *J. High Resolution Chromatogr.* 22, 43–46.

Kurzrock T. and Speer K. (1997a) Fatty acid esters of cafestol. *Proc. 17th ASIC Coll.*, pp. 133–140.

Kurzrock T. and Speer K. (1997b) Identification of cafestol fatty acid esters. In R. Amadò and R. Battaglia (eds), *Proc. Euro Food Chem. IX (Interlaken)*, Volume 3, pp. 659–663.

Kurzrock T. and Speer K. (2001a) Diterpenes and diterpene esters in coffee. *Food Rev. Int.* 17, 433–450.

Kurzrock T. and Speer K. (2001b) Identification of kahweol fatty acid esters in arabica coffee by means of LC/MS. *J. Sep. Sci.* 24, 843–848.

Labuza T.P., McNally L., Gallager D. *et al.* (1972) Stability of intermediate moisture foods. 1. Lipid oxidation. *J. Food Sci.* 37, 154–159.

Lack E. and Seidlitz H. (1993) Commercial scale decaffeination of coffee and tea using supercritical CO_2. In M.B. King and T.R. Bott (eds), *Extraction of Nature Products using Near Critical Solvents*. Glasgow: Blackie, pp. 101–139.

Lehmann G., Neunhoeffer O., Roselius W. and Vitzthum O. (1968) Antioxidants made from green coffee beans and their use for protecting autoxidizable foods. *German Patent* 1,668,236.

Levi C.P., Trenk H.L. and Mohr H.K. (1974) Study of the occurrence of ochratoxin A in green coffee beans. *J. Assoc. Official Analyt. Chemists* 57, 866–870.

Lipke U. (1999) Untersuchungen zur Charakterisierung von Rohkaffeepeptiden. Thesis, University of Dresden.

Lopez Garay C., Bautista Romero E. and Moreno Gonzales E. (1987) Use of gamma radiation for the preservation of coffee quality during storage. *Proc. 12th ASIC Coll.*, pp. 771–782.

Ludwig H., Obermann H. and Spiteller G. (1975) New diterpenes found in coffee. *Proc. 7th ASIC Coll.*, pp. 205–210.

Lüllmann C. and Silwar R. (1989) Investigation of mono- and disaccharide content of arabica and robusta green coffee using HPLC. *Lebensm. Gericht. Chem.* 43, 42–43.

Maier H.G. (1981) *Kaffee*. Berlin and Hamburg: Paul Parey.

Maier H.G. and Wewetzer H. (1978) Bestimmung von Diterpen-Glykosiden im Bohnenkaffee. *Z. Lebensm. Unters. Forsch.* 167, 105–107.

Mariani C. and Fedeli E. (1991) Sterols of coffee grain of arabica and robusta species. *Rivista Ital. Sostanze Grasse* 68, 111–115.

Martin H. (1982) Selective extraction of caffeine from green coffee beans and application of similar processes on other natural products. *Proc. 10th ASIC Coll.*, pp. 21–28.

Matiello J.B., Santinato R., Garcia A.W.R., Almeida S.R. and Fernandez D.R. (2002) Coltura do café no Brasil. Novo manual de recomendaçoes MAPA – SARC/Procafé – SPC/Decaf. São Paulo (Brazil), p. 357.

Mazzafera P. (1999) Chemical composition of defective coffee beans. *Food Chem.* 64, 547–554.

McGaw D., Comissiong E., Tripathi K., Maharaja A. and Paltoo V. (2001) The drying characteristics of coffee beans. *Proc. 19th ASIC Coll.*, CD-ROM.

Mello M., Fazuoli L.C., Teixeira A.A. and Amorim H.V. (1980) Alterações físicas, químicas e organolépticas em grãos de café armazenados. *Ciência e Cultura (São Paulo)* 32 (4), 467–472.

Moraes R.M., Angelucci E., Shirose T. and Medina J.C. (1973) Soluble solids determination in arabica and robusta coffees. *Coll. Inst. Techn. Alim. (Campinas)*, 5, 199–221.

Moret S., Grob K. and Conte L.S. (1997) Mineral oil polyaromatic hydrocarbons in foods, e.g. from jute bags, by on-line LC-solvent evaporation (SE)-LC-GC-FID. *Z. Lebensm. Unters. Forsch.* 204, 241–246.

Multon J.L., Poisson J., Cachagnier B. *et al.* (1973) Evolution de plusieurs caractéristiques d'un café arabica au cours d'un stockage expérimental effectué à 5 humidités relatives et 4 températures différentes. *Proc. 6th ASIC Coll.*, pp. 268–277.

Nagasampagi B.A., Rowe J.W., Simpson R. and Goad L.J. (1971) Sterols of coffee. *Phytochemistry* 10, 1101–1107.

Naish M., Clifford M.N. and Birch G.G. (1993) Sensory astringency of 5-O-caffeoylquinic acid, tannic acid and grapeseed tannin by a time-intensity procedure (1993). *J. Sci. Food Agric.* 61, 57–64.

Nakajima M., Tsubouchi H., Miyabe M. and Ueno Y. (1997) Survey of aflatoxin B_1 and ochratoxin A in commercial green coffee beans by high-performance liquid chromatography linked with immunoaffinity chromatography. *Food Agric. Immunol.* 9, 77–83.

Navellier P. and Brunin R. (1963) Suggestions pour l'expression des résultats des analyses de café. *Proc. 1st ASIC Coll.*, pp. 317–320.

Northmore J.M. (1969) Overfermented beans and stinkers as defectives of arabica coffee. *Proc. 4th ASIC Coll.*, pp. 47–54.

Noyes R.M. and Chu C.M. (1993) Material balance on free sugars in the production of instant coffee. *Proc. 14th ASIC Coll.*, pp. 202–210.

Obermann H. and Spiteller G. (1976) Die Strukturen der Kaffee-Atractyloside. *Chem. Ber.* 109, 3450–3461.

Ogawa M., Herai Y., Koizumi N., Kusano T. and Sano H. (2001) 7-Methylxanthine methyltransferase of coffee plants – gene isolation and enzymatic properties. *J. Biol. Chem.* 276, 8213–8218.

Ogawa M., Kamiya C. and Iida Y. (1989) Contents of tocopherols in coffee beans, coffee infusions and instant coffee. *Nippon Shokuhin Kogyo Gakkaishi* 36, 490–494.

Paneer S., Velmourougane K., Shanmukhappa J.R. and Naidu R. (2001) Studies of microflora association during harvesting and on-farm processing of coffee in India. *Proc. 19th ASIC Coll.*, CD-ROM.

Parliment T.H. (2000) An overview of coffee roasting. In T.H. Parliment, C-T. Ho, P. Schieberle (eds), Caffeinated Beverages, Health Benefits, Physiological Effects, and Chemistry. ACS symposium series No. 754, pp. 188–201.

Pettitt B.C. Jr. (1987) Identification of the diterpene esters in arabica and canephora coffees. *J. Agric. Food Chem.* 35, 549–551.

Picard H., Guyot B. and Vincent J.-C. (1984) Étude des composés stéroliques de l'huile de café *Coffea canephora*. *Café Cacao Thé* 28, 47–62.

Pitt J.I., Taniwaki M.H., Teixeira A.A. and Iamanaka B.T. (2001) Distribution of *Aspergillus ochraceus*, *A. niger* and *A. carbonarius* in coffee in four regions of Brazil. *Proc. 19th ASIC Coll.*, CD-ROM.

Pittet A., Tornare D., Huggett A. and Viani R. (1996) Liquid chromatographic determination of ochratoxin A in pure and adulterated soluble coffee using an immunoaffinity column cleanup procedure. *J. Agric. Food Chem.* 44, 3564–3569.

Pokorny J., Nguyen-Huy C., Smidralova E. and Janicek G. (1975) Nonenzymic browning. XII. Maillard reactions in green coffee beans on storage. *Z. Lebensm. Unters. Forsch.* 158, 87–92.

Prodolliet J., Baumgartner M., Martin Y.L. and Remaud G. (1998) Determination of the geographic origin of green coffee by stable isotope techniques. *Proc. 17th ASIC Coll.*, pp. 197–200.

Puzzi D. (1973) *Conservação dos Grãos Armazenados. Armazéns e Silos.* S. Paulo: Editora Agronômica Ceres.

Quijano Rico M. and Spettel B. (1975) Determinacion del contenido en varios elementos en muestras de cafes de diferentes variedades. *Proc. 7th ASIC Coll.*, pp. 165–173.

Rigitano A., Souza O.F. and Fava J.F.M. (1963) Coffee processing. In C.A. Krug (ed.), *Agricultural Practices and Fertilization of Coffee*. Instituto Brasilero Potassa (S. Paulo) (in Portuguese), pp. 215–259.

Roselius L. (1937) Die Erfindung des coffeinfreien Kaffes. *Chemiker Zeitung* 61 (1), 13.

Runge F. (1820) *Neueste phytochemische Entdeckungen* 1, 144–159.

Scholze A. and Maier H.G. (1983) Die Säuren des Kaffees. VII. Ameisen, Äpfel-, Citronen- und Essigsäure. *Kaffee Tee Markt* 33, 3–6.

Scholze A. and Maier H.G. (1984) Säuren des Kaffees. VIII. Glycol- und Phosphorsäure. *Z. Lebensm. Unters. Forsch.* 178, 5–8.

Spadone J.C., Takeoka G. and Liardon R. (1990) Analytical investigation of rio off-flavor in green coffee. *J. Agric. Food Chem.* 38, 226–233.

Speer K. (1989) 16-O-Methylcafestol – ein neues Diterpen im Kaffee – Methoden zur Bestimmung des 16-O-Methylcafestols in Rohkaffee und in behandelten Kaffees. *Z. Lebensm. Unters. Forsch.* 189, 326–330.

Speer K. (1991) 16-O-methylcafestol – a new diterpene in coffee; the fatty acid esters of 16-O-methylcafestol. In W. Baltes, T. Eklund, R. Fenwick, W. Pfannhauser, A. Ruiter and H.-P. Thier (eds), *Proc. Euro Food Chem. VI Hamburg, Germany*, Volume 1. Hamburg: Behr's Verlag, pp. 338–342.

Speer K. (1995) Fatty acid esters of 16-O-methylcafestol. *Proc. 16th ASIC Coll.*, pp. 224–231.

Speer K. and Mischnick P. (1989) 16-O-Methylcafestol – ein neues Diterpen im Kaffee – Entdeckung und Identifizierung. *Z. Lebensm. Unters. Forsch.* 189, 219–222.

Speer K. and Mischnick-Lübbecke P. (1989) 16-O-Methylcafestol – ein neues Diterpen im Kaffee. *Lebensmittelchemie* 43, 43.

Speer K. and Montag A. (1989) 16-O-Methylcafestol – ein neues Diterpen im Kaffee – Erste Ergebnisse: Gehalte in Roh- und Röstkaffees. *Dtsch. Lebensm.-Rundsch.* 85, 381–384.

Speer K., Hruschka A., Kurzrock T. and Kölling-Speer I. (2000) Diterpenes in coffee. In T.H. Parliment, C-T. Ho and P. Schieberle (eds), *Caffeinated Beverages, Health Benefits, Physiological Effects, and Chemistry*. ACS symposium series No. 754, pp. 241–251.

Speer K., Sehat N. and Montag A. (1993) Fatty acids in coffee. *Proc. 15th ASIC Coll.*, pp. 583–592.

Speer K., Steeg E., Horstmann P., Kühn T. and Montag A. (1990) Determination and distribution of polycyclic aromatic hydrocarbons in native vegetable oils, smoked fish products, mussels and oysters, and bream from the river Elbe. *J. High Resolution Chromatogr.* 13, 104–111.

Speer K., Tewis R. and Montag A. (1991a) 16-O-Methylcafestol – ein neues Diterpen im Kaffee – Freies und gebundenes 16-O-Methylcafestol. *Z. Lebensm. Unters. Forsch.* 192, 451–454.

Speer K., Tewis R. and Montag, A. (1991b) 16-O-methylcafestol – a quality indicator for coffee. *Proc. 14th ASIC Coll.*, pp. 237–244.

Speer K., Tewis R. and Montag A. (1991c) A new roasting component in coffee. *Proc. 14th ASIC Coll.*, pp. 615–621.

Stennert A. and Maier H.G. (1994) Trigonelline in coffee. II. Content of green, roasted and instant coffee. *Z. Lebensm. Unters. Forsch.* 199, 198–200.

Stiles J.I., Moisyadi I. and Neupane K.R. (2000) Purified proteins, recombinant DNA sequences and processes for producing caffeine free beverages. *US Patent 6,075,184.*

Suggi Liverani F. (1991) A tool for the classification of green coffee samples. *Proc. 14th ASIC Coll.*, pp. 657–665.

Suggi Liverani F. (1995) Green coffee grading using fuzzy classification. In G. Della Riccia, R. Kruse and R. Viertl (eds), *Mathematical and Statistical Methods in Artificial Intelligence.* New York: Springer, pp. 237–245.

Teixeira A.A. (1978) Estudo preliminar sobre a qualidade do café no estado de São Paulo safra 78/79. *6° Congresso Brasileiro de Pesquisas Cafeeiras* (Ribeirão Preto, SP), pp. 316–322.

Teixeira A.A. and Ferraz M.B. (1963) A caracterização da membrana prateada nos cafés despolpados. *A Rural (São Paulo)* 43, 28–29.

Teixeira A.A. and Figuereido J.P. (1985) Efeito do brunimento sobre a qualidade do café. *Biológico (São Paulo)* 51 (9), 233–237.

Teixeira A.A., Carvalho A., Mônaco L.C. and Fazuoli L.C. (1971) Graõs defeituosos em café colhido verde. *Bragantia* (Campinas) 30(8), 77–90.

Teixeira A.A., Gomes F.P., Pereira L.S.P., Moraes R.S. and Castilho A. (1969) A influência de grãos verdes em ligas com cafés de bebida mole. *Ciência e Cultura* 21, 355–356.

Teixeira A.A., Hashizume H., Nobre G.W., Cortez J.G. and Fazuoli L.C. (1982) Efeito da temperatura de secagem na caracterização dos efeitos provenientes de frutos colhidos verdes. *Proc. 10th ASIC Coll.*, pp. 73–80.

Teixeira A.A., Pereira L.S.P. and Pinto J.C.A. (1970) Classificação de café – noções gerais. Ministério da Indústria e do Comércio – Instituto Brasileiro do Café.

Teixeira A.A., Taniwaki M.H., Pitt J.I., Iamanaka B.T. and Martin C.P. (2001) The presence of ochratoxin A in coffee due to local conditions and processing in four regions in Brazil. *Proc. 19th ASIC Coll.*, CD-ROM.

Teixeira A.A., Toledo A.C.D., Toledo J.L.B., Inskava J.M. and Azevedo W.O. (1991) O prejuízo causado pelos grãos de café denominados defeitos verdes e preto verdes. *17° Congresso de Pesquisas Cafeeiras* (Varginha, MG), pp. 25–27.

Tiscornia E., Centi-Grossi M., Tassi-Micco C. and Evangelisti F. (1979) Sterol fractions of coffee seeds oil (*Coffea arabica* L.). *Rivista Ital. Sostanze Grasse* 56, 283–292.

Tosello A. (1946) Studies on the drying of agricultural products. *Bragantia* 6 (2), 39–107 (in Portuguese).

Trautwein E. (1987) Untersuchungen über den Gehalt an freien und gebundenen Aminosäuren in verschiedenen Kaffeesorten sowie über deren Verhalten während des Röstens.Thesis, University of Kiel.

Trouche M.-D., Derbesy M. and Estienne J. (1997) Identification of robusta and arabica species on the basis of 16-O-methylcafestol. *Ann. Fals. Exp. Chim.* 90, 121–132.

Tsubouchi H., Terada H., Yamamoto K., Hisada K. and Sakabe Y. (1988) Ochratoxin A found in commercial roast coffee. *J. Agric. Food Chem.* 36, 540–542.

Van der Stegen G. and Blanc M. (1999) Report on the workshop 'Enhancement of coffee quality by reduction of mould growth'. *Proc. 18th ASIC Coll.*, pp. 219–222, and following papers.

Van der Stegen G.H.D. and Noomen P.J. (1977) Mass-balance of carboxy-5-hydroxytryptamindes (C-5-HT) in regular and treated coffee. *Lebensmittelwiss. Technol.* 10, 321–323.

Van der Stegen G.H.D. and Van Duijn, J. (1987) Analysis of normal organic acids in coffee. *Proc. 12th ASIC Coll.*, pp. 238–246.

Van der Stegen G., Blanc M. and Viani R. (2001) Highlights of the workshop – Moisture management for mould prevention. *Proc. 19th ASIC Coll.*, CD-ROM.

Viani R. (1993) The composition of coffee. In S. Garattini (ed.), *Caffeine, Coffee, and Health.* New York: Raven Press, pp. 17–41.

Viani R. (2002) Effect of processing on ochratoxin A (OTA) content of coffee. *Adv. Med. Biol.* 504, 189–193.

Viani R. (2003) Coffee physiology. In *Encyclopedia of Food Science and Technology*, 2nd edn., Volume 3. London: Elsevier Science, pp. 1511–1516.

Vilela A.R.E., Chandra P.K. and Oliveira G.A. (2000) Efeito da temperatura e umidade relativa no branqueamento de grãos de café. *Ver. Brás. Viçosa Especial* 1, 31–37.

Vincent J.C. (1987) International standardization. In R.J. Clarke and R. Macrae (eds), *Coffee: Volume 1 – Technology.* London: Elsevier Applied Science, pp. 28–30.

Vitzthum O.G., Weisemann C., Becker R. and Köhler H.S. (1990) Identification of an aroma key compound in robusta coffees. *Café Cacao Thé* 34, 27–33.

Wajda P. and Walczyk D. (1978) Relationship between acid value of extracted fatty matter and age of green coffee beans. *J. Sci. Food Agric.* 29, 377–380.

Weidner M. and Maier H.G. (1999) Seltene Purinalkaloide in Röstkaffee. *Lebensmittelchemie* 53, 58.

Wettstein A., Spillmann M. and Miescher K. (1945) Zur Konstitution des Cafesterols 6. *Mitt. Helv. chim. Acta* 28, 1004–1013.

White D.R. (1995) Coffee adulteration and a multivariate approach to quality control. *Proc. 16th ASIC Coll.*, pp. 259–266.

Wilbaux R. (1967) Rapport sur les recherches en collaboration relatives aux méthodes de dosage de l'extrait soluble dans l'eau du café torréfié. *Proc. 3rd ASIC Coll.*, pp. 77–85.

Wöhrmann R., Hojabr-Kalali B. and Maier H.G. (1997) Volatile minor acids in coffee. I. Contents of green and roasted coffee. *Dtsche Lebensm.- Rundsch.* 93, 191–194.

Wootton A.E. (1971) The dry matter loss from parchment coffee during filed processing. *Proc. 5th ASIC Coll.*, pp. 316–324.

Zosel K. (1971) Verfahren zur Entcoffeinierung von Rohkaffee. *German Patent* 2,005,293.

CHAPTER 4

Roasting

B. Bonnländer, R. Eggers,
U.H. Engelhardt and H.G. Maier

4.1 THE PROCESS

R. Eggers

At first sight, roasting of coffee seems to be a well-known and simple process: It is just the application of heat to raw coffee beans. What is important is to generate and control the correct temperatures at the right moment, then stop the process when the aroma has fully developed and the colour of the coffee is homogeneous throughout the whole bean. However, on closer inspection, questions arise that have not yet received an answer: the dependency of the instationary (changing with the time) temperature distribution in the coffee bean on the parameters governing the process, such as roast gas temperature, fluid flow conditions and material properties of the coffee bean. The difficulty in seizing the whole process comes from the dramatic changes in nearly all the parameters related to the process: the temperatures, the material properties and the geometry of the beans. Figure 4.1 demonstrates the meaning of temperature distribution in a coffee bean during roasting.

The bean has a finite geometry of complicated shape; its inner structure is heterogeneous; by heat admission its volume swells and the inner structure changes. Mathematically, temperature becomes a three-dimensional function of non-steady-state character with unknown moving boundaries. From a chemical-engineering point of view the roasting process consists of a combined heat and mass transport superposed by endothermic and exothermic reactions. Thus the application of heat to the coffee beans not only generates a temperature field, it also causes inner pressures and a re-distribution of moisture depending on time and location.

These effects are illustrated in Figure 4.2: Heat energy is admitted to the surface of the whole green bean, mainly by external hot gas flow, with

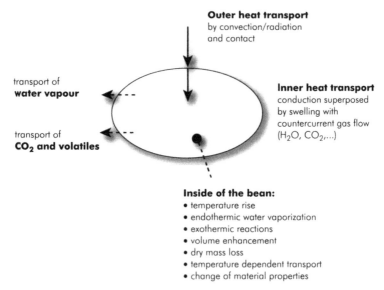

Outer heat transport
by convection/radiation
and contact

transport of
water vapour

transport of
CO₂ and volatiles

Inner heat transport
conduction superposed
by swelling with
countercurrent gas flow
(H_2O, CO_2,...)

Inside of the bean:
- temperature rise
- endothermic water vaporization
- exothermic reactions
- volume enhancement
- dry mass loss
- temperature dependent transport
- change of material properties

Figure 4.1   Roasting of coffee beans – main aspects

additional radiation and contact heat transfer depending on the type of roaster.

The temperature of the bean surface increases – with heat conduction into the porous material – due to the temperature gradient. When the local temperature reaches the evaporation temperature of the bean moisture, a front of evaporation starts moving towards the centre of the bean. During this first part of the roasting process the walls of the whole bean are still relatively firm. Thus the vapour that has been generated cannot permeate and the pressure build-up makes the bean volume expand. Evaporation of the bean moisture, being an endothermic operation, needs latent energy leading to a slowed down kinetic in the temperature rise within the bean. The swelling and drying result in a strong decrease in heat conductivity within the section between the vaporization front and the outer surface of the bean. As a consequence the temperature gradient is steeper in the dried region of the bean because of the enhanced resistance to heat transfer. Mechanical and thermal stresses moving towards the centre of the bean are created, which make the beans crack or even burst if the superposed stresses overcome the tensile strength of the bean.

Roasting reactions – browning with formation of flavour compounds at elevated inner pressures – begin at higher temperatures (T>160 °C) –

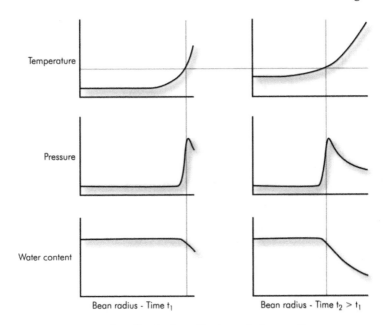

Figure 4.2   Assumed profiles in the coffee bean during roasting

starting at the bean surface and moving towards the inner dry pre-expanded structure of the bean. This second front of moving latent heat is exothermic. Gaseous reaction products – mainly carbon dioxide – are generated and entrapped within the cell structure, increasing inner pressure until they permeate through the walls that are weakened and partly destructed by the high temperatures as well (Perren *et al.*, 2001). The roasting process is then a counter-current process with heat transport inside and mass outside. The phenomenological temperature profile (Figure 4.2) shows that the distance between the two moving zones of evaporation and roasting depends on the outer heat transfer, given by the roasting technique applied, and on the structure of the bean. The latter is linked with the geographical origin of the coffee to be roasted (Figure 4.3).

In order to achieve a roasting profile as homogeneous as possible, the process must be precisely controlled, aiming at small temperature gradients throughout the bean; on the contrary, fast roasting leads to an overlapping of the evaporation and roasting steps and to an inhomogeneous profile.

When the desired degree of roast is reached, the beans have to be cooled down rapidly by water quenching or cold air in order to stop further changes in colour, flavour and volume. The roasting loss is either

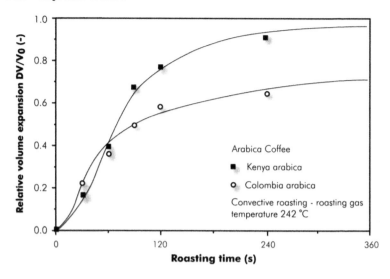

Figure 4.3   Volume expansion for coffees of different origin

measured by the dry mass loss (the weight loss based on dry green beans) or by the total mass loss (moisture plus organic matter loss). Table 4.1 shows the correspondence between roasting degree and dry mass loss (Clarke, 1987). During roasting the density of the beans decreases to nearly half its initial value.

The shape of a coffee bean can be defined as half an ellipsoidal body. Attempts at modelling the temperature field of a coffee bean during roasting have been made. The influence of the endothermic vaporization enthalpy as well as of the exothermic reaction enthalpy is shown in Figure 4.4. In the comparison the temperature increase has been

Table 4.1   Material data of arabica coffee beans

Coffee	Mass	Moisture	Roasting loss	Organic loss	Density	Volume	Radius*	Porosity
	(g)	(wt%)	(wt%)	(wt%)	(g/ml)	(ml)	(mm)	(–)
Green	0.15	10–12	0	0	1.2–1.4	0.11–0.13	3	<0.1
Medium to dark roast	0.13	2–3	15–18	5–8	0.7–0.8	0.16–0.19	3.5	0.5

*Of a sphere with the same volume.

Without latent energies   With latent energies

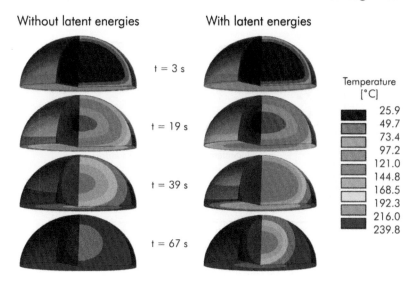

t = 3 s

Temperature
[°C]

t = 19 s

25.9
49.7
73.4
97.2
121.0
144.8
168.5
192.3
216.0
239.8

t = 39 s

t = 67 s

Figure 4.4   Temperature calculation of a roasting bean

calculated without taking into consideration latent energies. The results, when latent energies are introduced, indicate a delayed temperature rise up to 150 °C caused by vaporization followed by an accelerated exothermic temperature increase above 180 °C. Additional measurements of three temperatures – at the surface, between core and surface and in the core itself – show some remarkable aspects of roasting in a roasting experiment lasting 600 seconds (Figure 4.5).

Starting at ambient conditions all three temperatures increase with declining slope until nearly constant temperature has been reached. During the first period (0–50 sec) of heating up the bean the difference between core and surface temperatures passes a maximum of at least 70 degrees, whereas the temperature difference between core and half distance to the surface of the bean slowly increases up to a maximum of approximately 10 degrees throughout nearly the whole roasting time. Obviously, the roasting process is much faster in the outer sections of the bean compared to the core, probably leading to high thermal stresses in the outer parts of the bean.

A small temperature drop occurs suddenly after 200 seconds' roasting time. At this time the inner temperature differences are clearly higher in comparison to the outer temperature differences. However, the temperature course drops slightly simultaneously throughout the whole bean. After this unsteady phenomenon, temperatures increase more rapidly than before! This could be explained by a pressure built up in the centre

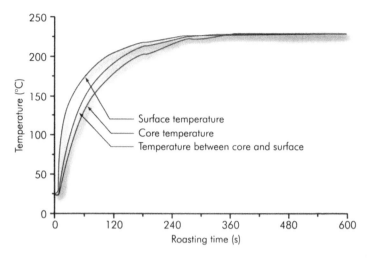

Figure 4.5   Experimental temperature profiles in the bean

of the bean due to water evaporation starting from the surface towards the central region, but proceeding with decreasing velocity because the driving force of heat transfer – the temperature difference to the ambient – is diminishing all the time. So the pressure inside the bean increases until the temperature overcomes the vaporization temperature, which is higher at elevated pressure. Thus the remaining water in the centre of the bean evaporates spontaneously and an endothermic flash creates the slight temperature drop. Immediately after flashing the full exothermic roast reaction is enabled to run to the core of the bean.

In order to roast the bean as homogeneously as possible, a stepwise process with a slow increase of the temperature of the heating gas seems to be advantageous. Further, a build-up of pressure in the bean is important for the generation of sufficient aroma. Thus, temperature control of the heating gas, permitting not only temperature profiles with moderate differences, but also a sufficient pressure built up inside the bean, has to be the objective of roasting.

4.2   ROASTING TECHNIQUES

R. Eggers

Conventional roasting techniques have been described by Sivetz and Desrosier (1979), Rothfos (1984) and Clarke (1987). Recently, newer developments have been summarized by Clarke and Vitzthum (2001). Documentation is also available from manufacturers of roasting equipment, e.g. Probat Werke at Emmerich and Neuhaus Neotec at Reinbek, both in Germany.

From an engineering standpoint, the principles of roasting can be described from a mechanical, thermal and operational point of view (Figure 4.6).

In the case of traditional roasters, a tendency towards a larger variety in coffee products has led to smaller uniform charges and therefore the demand for large continuous roasters is decreasing in the roast and ground market. Due to modern control techniques, very consistent products can be obtained with batch roasters, where the heat input can be varied over time. Roasters operating under pressure and steam roasters are unavailable commercially, in spite of ongoing research.

Table 4.2 summarizes the basic principles of modern roasting technology: a forced convective flow of hot gases passes through a moving bed of coffee beans. The motion of the beans is either produced by rotation or by the flow of roasting gases. Hot gas is either produced by gas or by oil burners.

Due to increasing cost of energy and environmental considerations, modern roasting equipment usually includes re-circulation of the exhaust gas, after removal of the solid particles it carries (chaff, dust) by retention on a cyclone. The gas is either brought back to the burner, or to a thermal afterburner, operating at temperatures between 400 and 600 °C, to burn

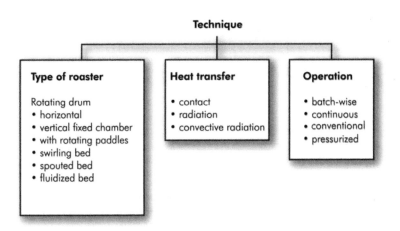

Figure 4.6 Principal roasting techniques

Table 4.2   Basic principles in modern roasting technology

Type	Principle	Characteristics
Rotating cylinder		Horizontal/vertical With/without perforated walls Direct heating by convective flow of hot gases Indirect heating by hot drum walls Batch-operated Continuously operated by an inner conveyer Gas temperatures: 400–550 °C Roasting times: 8.5–20 min
Bowl		Direct heating by convective flow of hot gases Continuously operated across the gas stream; rotating Gas temperatures: 480–550 °C Roasting times: 3–6 min
Fixed drum		Direct heating by convective flow of hot gases Batch operated Gas temperatures: 400–450 °C Roasting times: 3–6 min
Fluidized bed		Direct heating by fluidizing gas Batch operated Gas temperatures: 240–270 °C Roasting times: 5 min
Spouted bed	symmetric asymmetric	Direct heating by fluidizing gas Batch operated Fast roasting: Gas temperatures: 310–360 °C Roasting times: 1.5–6 min Slow roasting: Gas temperatures: 230–275 °C Roasting times: 10–20 min

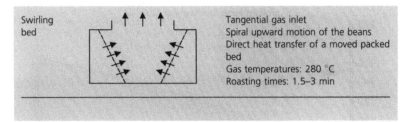

Swirling bed	Tangential gas inlet
	Spiral upward motion of the beans
	Direct heat transfer of a moved packed bed
	Gas temperatures: 280 °C
	Roasting times: 1.5–3 min

off both the residual particles from above 1600 to well below 50 mg/Nm³ and the volatile organic matter (aerosols, condensate, etc.) present, as required by law in some countries. Its residual thermal energy can be used to preheat incoming fresh air (Illy and Viani, 1995, p. 94).

4.2.1   Energy balance and heat transfer

The investigation of the modes of heat transport from the heating gas to the surface and inside the bean is rather complex, hence only approximate calculations are possible. Nevertheless, the heat transfer theory helps one to understand the roasting process. Assuming spherical coffee beans with an average and constant diameter of 6 mm, the tendency of the convective heat transfer can be calculated for a given gas temperature. The heat transfer coefficient α is shown versus the Reynolds number Re in Figure 4.7. In a motionless system ($v_{rel} = 0$), the heat transfer coefficient is around 14 W/m²K and a minimum fluidization velocity (e.g. Re = 300) causes α-values in the range of 75 W/m²K.

The heat transfer to single beans is different from the heat transfer to a bed of beans. The calculated values are higher than those for single spheres at the same superficial velocity and the actual roasting process is expected to show heat transfer values in the region between the two curves. Two important consequences arise: on the one hand, only a slight increase in heat transfer is possible at higher velocities; on the other hand, the improvement of the outer heat transfer is limited due to the heat transfer resistance of the inner bean structure.

The amount of heat energy Q (KJ) transferred to the bean can be calculated with the following equation:

$$Q = \alpha \int A_{cs(t)} (T_g - T_{cs(t)}) dt \qquad (4.1)$$

where:

$A_{cs(t)}$ surface area of the coffee bean (increasing during roasting)

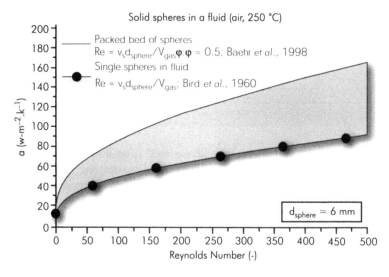

Figure 4.7 Heat transfer of coffee beans during roasting

$T_{cs\,(t)}$	surface temperature of the coffee bean (increasing during roasting)
α	convective heat transfer coefficient

Although the packed bed of coffee beans reveals the highest heat transfer coefficient, the advantage of roasting using fluidizing gases becomes obvious from equation 4.1. The surface area A_{cs} of the fluidized beans is in intimate contact with the hot gases and, as a result, the heat transfer is more effective and homogeneous.

Figure 4.8 shows the balance of energy flow of a roasting section operated continuously, where:

$\dot{M}$	mass flow
c	heat capacity
T	temperature
Index g	gas
Index c	coffee
Index in	input
Index out	output

The enthalpies of vaporization, Δh_v, and exothermic reaction, Δh_{ex}, have to be considered respectively as heat sinks and sources, in balancing

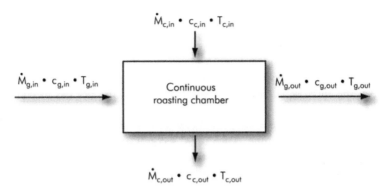

Figure 4.8 Balance of energy flow

the energy flows. There is a further heat loss $\dot{Q}_{loss}$ from the roaster to the atmosphere.

$$
\begin{aligned}
\dot{M}_{g,in} \cdot c_{g,in} \cdot T_{g,in} + \dot{M}_{c,in} \cdot \left(c_{c,in} \cdot T_{c,in} + \Delta h_{ex} - \Delta h_v \right) \\
= \dot{M}_{g,out} \cdot c_{g,out} \cdot T_{g,out} + \dot{M}_{c,out} \cdot T_{c,out} + \dot{Q}_{loss}
\end{aligned}
\tag{4.2}
$$

The figures for latent energy are not yet very well known.

By neglecting the difference between $\dot{M}_{g,in}$ and $\dot{M}_{g,out}$ that is due to the development of roast gases and flavour, and keeping the specific heat of gas constant, the specific energy demand $q[kJ/kg\,roastcoffee]$ becomes:

$$
q = \frac{\dot{M}_g \cdot c_g \left[\left(T_{g,in} - T_{g,out} \right) + \dot{Q}_{loss} \right]}{\dot{M}_{c,out}}
\tag{4.3}
$$

In the case of batch roasting, the temperature of the gas at the outlet of the roaster increases with time and the specific roasting energy needed can be calculated by integration:

$$
q = \frac{\dot{M}_g c_g \cdot \int \left[\left(T_{g,in} - T_{g,out(t)} \right) + \dot{Q}_{loss} \right] dt}{\dot{M}_{c,out}}
\tag{4.4}
$$

A comparison of the equation representing the energy demand with the heat transfer equation indicates that the roasting process to be energetically optimized must be operated with small temperature differences and high heat transfer coefficients, meaning low temperatures and high velocities of the hot gases.

4.2.2   High yield or fast roasting

Although processing at low temperatures has the advantage of a homogeneous roast in favourable energetic conditions, there is an alternative for enhanced heat transfer related to equations (4.1) and (4.4): the increase of the gas temperature T_g and the hot gas to coffee ratio.

Fast roasting, i.e., providing the thermal energy required for roasting in a shortened period of time, down to 90 seconds or even less (60 seconds has been indicated as the shortest roasting time achievable now), has been made possible by the development of forced convection roasting at temperatures of 300–400 °C. Because of the poor thermal conductivity of the bean, there is a roasting gradient within the bean (particularly noticeable at low roast temperatures), bean volume is increased with a characteristic puffing by 10–15% with respect to conventional roasting, bulk density is reduced below 300 g/l and extraction yield at brewing is increased by 20%.

Summarizing, the profile of bean temperature variation during roasting is an important parameter.

Both bean temperature and roasting time depend strongly on heat transfer and therefore on the technology applied. The results of the LTLT (Low Temperature Low Time) and HTST (High Temperature Short Time) processes integrate into a slightly increasing function. Personal measurements (Clarke and Vitzthum, 2001, p. 93) are in good agreement with the literature data. Data on the variation of bean temperature during roasting have been reported by Da Porto et al. (1991) and by Schenker et al. (1999). The latter authors roasted 100 g of arabica beans as a fast fluidized bed in a hot air flow of 0.01885 m³/s in two ways: first the so-called LTLT roasting with a hot gas temperature of 220 °C and a roasting time between 9 and 12 minutes, then HTST roasting at a gas temperature of 260 °C and roasting time between 2.6 and 3 minutes. In the case of LTLT roasting, the bean temperature and other parameters were given as a function of roasting time: the temperature rose continuously from 20 to 190 °C in 2 minutes; this increase corresponds to 90% of the final increase (211 °C after 14 minutes). The bean temperature data published by Da Porto et al. (1991) were obtained with a laboratory roaster (HTLT) and Brazilian Santos coffee. It can be seen that, in comparison with this traditional laboratory roaster, a fluidized bed roaster produces a significantly faster increase in bean temperature. However, because of the poor thermal conductivity of the bean, there is a gradient of roasting level within the bean, particularly noticeable at low roast; the volume is increased by 10–15% with respect to conventional roasting with a

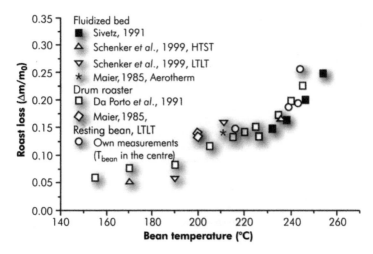

Figure 4.9   Roasting loss versus bean temperature – different processes

characteristic puffing of the beans and a bulk density lower than below 300 g/l, and extraction yield at brewing 20% higher.

High yield roasting is not considered optimal for Italian espresso, owing to the high residual content of chlorogenic acids, which carries into the cup an astringent sour note (Maier, 1985; Illy and Viani, 1995, p. 96).

4.3   CHANGES PRODUCED BY ROASTING

H. G. Maier

4.3.1   Physical changes

The main transformations occurring in the bean with increasing temperature are given in Table 4.3, while the most important macroscopic changes produced by roasting are indicated in Table 4.4 (Pittia *et al.*, 2001). The roasted coffee bean becomes brown to black, very brittle, with an increase in volume up to 100% (dark roasting) and a corresponding decrease of density (300–450 g/l in contrast to 550–700 g/l in raw beans) (see Figure 4.10), with many macro- and micropores (Schenker *et al.*, 2000), the cellular structure ruptured, especially upon espresso roasting. The water content should be near 1%, if no water quench is applied.

Table 4.3   Macroscopic changes during roasting

Temperature change within the bean (°C)	Effect
20–130	Liquid-vapour transition of water (bean drying). Colour fades
130–140	First endothermic maximum. Yellow colouring and swelling of bean with beginning of non-enzymatic browning. Roast gases are formed and begin to evaporate
140–160	Complex series of endothermic and exothermic peaks. Colour changes to light brown. Large increase in bean volume and micropores. Rests of silverskin are removed. Bean is very brittle. Some little fissures at the surface occur. Aroma formation starts
160–190	Roasting reactions move towards the inner dry structure of the bean (see also Figures 4.2 and 4.4)
190–220	Micro fissures inside the bean. Smoke escapes. Large volumes of carbon dioxide escape and leave the bean very porous. Typical flavour of roasted coffee appears

4.3.2   Chemical changes

4.3.2.1   Overall changes

As shown in Table 4.3, during roasting, water (the moisture water of the green beans and that generated by reactions) and carbon dioxide escape. They are accompanied by some carbon monoxide and organic volatiles. Water and carbon dioxide are generated by the very important Maillard reaction, which leads to the coloured products, the melanoidins, and to the main part of the organic volatiles. In addition, water and carbon dioxide are produced by numerous other pyrolytic reactions. Table 4.4 summarizes the macroscopic changes occurring throughout the roasting operation, and Figure 4.11 gives an idea of the effect on the contents of the main groups of coffee constituents. The values for roasted coffee are calculated on green coffee dry basis, in order to make a better comparison of the changes. The changes within the main groups will be dealt with in the following paragraphs.

4.3.2.2   Carbohydrates

Of the mono- and disaccharides of green coffee, after roasting only traces of free sugars remain. Sucrose is supposed to be partially hydrolysed, the rest being pyrolysed (caramelized). From the reducing sugars or their

Table 4.4 Macroscopic changes brought about by roasting

Parameter	Comment
Colour of bean	It fades to whitish-yellow at the beginning of roasting, it darkens regularly with increasing temperature (see Figure 4.10); natural arabicas and robustas need a higher final temperature to reach the same colour as washed arabicas. There is a dark to light colour gradient from the surface within the bean, particularly noticeable with low-roasted fast-roasted coffees
Surface of bean	Oil sweats to the surface rendering it brilliant, particularly at high roasting levels
Structure of the bean	The large volumes of carbon dioxide freed render the bean porous (see Figure 4.10c)
Brittleness	It increases with roasting to a maximum, with an important modification of the internal texture of the bean, which under the microscope looks like a volcanic land covered with an amalgam of the original constituents (see Figure 5.1)
Density	Decreases regularly from 550 to $700\,g\,l^{-1}$ in raw beans to $300–450\,g\,l^{-1}$ in roasted beans (the lowest figures are attained with fast-roasted coffee)
Hot water extract	It decreases slightly from green to light-roasted beans, then increases again slightly with increasing roasting degree (highest with fast-roasted coffee)
Moisture	Both free water, which decreases regularly all through roasting, and chemically bound water, liberated above 100 °C, decrease to below 1% unless water quenching is applied (moisture release is less efficient with fast-roasted coffee)
Organic losses	Destruction of carbohydrates, chlorogenic acids, trigonelline, amino acids becomes important above 160 °C, varying between 1 and 5% at the lightest commercial roasts, 5–8% for medium-roasted coffees, to above 12% for dark-roasted ones (lower for fast-roasted coffee) Release of CO_2 continues for several days
Volatile constituents	Aroma content reaches a maximum at low roasting, while destruction becomes more important than generation at medium roasting losses (aroma generation is higher in fast-roasted coffees, where a maximum is reached at medium roast)
pH of beverage	In washed arabicas it increases from 4.9 for low-roasted coffees to 5.4 for dark-roasted one; higher for dry-processed coffees (lower values for fast-roasted coffee)

fragmentation products, many volatiles (aroma compounds, volatile acids) and non-volatiles (melanoidins and their precursors, acids) are formed by the Maillard reaction and, to a lesser extend, by caramelization. The Maillard reaction has lower activation energy and is therefore favoured if reactive nitrogen compounds (amino acids, free amino groups in proteins and peptides) are present.

Figure 4.10   Stereo microscope section of a bean: (a) green; (b) toasted to c.70 °C; (c) roasted, showing the porous structure

The polysaccharides except cellulose are partially solubilized (Bradbury, 2001; Redgwell *et al.*, 2002). Nevertheless the sum of soluble carbohydrates is lower in roasted coffee. In espresso, the foam stability is related to the amount of galactomannan and arabinogalactan and a function of the degree of roast (Nunes *et al.*, 1997). The carboxyatractyloglycosides are decarboxylated quantitatively and the so generated atractyloglycosides are degraded about 50% under 'normal' roasting conditions (Bradbury, 2001).

4.3.2.3   Non-volatile lipids

Overall, there are only slight changes in the lipid fraction upon roasting. The sterols and most of the triglycerides remain unchanged. The level of

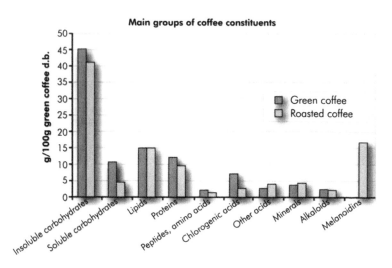

Figure 4.11   Main families of coffee constituents (average of arabica coffees)

trans fatty acids increases, especially the contents of $C_{18:2ct}$ and $C_{18:2tc}$. The linoleic acid content decreases slightly with roasting temperature. The diterpenes, cafestol and kahweol, are decomposed to some extent. Increasing with roasting temperature, dehydrocafestol, dehydrokahweol, cafestal and kahweal are formed (0.5–2.5% of each educt). 16-O-methylcafestol (in robusta beans) is less affected. Up to 20% of the tocopherols and 25–50% of the carbonic acid 5-hydroxytryptamides are destroyed (Wurziger and Harms, 1969; Wurziger, 1972; Speer and Kölling-Speer, 2001; Kurt and Speer, 2002).

4.3.2.4   Proteins, peptides and amino acids

Crude protein content (Kjeldahl nitrogen multiplied by 6.25) changes only very slightly upon roasting, but the nitrogenous components are thoroughly changed. There is a loss of amino nitrogen of 20–40%, in dark roasted espresso, about 50% (Thaler and Gaigl, 1963; Macrae, 1985). Essentially all of the protein in the green coffee is denatured (Macrae, 1985). Some cross-bonds between the amino acid residues of the proteins are formed (Homma, 2001). Parts of the L-amino acids are isomerized to form D-amino acids, in commercial blends 7–70%, depending on the individual acid (Nehring, 1991). The amino acid composition is changed too: the more stable acids, e.g. glutamic acid, survive while others, e.g. cysteine or arginine, decrease or are completely destroyed (Macrae, 1985). In espresso, the foamability depends on the amount of protein in the infusion and on the degree of roast (Nunes et al., 1997). Of the free amino acids, only traces are left after roasting. The reaction products are Maillard products (melanoidins, their precursors and volatiles) and dioxopiperazines, of which the proline-containing ones are bitter (Ginz, 2001). Some intact amino acid residues are incorporated into the melanoidins (Maier and Buttle, 1973).

4.3.2.5   Chlorogenic acids

Since chlorogenic acids are the most important group of acids in green coffee, they will be treated separately. These acids are for the most part destroyed during roasting (Figure 4.12). In addition, some intact chlorogenic acids may be bound by melanoidins (Heinrich and Baltes, 1987). It is assumed that the greatest part of chlorogenic acid is hydrolysed. In addition, isomerization and, to a smaller extent, lactonization takes place. The 6 lactones of the mono-hydroxycinnamoyl-quinic acids are assumed to contribute to the bitter taste of roasted coffee (Ginz,

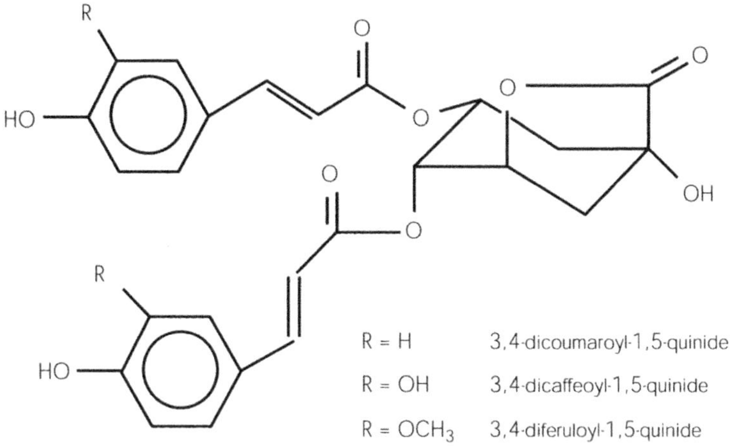

R = H   3,4-dicoumaroyl-1,5-quinide

R = OH   3,4-dicaffeoyl-1,5-quinide

R = OCH₃   3,4-diferuloyl-1,5-quinide

Figure 4.12   Quinides formed during roasting

2001). The 3 di-hydroxycinnamoyl-quinides are supposed to modulate the effect of caffeine (Figure 4.12) (Martin *et al.*, 2001) (see Chapter 10).

The free quinic acid is also isomerized (5 isomers), forming 4 γ-lactones and 3 δ-lactones (Scholz-Böttcher *et al.*, 1991; Homma, 2001). A little part of this acid is decomposed to form simple phenols like hydroquinone. The greatest part, however, remains unchanged, so that in dark roasts, quinic acid is the most prevailing of the individual low molecular acids.

Little is known about the fate of caffeic acid. A minor part degrades to form simple phenols. Two tetra-oxygenated phenylindans have been found in model roasting of caffeic acid as well as in coffee at levels of 10–15 mg/kg, and neolignans (caffeicins) only in model roastings (Clifford, 1997). The rest is assumed to be incorporated into the melanoidins (see 4.5).

4.3.2.6   *Other acids*

Most of the organic acids present in green coffee behave like the chlorogenic acids: they are partially decomposed, e.g. citric acid to form citraconic, itaconic, mesaconic, succinic, glutaric acids and eventually some minor acids, malic acid to form fumaric and maleic acids (Scholz-Böttcher, 1991; Bähre and Maier, 1999). It is likely that esters of these acids are formed upon roasting, because the acids are liberated on heating the brew (Maier *et al.*, 1984). Scholz-Böttcher (1991) detected a fumaric acid/malic acid ester and another malic acid monoester after roasting of malic acid. Phosphoric acid is stable, and its content increases by hydrolysis of the inositol phosphates (Franz and Maier, 1994). A great number of acids is generated by Maillard reaction or caramelization. The most prominent are formic and acetic acids. Their contents reach a maximum at medium roast; upon darker roasting volatilization prevails over formation. Moreover (in decreasing order of the contents in espresso), glycolic, lactic, metasaccharinic, threo-and erythro-3-deoxy-pentonic, glyceric, 2-furanoic and several minor acids are formed (Bähre and Maier, 1999).

4.3.2.7 Minerals

With the exception of phosphoric acid (see 4.3.2.6), the mineral content does not change upon roasting.

4.3.2.8 Alkaloids

Caffeine is stable upon roasting, but a small part is lost by sublimation. Nevertheless this is often overcompensated by the roasting loss. Trigonelline is partially decomposed. The remaining content is about 50% in a light roast, traces in a very dark roast (Viani and Horman, 1975; Stennert and Maier, 1996). The products of model roastings are nicotinic acid, N-methyl nicotinamide, methyl nicotinate and numerous volatile nitrogen compounds, e.g. 46% pyridines and 3% pyrroles (Viani and Horman, 1974). Most are found in coffee, especially nicotinic acid, which constitutes only 1.5% of the degraded trigonelline, but nevertheless acts as an important source of vitamin. The trigonelline/nicotinic acid ratio can be used for a rough estimation of the degree of roast (Stennert and Maier, 1996).

4.4 VOLATILE AROMA COMPOUNDS

B. Bonnländer

Green coffee contains about 300 volatiles (Flament, 2001). The content of some of them, e.g. 3-isobutyl-2-methoxypyrazine, is not changed by roasting. The content of others, e.g. ethyl-3-methylbutyrate, is diminished, but most of the volatiles increase upon roasting (Czerny and Grosch, 2000; Flament, 2001). In addition, about 650 new volatiles have been identified, bringing to more than 850 the number of volatiles identified in roasted coffee.

4.4.1 Generation of roast aroma

Most of the delightful aromatic character of coffee is the result of the roasting process. Green coffee shows a typical green bell pepper-like aroma, where isobutylmethoxypyrazine (MIBP) could be identified as the character impact compound with a 'peasy' smell (Vitzthum et al., 1976). Only part of the more than 300 additional green coffee aroma compounds identified so far (Flament, 2001) survive the roasting process. The high temperature (usually 170–230 °C for 10–15 min) and the elevated pressure inside the bean (up to 25 atm) trigger a vast number of chemical reactions leading to dark colour and more than 1000 volatile and non-volatile compounds identified in roasted coffee (Stadler et al., 2002b). In contrast with green coffee, no single compound but a mixture of approx. 25 very potent compounds represents the significant impression of coffee. The aroma compounds (approx. 1 g/kg) are concentrated in the coffee oil. The concentrations of the most potent ones are in the lower ppm (part per million) or even ppt (part per trillion) range.

4.4.2 Precursors of aroma compounds

The non-volatile constituents play an important role as precursors in the formation of the volatile compounds in roast coffee. Along with the rising temperature, and the subsequent drying and swelling of the bean, the following reactions take place under elevated pressure conditions:

- caramelization and degradation of carbohydrates forming mainly aldehydes and volatile acids (Yeretzian et al., 2002);
- denaturation of proteins and reaction of free amino acids with carbohydrates and their degradation products (Maillard reactions);
- production of phenols and taste active compounds from chlorogenic acids (Pypker and Brouwer, 1970);
- degradation of trigonelline (Stadler et al., 2002a).

According to the different precursor compositions of the two main coffee species, also a different aroma is observed. The aroma of *Coffea canephora* (robusta) is characterized by higher amounts of phenolic compounds (guaiacol, vinylguaiacol), perceived as harsh earthy notes, originating from the chlorogenic acids (Vitzthum et al., 1990). Another interesting class of aroma precursors is that of the glycosides. They can liberate the bound aroma compounds enzymatically, during post-harvest treatment or roasting (Weckerle et al., 2002), and are of special interest also in other foods as aroma storage forms. The common precursor prenylalcohol is a source for sulphur odorants as shown by model reactions with sulphur-containing amino acids (Holscher and Steinhart, 1992).

4.4.3 Identification and characterization of aroma compounds

Pioneer research on coffee aroma was already performed before 1926 when the Nobel Prize winners Reichstein and Staudinger succeeded in identifying the first aroma compounds like furfurylthiol and the guaiacols well before the discovery of gas chromatography (GC). After a period of identifying more and more compounds by GC-mass spectrometry (GC-MS) the current trend in research is to identify the active-smelling ones by GC-olfactometry (GC-O) in combination with the chemical structural information obtained from GC-MS. Generally, the procedure requires extraction and enrichment of the aroma compounds from the natural material, followed by chromatographic fractionation, and finally the authentic identification.

4.4.3.1 Extraction procedures

The extraction procedure has to separate the volatile compounds from the coffee matrix. A very important question is how representative is the extract, or, in other words, how to avoid artefact formation during extraction and enrichment (Sarrazin et al., 2000). Several techniques are widely known in literature, such as simultaneous distillation extraction (SDE) and high vacuum distillation (HVD) or solvent-assisted flavour extraction (SAFE, Engel et al., 1999). In addition, the volatile compounds in the headspace above ground or brewed coffee can be analysed by headspace (HS) techniques either statically or in dynamic mode. Solid phase micro extraction (SPME) can be operated without solvent in fully automated manner either from the brew (IS–SPME) or from the headspace (HS–SPME). Increasing the amount of stationary phase dramatically improves the detection limits and is used as stir bar sorptive extraction (SBSE) in the brew or in the headspace (HSSE), reported by Bicchi et al. (2002). As the aroma compounds tend to be very volatile, extraction losses are compensated by adding stable isotope-labelled reference compounds.

4.4.3.2 Instrumental sensory analysis

GC in combination with olfactometric detection (GC-O) helps to detect potent odorants without knowing their chemical structure. The effluent of a gas chromatograph is split to a conventional detector and a heated sniffing port, where trained people evaluate and record the sensory impression of individual compounds (Holscher *et al.*, 1990). Dilution techniques are used to determine the so-called flavour dilution (FD) factors. By a stepwise dilution of the aroma extract (1:1 by volume) followed by GC-O analysis the most important contributors can be smelled at the highest dilution, thus getting the highest FD factors. This technique, developed by Ullrich and Grosch (1987), is known as aroma extract dilution analysis (AEDA). The odour activity value (OAV) can be expressed as the concentration divided by its threshold only if the structure and the odour threshold of a substance are known. A method for the prediction of OAV from FD factors is reported as well as the precision and the optimal design of AEDA (Ferreira *et al.*, 2001).

4.4.4   Aroma impact compounds in roasted coffee

In arabica coffee the most important contributors to the aroma of roast and ground coffee are determined by the techniques mentioned above. 3-Mercapto-3-methylbutylformate (MMBF), 2-furfurylthiol, methional, β-damascenone as well as two pyrazines and furanones show the highest FD factors (Holscher *et al.*, 1992). A comparison of the results of the major research groups in coffee aroma is given in Table 4.5, together with the aroma impressions of the substance.

The absolute amounts of the aroma compounds shown in the last column differ by two orders of magnitude. The potency, especially that of the sulphur compounds, is demonstrated by their low concentrations. Because of a very low odour threshold (in air), only 130 µg/kg of MMBF are sufficient to generate the strongest odour impression, which is more than 10 000 times over its threshold of 0.003 ng/l of air (OAV of 37 000).

4.4.5   Effects on cup impression

Everybody who has ever had a really bad cup of coffee in direct comparison to a perfect one knows how big the difference can be. The perceived quality depends on objective criteria such as quality of green coffee, but also on subjective or cultural preferences like type of preparation.

Table 4.5 Aroma impact compounds in roasted coffee powder

Odorant	Odour impression	Holscher[1] (FD)	Grosch[2] (FD)	Schenker[3] (FD)	Mayer[4] (μg/kg)
Methanethiol	Putrid, cabbage-like	25	–	–	4500
2-Methylpropanal	Pungent, malty	100	–	–	24000
2-Methylbutanal	Pungent, fermented	100	–	16–128	28600
2,3-Butanedione	Buttery	200	–	256–1024	55700
2,3-Pentanedione	Buttery	100	–	4–128	28300
3-Methyl-2-buten-1-thiol	Animal-like, skunky	200	128	64–256	13
2-Methyl-3-furanthiol	Roasted meat-like	500	–	32	60
Mercaptopentanone	Sweaty, catty	100	–	–	–
2,3,5-Trimethylpyrazine	Roasty, musty	200	64	16–32	6000
2-Furfurylthiol	Roasty, coffee-like	500	256	1024	1350
2-Isopropyl-3-methoxy-pyrazine	Peasy	100	128	–	–
Acetic acid	Vinegar-like	100	–	–	–
Methional	Cooked potato	500	128	1024	148
2-Ethyl-3,5-dimethyl-pyrazine	Roasty, musty	200	2048	1024	55
(E)-2-Nonenal	Fatty	5	64	–	100
2-Vinyl-5-methyl-pyrazine	Roasty, musty	200	–	–	53
3-Mercapto-3-methyl-butylformate	Catty, roasted coffee-like	500	2048	1024	130
2-Isobutyl-3-methoxy-pyrazine	Paprika-like	500	512	4–64	84
5-Methyl-5H-6,7-dihydro-cyclopentapyrazine	Peanut-like	50	–	–	–
2-Phenylacetaldehyde	Honey-like	25	64	–	2500
3-Mercapto-3-methylbutanol	Soup-like	100	32	–	–
2/3-Methylbutanoic acid	Sweaty	500	64	1024	25000
(E)-β-Damascenone	Honey-like, fruity	500	1024	16–128	258
Guaiacol	Phenolic, burnt	200	–	512–1024	3420
4-Ethylguaiacol	Clove-like	25	256	–	1780
4-Vinylguaiacol	Clove-like	200	512	256	45100
Vanilline	Vanilla-like	–	32	–	4050

Peaks sorted by retention time on DB-wax column. Compounds with FD ≥ 32 are reported.
Sources: [1]Holscher et al. (1990); [2]Czerny and Grosch (2000); [3]Schenker et al. (2002), range according to roasting isothermal high/low temperature; [4]Mayer and Grosch (2001)

4.4.5.1   Green coffee quality

Correlations of geographical origins to the cup impression can be recognized by a trained cup taster, but chemical mapping is also possible. Modern statistical methods (e.g. analysis of variance, ANOVA) help to map the composition of the aroma against the growing region (Freitas and Mosca, 1999). Even a mapping against different growing conditions (e.g. shade or sun) is possible (Bonnlaender, Unpublished results). Next in the chain of production are the harvest and the post-harvest treatments of the cherries. Errors in handling can lead to mould growth or other defects (see 2.3.6 and 3.7).

4.4.5.2   Roasting

Older studies compared the composition of the volatiles extracted at different degrees of roasting (Gretsch et al., 2000). The use of modern fast analysis tools, such as resonance enhanced multiphoton ionization time-of-flight mass spectrometry (REMPI/TOFMS), makes online monitoring of the roasting process possible; a continuous monitoring with sampling rates up to one per second can be performed showing the evolution characteristics of specific compounds (Yeretzian et al., 2002). Generally, most of the aroma compounds are generated at medium roast. Some aroma compounds degrade at higher temperatures, whereas others like the guaiacols, furfurylthiol and pyridine, show an increase up to very high roasts; they contribute particularly to the aroma of dark roasted coffees (Mayer et al., 1999).

4.4.5.3   Preparation

During the preparation of the beverage the flavour compounds are extracted from roast and ground coffee as a function of their solubility in water. Whereas large amounts of 2,3-butandione, 2,3-pentandione and the furanones get extracted from coffee, the yield of less polar compounds like furfurylthiol is reduced (Mayer and Grosch, 2001). Figure 4.13 shows the headspace stir bar sorptive extraction (HSSE) GC-MS analysis of an espresso versus a drip cup to illustrate differences between the important Italian beverage espresso and the drip preparation used worldwide.

Addition of milk to the beverage generally reduces the perceived coffee aroma, increasing the creamy impression. Bucking and Steinhart (2002) found a characteristic change in sensorial impression, depending on the type of milk used (skim milk, coffee whitener).

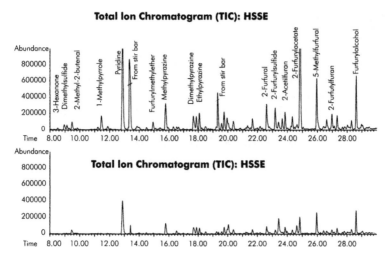

Figure 4.13   Differences in preparation techniques. HSSE of an espresso (upper total ion chromatogram) versus a drip coffee (lower chromatogram) analysed by thermo desorption gas chromatography–mass spectrometry (TDS GC–MS EI)

4.4.5.4   Staling of coffee (see also 6.1.2 and 6.1.3)

The delicious aroma of fresh roast and ground coffee only lasts for a short time. Soon after grinding the very fresh, mild, roasty notes diminish and strong spicy notes appear in the flavour (Mayer and Grosch, 2001). In particular, the very potent sulphur compounds deteriorate in contact with air and the ground coffee starts smelling stale after about 10 days. The structure of the whole bean provides protection to a certain degree by keeping CO_2 gas in its pores. Analytically the ratio of 2-methylfuran to 2-butanone (M/B ratio) can be used as a good freshness indicator, before lipid oxidation leads to rancid products after several weeks (Arackal and Lehmann, 1979). Packaging under inert gas (CO_2 or N_2) with a minimum of residual oxygen guarantees long shelf life. Freezing also helps, slowing down staling reactions. After the package has been opened, it is recommended to store coffee in cool and dry conditions in the dark. The staling reactions have not been completely explained. Hofmann and Schieberle (2002) found reduction of sulphur compounds (MMBF, furfurylthiol) in roast and ground and in liquid coffees by binding to melanoidins. A radical mechanism is suggested by Blank et al. (2002). Hydrolysis of chlorogenic quinic lactone leads to liberation in the brew of the organoleptically undesirable quinic acid.

4.5 MELANOIDINS

U.H. Engelhardt

Melanoidins are coloured (brown) pigments developed during thermal treatment of foods via the Maillard reaction or by dehydration (caramelization) reactions of carbohydrates followed by polymerization (Bradbury, 2001). Basically, melanoidins are formed by interactions between carbohydrates and compounds, which possess a free amino group, such as amino acids or peptides. Recently, melanoidins – not only those from coffee – have attracted lots of interest as regards their occurrence in foods and the corresponding impact on human health. This is documented by the fact that the European Community set up a COST action (COST Action 919: 'Melanoidins in food and health') starting in 1999 and ending in 2004. This action includes research on the separation and characterization of melanoidins and related macromolecules, the flavour binding, colour, texture and antioxidant properties of melanoidins and investigation of the physiological effects and fate of melanoidins (COST, 2002). The foods of relevance specified are coffee, malt, beer, breakfast cereals and bread.

4.5.1 Chemistry

It can be anticipated that melanoidins are a very heterogeneous group of compounds as regards molecular mass as well as the chemical and biological properties. Taking this into account, one may argue that there is no justification for referring to all those compounds as melanoidins. On the other hand, is there any alternative? This situation is comparable with other polymers in foods and beverages, e.g. the thearubigins in tea or phlobaphenes in cocoa. It is agreed that those are constituents of the beverage, and contents can be found in tables, but nobody really knows what the properties and the structures of the compounds are.

What do we know about the formation of melanoidins? Most of the free mono- and disaccharides are lost during the roasting process; the same is true for free amino acids (Trautwein, 1987; Bradbury, 2001). They are at least in part converted into melanoidins. Another part of the sugars undergoes caramelization reactions, also yielding melanoidin-like pigments (Bradbury, 2001) or is degraded yielding acidic compounds (Ginz et al., 2000). A small proportion of the amino acids could be converted into diketopiperazines (cyclic dehydration products of dipeptides), but, according to Ginz (2001), it seems to be more likely that the diketopiperazines are formed via protein degradation. Not much is yet known about the structures of melanoidins. As already mentioned above, melanoidins are formed during the Maillard reaction along with a variety of flavour compounds. The Maillard reaction in a food matrix is a complex process because of the multitude of compounds present. Consequently, information on the Maillard reaction is best generated by model experiments with only a limited number of educts present; which, however, still give rise to a considerable number of volatile and non-volatile products. Reviews of the Maillard reaction can be found in Flament (2001, p. 39).

Coffee melanoidins contain phenolic moieties (Heinrich and Baltes, 1987; Homma, 2001), as indicated by experiments using Curie point pyrolysis high-resolution gas chromatography/mass spectrometry (HRGC/MS), which gave 97 products, 33 of which were phenols (Heinrich and Baltes, 1987). The molecular mass of the melanoidin fractions is between 3000 and 100 000 Dalton according to studies using gel chromatography or HPLC as analytical tools (Homma, 2001). There seems to be more high molecular weight material in robusta coffee compared to arabica (Steinhart et al., 1989). Contents of melanoidins in coffee up to 30% (Bradbury, 2001) or 29.4% (Belitz et al., 2001) are reported in the literature – in the latter referred to as unknown constituents (colorants, bitter compounds), while other authors give 15% for brewed coffee and about 23% for roasted coffee on a dry weight base (Parliment, 2000). However, it should be stressed that this is just a determination of the melanoidins by difference and not at all a determination of the compounds themselves. Regardless what the figures might be, melanoidins make a relevant contribution in the cup.

Gel filtration chromatography was usually employed to purify the melanoidin fractions according to the procedure of Maier et al. (1968) and Maier and Buttle (1973). This fractionation includes extraction with diethyl ether and adsorption of the water-soluble compounds on polyamide. The compounds not adsorbed on polyamide were separated into three fractions using Sephadex® G-25. The amount adsorbed gave two fractions, one of which gave three subfractions. Details can be found in papers by Maier et al. (1968), Maier (1981) and Macrae (1985). More recent work usually also employs gel chromatography, but in most cases without the polyamide step (Steinhart et al., 1989; Nunes and Coimbra, 2001). Basically, coffee is prepared by hot water extraction and defatted using dichloromethane. The freeze-dried extract is either treated by ultrafiltration or separated by gel chromatography (Sephadex® G-25; eluent: water) and again freeze-dried (Hofmann et al., 2001; Steinhart et al., 2001). Hydrolysis was often employed to detect information about the moieties present in the polymers and sugars, amino acids and phenols.

An interesting intermediate involved in melanoidin formation has been recently detected by Hofmann *et al.* (1999) using EPR and LC-MS. The compounds are named CROSSPY – (1,4-bis-5-amino-5-carboxy-1-pentyl)pyrazinium radical cations – which are oxidized yielding diquaternary pyrazinium ions. Figure 4.14 shows the reaction pathway proposed by Hofmann and Schieberle (2002), and Hofmann *et al.* (2001).

Even though the formation of melanoidins is far from being fully understood, these findings are an important step in gaining knowledge of

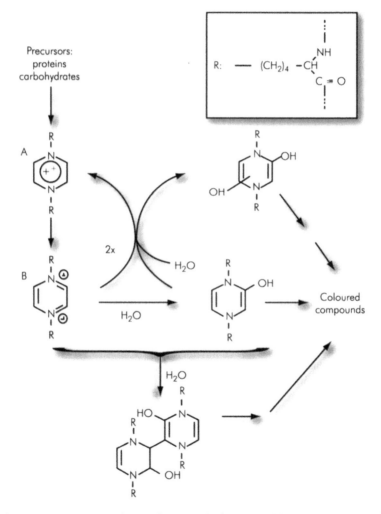

Figure 4.14 Reaction scheme of melanoidin formation: (a) via CROSSPY radical; (b) via diquaternary pyrazinium ions

the mechanisms involved. The amounts of high-molecular mass melanoidins increased with increasing roasting loss (Ottinger and Hofmann, 2001).

In a study on the interaction of isolated coffee melanoidins and selected aroma compounds in model experiments Hofmann et al. (2001) detected a decrease in free furfurylthiol, 3-methyl-2-butene-1-thiol and mercapto-3-methylbutylformate, which was most significant when the 1500–3000 Da coffee melanoidins were used. In another experiment by Hofmann and Schieberle (2002) with (^{2}H$_2$)-2-furfurylthiol the spectroscopic data (^{2}H-NMR, LC-MS) indicated that furfurylthiol reacts with oxidation products of CROSSPY. The authors proposed that a covalent bond is formed between these compounds leading to a decrease of the odour quality of coffee right after brewing (especially as regards the sulphury-roasty odour quality). Other useful techniques used in melanoidin/Maillard research are capillary electrophoresis (Borelli et al., 2002; del Castillo et al., 2002; Re et al., 2002) and matrix-assisted laser desorption ionization time-of-flight mass spectrometry (MALDI/TOFMS) (Borelli et al., 2002; Kislinger et al., 2002). Capillary electrophoresis (after a fractionation by ultrafiltration with a 5000 Da cut-off dialysis membrane) has been used by Ames et al. (2000) to monitor the development of coloured compounds in coffee. It can be expected that both techniques will be used in the characterization of coffee melanoidins.

4.5.2 Physiological effects of melanoidins (see also Chapter 10)

Maillard reaction products in general may have negative and positive (among those antioxidant activity) effects (Friedman, 1996). Claims have been made that the melanoidins do contribute to the antioxidant activity of coffee brews. Daglia et al. (2000) studied the in vitro antioxidant activity and the ex vivo protective activity of coffee constituents and stated that the higher molecular mass fractions had an antioxidant activity while the lower molecular mass fractions had a protective activity. The antioxidant properties are related to the degree of roast. Medium roasted coffee had the most pronounced antioxidative effect (Steinhart et al., 2001). Del Castillo et al. (2002) used the ABTS·+ assay to assess the contribution of high and low molecular compounds to antioxidant activity of coffee brews. The ABTS test is a spectrophotometric method which is based on the different colour of 2,2′-azino-di-(3-ethylbenzthiazoline-6-sulfonic acid and the corresponding radical cation, by monitoring at 734 nm the decrease in ABTS·+concentration after addition of the antioxidant; Trolox

(6-hydroxy-2,5,7,8-tetramethylchroman-2-carbon acid) serves as reference. Again, it was observed that medium-roasted coffee had the highest
antioxidant activity *in vitro* and that the low molecular mass compounds
had a greater contribution to the antioxidant activity compared to the
high molecular mass compounds. The same group analysed the antioxidant
activity of a coffee model system (consisting of chlorogenic acid, N-acetyl-
l-arginine, sucrose, and cellulose) and found that there was no relationship
between antioxidant activity and colour generation (Charurin *et al.*,
2002). Morales and Babbel (2002) also used an *in vitro* system (bleaching of
DMPD·+) and found a significant lower antioxidant activity of melanoidins compared with classical antioxidants such as caffeic or gallic acid. As
with other polymers in foods, there is practically no information available
on the bioavailability of the melanoidins. It seems reasonable to anticipate
that the fractions with different molecular mass also will have a different
bioavailability.

Claims have been made that melanoidins suppress the formation of
n-nitroso amines (Wuerzner *et al.*, 1989). Among other components of
coffee, such as trigonelline, nicotinic and chlorogenic acids, a low
molecular weight coffee melanoidin fraction (1000–3500 Da) contributes
to anti-adhesive properties against *Streptococcus mutans* adhesive activity
on saliva-coated hydroxyapatite beads, which might contribute to an
anticaries effect (Daglia *et al.*, 2002). As animal data are missing
the authors recommend a careful interpretation of their findings.
Wijewickreme and Kitts (1998) stated a modulation of metal-induced
cytotoxicity by Maillard reaction products from coffee brews.

4.5.3 Summary and outlook

Our knowledge on the chemistry and physiology of coffee melanoidins
and of their impact on the taste of coffee are still fragmentary and far away
from a real structural elucidation. Recently, new information has become
available on one type of melanoidins containing CROSSPY and on their
interaction with sulphur-containing odorants. It might be expected that
our knowledge on melanoidins in general and coffee melanoidins in
particular will increase within the next few years as concentrated efforts
are made in this field of research (e.g. COST action). As soon as
individual structures of melanoidins are known, biological testing of the
compounds becomes possible as well as a more precise evaluation of the
impact of the melanoidins on coffee taste. These results might then
enable roasters to control the formation of melanoidins, which is very
important for the aroma and the overall taste of the coffee beverage.

4.6 CONTAMINANTS

H.G. Maier

4.6.1   Mycotoxins (see also 3.11.11.1)

The mycotoxins are largely destroyed upon roasting (Maier, 1991). For instance, the destruction of aflatoxin B1 in artificially contaminated green coffee beans during roasting ranges from 90 to 100% (Micco et al., 1991). Beyond that, in non-mouldy beans, only ochratoxin A has been found (Maier, 1989). In all the experiments where naturally contaminated beans, a sampling procedure adapted to mycotoxin inhomogeinity and roasting conditions within the range of actual practice were employed, half to almost all of the ochratoxin A disappeared during roasting (Blanc et al., 1998; Van der Stegen et al., 2001; Viani, 2002).

4.6.2   Polycyclic aromatic hydrocarbons (see also 3.11.11.2)

In most cases, benzo[a]pyrene (BP) has been determined as a leading substance for all polycyclic aromatic hydrocarbons. After normal roast the content is often reduced in respect to green coffee. No differences between roasting by direct firing and by indirect air heating could be detected. More BP is formed by heat transfer to the beans from contact with hot surfaces than by transfer from hot gases. The transfer of BP to the coffee brew depends on the concentration of benzo[a]pyrene in the roasted coffee and on the water-to-coffee ratio, amounting on average to approximately 5%, so that the content in coffee brews and extracts is insignificant (Maier, 1991).

REFERENCES

Ames J.M., Royle L. and Bradbury A.G.W. (2000) Capillary electrophoresis of roasted coffees. In T.H. Parliment, C.-T. Ho, P. Schieberle (eds), Caffeinated Beverages. ACS Symposium Series 754, pp. 364–373.

Arackal A. and Lehmann G. (1979) Messung des Quotienten 2-Methylfuran/2-Butanon von ungemahlenem Röstkaffee während der Lagerung unter Luftausschluß. Chem. Mikrobiol. Technol. Lebensm. 6, 43–47.

Bähre F. and Maier H.G. (1999) New non-volatile acids in coffee. Dtsch. Lebensm.-Rundsch. 95, 399–402.

Belitz H.D., Grosch W. and Schieberle P. (2001) Lehrbuch der Lebensmittelchemie, 5th edn. Heidelberg: Springer, p. 934.

Bicchi C., Iori C., Rubiolo P. and Sandra P. (2002) Headspace sorptive extraction (HSSE), stir bar sorptive extraction (SBSE), and solid phase microextraction (SPME) applied to the analysis of roasted arabica coffee and coffee brew. *J. Agric. Food Chem.* 50, 449–459.

Blanc M., Pittet A., Muñoz-Box R. and Viani R. (1998) Behavior of ochratoxin A during green coffee roasting and soluble coffee manufacture. *J. Agric. Food Chem.* 46, 673–675.

Blank I., Pascaul E., Devaud S., Fay L., Stadler R., Yeretzian C. and Goodman B. (2002) Degradation of the coffee flavor compound furfuryl mercaptan in model fenton-type reaction systems. *J. Agric. Food Chem.* 50, 2356–2364.

Borelli R.C., Fogliano V., Monti S.M. and Ames J.M. (2002) Characterization of melanoidins from a glucose-glycine model system (DOI 10.1007/s00217-002-0531-0, 26 June 2002).

Bradbury A.G.W. (2001) Carbohydrates. In R. J. Clarke and O. G. Vitzthum (eds), *Coffee – Recent Developments*. Oxford: Blackwell Science, pp. 1–17.

Bucking M. and Steinhart H. (2002) Headspace GC and sensory analysis characterization of the influence of different milk additives on the flavor release of coffee beverages. *J. Agric. Food Chem.* 50, 1529–1534.

Charurin P., Ames J.M. and del Castillo M.D. (2002) Antioxidant activity of coffee model systems. *J. Agric. Food. Chem.* 50 (13), 3751–3756.

Clarke R. (1987) Roasting and grinding. In R.J. Clarke and R. Macrae (eds), *Coffee: Volume 2 – Technology*. London: Elsevier.

Clarke R.J. and Vitzthum O. (2001) *Coffee – Recent Developments*. London: Blackwell Science.

Clifford M.N. (1997) The nature of chlorogenic acids. Are they advantageous compounds in coffee? *Proc. 17th ASIC Coll.*, pp. 79–91.

COST (2002) see under http://www.cifa.ucl.ac.be/COST/.

Czerny M. and Grosch W. (2000) Potent odorants of raw arabica coffee. Their changes during roasting. *J. Agric. Food Chem.* 48, 868–872.

Da Porto C., Nicoli M.C., Severini C., Sensidoni A. and Lerici C.R. (1991) Study on physical and physiochemical changes in coffee beans during roasting, Note 2. *Ital. J. Food Sci.*, pp. 197–207.

Daglia M., Papetti A., Gregotti C., Bertè F. and Gazzani G. (2000) In vitro antioxidant and ex vivo protective activities of green and roasted coffee. *J. Agric. Food Chem.* 48, 1449–1454.

Daglia M., Tarsi R., Papetti A., Grisoli P., Dacarro C., Pruzzo C. and Gazzani G. (2002) Antiadhesive effect of green and roasted coffee on *Streptococcus mutans*' adhesive properties on saliva-coated hydroxyapatite beads. *J. Agric. Food Chem.* 50, 1225–1229.

del Castillo M.D., Ames J.M. and Gordon M.H. (2002) Effect of roasting on the antioxidant activity of coffee brews. *J. Agric. Food Chem.* 50, 3698–3703.

Engel W., Bahr W. and Schieberle P. (1999) Solvent assisted flavour evaporation – a new and versatile technique for the careful and direct isolation of aroma compounds from complex food matrices. *Eur. Food Res. Technol.* 209, 237–241.

Ferreira V., Pet'ka J. and Aznar M. (2001) Aroma extract dilution analysis. Precision and optimal experimental design. *J. Agric. Food Chem.* 50, 1508–1514.

Flament I. (2001) The volatile compounds identified in green coffee beans. In *Coffee Flavor Chemistry*. Chichester: J. Wiley & Sons, pp. 29–34.

Franz H. and Maier H.G. (1994) Inositolphosphate in Kaffee und Kaffeemitteln. II. Bohnenkaffee. *Dtsch. Lebensm.-Rundsch.* 90, 345–349.

Freitas A. and Mosca A. (1999) Coffee geographic origin – an aid to coffee different. *Food Res. Int.* 32, 565–573.

Friedman M (1996) Food browning and its prevention: an overview. *J. Agric. Food Chem.* 44, 631–653.

Ginz M. (2001) Bittere Diketopiperazine und Chlorogensäurederivate in Röstkaffee. Thesis, TU Braunschweig, Germany.

Ginz M., Balzer H.H., Bradbury A.G.W. and Maier H.G. (2000) Formation of aliphatic acids by carbohydrate degradation during the roasting of coffee. *Eur. Food Res. Technol.* 211, 404–410.

Gretsch C., Sarrazin C. and Liardon R. (2000) Evolution of coffee aroma characteristics during roasting. *Proc. 18th ASIC Coll.*, pp. 27–34.

Heinrich L. and Baltes W. (1987) Vorkommen von Phenolen in Kaffee-Melanoidinen. *Z. Lebensm. Unters. Forsch.* 185, 366–370.

Hofmann T. and Schieberle P. (2002) Chemical interactions between odor-active thiols and melanoidins involved in the aroma staling of coffee beverages. *J. Agric. Food Chem.* 50, 319–326.

Hofmann T., Bors W. and Stettmaier K. (1999) Radical-assisted melanoidin formation during thermal processing of foods as well as under physiological conditions. *J. Agric. Food Chem.* 47, 391–396.

Hofmann T., Czerny M., Calligaris S. and Schieberle P. (2001) Model studies on the influence of coffee melanoidins on flavor volatiles of coffee beverages. *J. Agric. Food Chem.* 49, 2382–2386.

Holscher W. and Steinhart H. (1992) New sulfur-containing aroma-impact compounds in roasted coffee. *Proc. 14th ASIC Coll.*, pp. 130–136.

Holscher W., Vitzthum O. and Steinhart H. (1990) Identification and sensorial evaluation of aroma-impact-compounds in roasted Colombian coffee. *Café Cacao Thé* 34, 205–212.

Holscher W., Vitzthum O. and Steinhart H. (1992) Prenyl alcohol – source for odorants in roasted coffee. *J. Agric. Food Chem.* 40, 655–658.

Homma S. (2001) Non-volatile compounds, Part II. In R.J. Clarke and O.G. Vitzthum (eds), *Coffee – Recent Developments*. Oxford: Blackwell Science, pp. 50–67.

Illy A. and Viani R. (eds) (1995) *Espresso Coffee*. London: Academic Press.

Kislinger T., Humeny A., Seeber S., Becker C.-M. and Pischetsrieder M. (2002) Qualitative determination of early Maillard-products by MALDI-TOF mass spectrometry peptide mapping. *Eur. Food Res. Technol.* 215, 65–71.

Kurt A. and Speer K. (2002) Untersuchungen zum Einfluss der Dämpfungsparameter auf die Diterpengehalte von Arabica Roh- und Röstkaffees. *Dtsch. Lebensm. Rdsch.* 98, 1–4.

Macrae R. (1985) Nitrogenous compounds. In R. J. Clarke and R. Macrae (eds), *Coffee: Volume 1 – Chemistry*. Barking: Elsevier Applied Science, pp. 115–152.

Maier H.G. (1981) *Kaffee*. Berlin: Paul Parey, pp. 64–67.

Maier H.G. (1985) Zur Zusammensetzung kurzzeitgerösteter Kaffees. *Lebensmittelchem. Gerichtl. Chem.* 39, 25–29.

Maier H.G. (1989) Zum Stand der Forschungen über Kaffee. *Lebensmittelchem. Gerichtl. Chem.* 43, 25–33.

Maier H.G. (1991) Teneur en composés cancérigènes du café en grains. *Café Cacao Thé* 35, 133–142.

Maier H.G. and Buttle H. (1973) Zur Isolierung und Charakterisierung der braunen Kaffeeröststoffe. II. Mitteilung. *Z. Lebensm. Unters. Forsch.* 150, 331–334.

Maier H.G., Diemair W. and Ganssmann J. (1968) Zur Isolierung und Charakterisierung der braunen Kaffeeröststoffe. *Z. Lebensm. Unters.-Forsch.* 137, 282–292.

Maier H.G., Engelhardt U.H. and Scholze A. (1984) Säuren des Kaffees. IX. Zunahme beim Warmhalten des Getränks. *Dtsch. Lebensm. Rdschau* 80, 265–268.

Martin P.R., Depaulis T. and Lovinger D.M. (2001) Non-caffeine dicinnamoylquinide constituents of roasted coffee inhibit the human adenosine transporter. *Proc. 19th ASIC Coll.*, CD-ROM.

Mayer F. and Grosch W. (2001) Aroma simulation on the basis of the odorant composition of roasted coffee headspace. *Flavour Fragr. J.* 16, 180–190.

Mayer F., Czerny M. and Grosch W. (1999) Influence of the provenance and roast degree on the composition of potent odorants in Arabica coffees. *Eur. Food Res. Technol.* 209, 242–250.

Micco C., Miraglia M., Brera C., Desiderio C. and Masci V. (1991) The effect of roasting on the fate of aflatoxin B1 in artificially contaminated green coffee beans. *Proc. 14th ASIC Coll.*, 183–189.

Morales F.J. and Babbel M.B. (2002) Melanoidins exert a weak antiradical activity in watery fluids. *J. Agric. Food Chem.* 50, 4657–4661.

Nehring U. (1991) D-Aminosäuren in Röstkaffee. Thesis, TU Braunschweig, Germany.

Nunes F.M. and Coimbra M.A. (2001) Chemical characterization of the high molecular weight material extracted with hot water from green and roasted arabica coffee. *J. Agric. Food Chem.* 49, 17773–1782.

Nunes F.M., Coimbra M.A., Duarte A.C. and Delgadillo I. (1997) Foamability, foam stability, and chemical composition of espresso coffee as affected by the degree of roast. *J. Agric. Food Chem.* 45, 3238–3243.

Ottinger H. and Hofmann T. (2001) Influence of roasting on the melanoidin spectrum in coffee beans and instant coffee. In: *Proceedings of the COST Action 919*, Volume 2, pp. 119–125.

Parliment T.H. (2000) An overview of coffee roasting. In T.H. Parliment, C.-T. Ho and P. Schieberle (eds), *Caffeinated Beverages*. ACS Symposium Series 754, pp. 188–201.

Perren R., Geiger R. and Escher F. (2001) Mechanism of volume expansion in coffee beans during roasting. *Proc. 19th ASIC Coll.*, CD-ROM.

Pittia P., Manzocco L. and Nicoli M.C. (2001) Thermophysical properties of coffee as affected by processing. *Proc. 19th ASIC Coll.*, CD-ROM.

Re R., Pellegrini N., Proteggente A., Pannala A., Yang M. and Rice-Evans C. (1999) Antioxidant activity applying an improved ABTS radical cation decolorization assay. *Free Radic. Biol. Med.* 26, 1231–1237.

Redgwell R.J., Trovato V., Curti D. and Fischer M. (2002) Effect of roasting on degradation and structural features of polysaccharides in arabica coffee beans. *Carbohydrate Research* 337, 421–431.

Rothfos B. (1984) *Kaffee.* Hamburg: Gordian.

Sarrazin C., Le Quere J.-L., Gretsch C. and Liardon R. (2000) Representativeness of coffee aroma extracts – a comparison of different extraction methods. *Food Chem.* 70, 99–106.

Schenker S., Handschin S., Frey B., Perren R. and Escher F. (1999) Structural properties of coffee beans as influenced by roasting conditions. *Proc. 18th ASIC Coll.*, pp. 127–135.

Schenker S., Handschin S., Frey B., Perren R. and Escher F. (2000) Pore structure of coffee beans affected by roasting conditions. *J. Food Sci.* 65, 452–457.

Schenker S., Heinemann C., Huber M., Pompizzi R., Perren R. and Escher F. (2002) Impact of roasting conditions on the formation of aroma compounds in coffee beans. *J. Food Sci.* 67, 60–66.

Scholz-Böttcher B. (1991) Bildung von Säuren und Lactonen, insbesondere aus Chlorogensäuren, beim Rösten von Kaffee. Thesis, TU Braunschweig, Germany.

Scholz-Böttcher B.M., Ernst L. and Maier H.G. (1991) New stereoisomers of quinic acid and their lactones. *Liebigs Ann. Chem.* 1991, 1029–1036.

Sivetz M. and Desrosier N.W. (1979) *Coffee Technology.* Westport, CT: AVI.

Speer K. and Kölling-Speer I. (2001) Lipids. In R.J. Clarke and O.G. Vitzthum (eds), *Coffee – Recent Developments.* Oxford: Blackwell Science, pp. 33–49.

Stadler R., Varga N., Hau J., Vera F. and Welti D. (2002a) Alkylpyridiniums. 1. Formation in model systems via thermal degradation of trigonelline. *J. Agric. Food Chem.* 50, 1192–1199.

Stadler R., Varga N., Milo C., Schilter B., Vera F. and Welti D. (2002b) Alkylpyridiniums. 2. Isolation and quantification in roasted and ground coffees. *J. Agric. Food Chem.* 50, 1200–1206.

Steinhart H., Luger A. and Piost J. (2001) Antioxidative effect of coffee melanoidins. *Proc. 19th ASIC Coll.*, CD-ROM.

Steinhart H., Möller A. and Kletschkus H. (1989) New aspects in the analysis of melanoidins in coffee with liquid chromatography. *Proc. 13th ASIC Coll.*, pp. 197–205.

Stennert A. and Maier H.G. (1996) Trigonelline in coffee. Part 3. Calculation of the degree of roast by the trigonelline/nicotinic acid ratio. New gas chromatographic method for nicotinic acid. *Z. Lebensm. Unters. Forsch.* 202, 45–47.

214 Espresso Coffee

Thaler H. and Gaigl R. (1963) Untersuchungen an Kaffee und Kaffee-Ersatz. VIII. Das Verhalten der Stickstoffsubstanzen beim Rösten von Kaffee. Z. Lebensm. Unters. Forsch. 120, pp. 357–363.

Trautwein E. (1987) Untersuchungen über den Gehalt an freien und gebundenen Aminosäuren in verschiedenen Kaffee-Sorten sowie über das Verhalten während des Röstens. Thesis University of Kiel, Germany.

Ullrich F. and Grosch W. (1987) Identification of the most intensive volatile flavor compounds formed during autoxidation of linoleic acid. Z. Lebensm. Unters. Forsch. 184, 277–282.

Van der Stegen G.H.D., Essens P.J.M. and van der Lijn J. (2001) Effect of roasting conditions on reduction of ochratoxin A in coffee. J. Agric. Food Chem. 49, 4713–4715.

Viani R. (2002) Effect of processing on ochratoxin A (OTA) content of coffee. Adv. Exp. Med. Biol. 504, 89–93.

Viani R. and Horman I. (1974) Thermal behaviour of trigonelline. J. Food Sci. 39, 1216–1217.

Viani R. and Horman I. (1975) Determination of trigonelline in coffee. Proc. 7th ASIC Coll., 273–278.

Vitzthum O., Weissmann C., Becker R. and Köhler H. (1990) Identification of an aroma key compound in robusta coffees. Café Cacao Thé 34, 27–36.

Vitzthum O., Werkhoff P. and Ablanque E. (1976) Fluechtige Inhaltsstoffe des Rohkaffees. Proc. 7th ASIC Coll., pp. 115–123.

Weckerle B., Gati T., Toth G. and Schreier P. (2002) 3-Methylbutanoyl and 3-methylbut-2-enoyl disaccharides from green coffee beans (Coffea arabica). Phytochem. 60, 409–414.

Wijewickreme A.N. and Kitts D.D. (1998) Modulation of metal-induced cytotoxicity by Maillard reaction products from coffee brew. Toxicol. Environm. Health, Part A 55, 151–168.

Wuerzner H.P., Jaccaud E. and Aeschbacher U. (1989) In vivo inhibition of nitrosamide formation by coffee and coffee constituents. Proc. 13th ASIC Coll., pp. 73–81.

Wurziger J. (1972) Carbonsäurehydroxytrpyptamide oder ätherlösliche Extraktstoffe zum Nachweis und zur Beurteilung von bearbeiteten bekömmlichen Röstkaffees. Kaffee- und Tee-Markt 22 (14), 3–11.

Wurziger J. and Harms U. (1969) Beiträge zum Genußwert und zur Bekömmlichkeit von Rohkaffee. IV. Hydroxytryptamide in Röstkaffees aus normalen Handelssorten sowie aus in verschiedener Weise bearbeiteten Rohkaffees. Kaffee- und Tee-Markt 19 (18), 26–29.

Yeretzian C., Jordan A., Badoud R. and Lindinger W. (2002) From the green bean to the cup of coffee: investigating coffee roasting by on-line monitoring of volatiles. Eur. Food Res. Technol. 214, 92–104.

CHAPTER 5

Grinding

M. Petracco

The classical representation of coffee usually pictures it in beans, either green or roasted. But for coffee to be consumed as food, it is necessary to destroy the beautiful form nature has given the seed by transforming it into a powder. The operation which effects this conversion is called, in engineering language 'comminution', which means breaking down particles into smaller fragments, while in coffee industry jargon it is commonly known as grinding.

Grinding is performed by a grinder, or coffee-mill, and the resulting product is ground coffee. The term 'grind' applies to an important variable in espresso coffee preparation, as we shall see later: it refers to the degree of comminution associated with the concept of fineness of ground coffee.

The main objective of grinding is to increase the specific extraction surface, or rather, to increase the extent of the interface between water and the solid per unit weight of coffee, so as to facilitate the transfer of soluble and emulsifiable substances into the brew. Each time a solid body is broken up, additional surface is generated that comes into contact with the surrounding environment, in this case the extraction water.

Two apparently contradictory needs must be satisfied to prepare a good cup of espresso: on the one hand a short percolation time is required, while, on the other hand, high concentration of soluble solids must be reached. Both requirements can only be attained if a close contact between solid particles and extraction water can be achieved. Thus, espresso percolation needs a plurimodal particle size distribution, where the finer particles enhance the exposed extraction surface (chemical need) and the coarser ones allow the water flow (physical need).

5.1 THEORY OF FRACTURE MECHANICS

Before dealing with the various types of devices employed for grinding, it is useful to give a qualitative notion of the fundamental laws of physics acting in comminution. To reduce solid blocks into smaller sizes, it is necessary to produce fractures by applying a set of forces. A quantitative theoretical analysis of this process would only be possible for very simple geometric forms; nevertheless, a general outline may be useful to understand the machine's application of complex stress configurations on the material (Fayed and Otten, 1984).

Stresses exerted on an elementary object fall into two categories: compressive stresses and shear stresses. Both produce strains that may follow either simple laws, like Hooke's law in the case of elastic materials ($\varepsilon = \sigma E$, where ε stands for strain caused by stress σ, and E represents the modulus of elasticity or Young's modulus), or more complex laws for materials displaying a plastic or viscous-plastic behaviour, which is not the case of roasted coffee. A fracture results when the stress exceeds the limit given by Hooke's law, that is, when it reaches the so-called 'breaking load' (Pittia and Lerici, 2001).

In reality, the stress is not distributed uniformly inside an object, not even in the ideal case of an object having a regular form and consisting of a homogeneous and isotropic material. As a matter of fact, the interior of an object always displays slight discontinuities, or flaws, like the defects in the crystal lattice of a mineral, or inhomogeneities in animal or plant tissues.

Lines of force are concentrated along the rim of discontinuities, and may easily cause the stresses to reach very high values, which lead to local micro-fractures that propagate rapidly. In the case of shear stresses, the relevant flaws tend to lie on the plane of shear, and cause the formation of a roughly uniform surface of fracture (depending on the nature of the material), which generally splits the object into two. In the case of compression forces, the level of stresses is more uniform, and the flaws may be randomly scattered within the object: a concurrent formation of fracture lines in various points causes the object to burst, creating fragments of different random sizes. In an object that does not possess an elementary form, but has an articulated or compound geometry, the most stressed points susceptible to failure can be determined by the use of the design techniques employed in construction science.

The geometry of a coffee bean is anything but simple, as is often the case with the works of nature. The coffee bean displays two distinct levels of complexity:

■ At a macroscopic level, the bean is made of a layer of tissue folded together (interposed with a thin layer of a different tissue, the perisperm, also called 'silverskin'), forming a solid resembling an ellipsoid of revolution, as was shown in Figure 4.10c.

■ At a microscopic level, the tissue is composed of cells of different shapes filled with a mass of sugars, proteins and lipids, enclosed by walls made of polysaccharides (see Figure 3.21b). This content can be observed only in green coffee, while it apparently disappears after roasting due to transformation into polymeric material and volatile compounds (Figure 5.1).

A study of the cellular geometry of roasted coffee, performed by computer simulation employing the 'finite element method', shows that the application of shear to a reticular structure tends to produce a comminution into many small particles and a few larger ones of different size, as shown in Figure 5.2 (Petracco and Marega, 1991).

Thus, two phases of comminution must be taken into consideration in the design of a grinder:

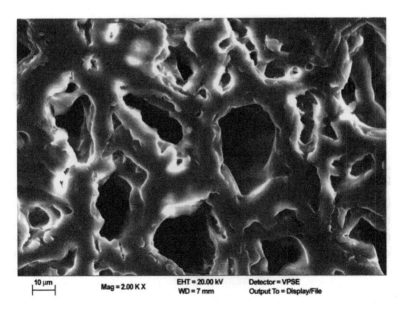

| 10 μm | Mag = 2.00 K X | EHT = 20.00 kV | Detector = VPSE |
| | | WD = 7 mm | Output To = Display/File |

Figure 5.1   SEM structure of roasted coffee bean, where the cell walls are no longer visible, as they are covered by oil

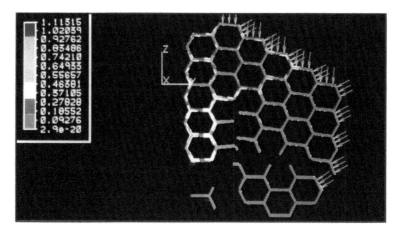

Figure 5.2   Simulation of sub-particle formation

■ a crushing phase, where the convolute structure of the brittle roasted
  bean is broken down into fragments of approximately one millimetre;
  and
■ a second phase, properly called grinding, in which such fragments are
  submitted to shearing forces.

This grinding process produces a particle distribution that, as proposed in
section 7.5, has a favourable if not indispensable effect on espresso
brewing, though theoretically it does not contribute to maximization of
the surface.

5.2 COFFEE GRINDERS

Roasted coffee possesses a particular feature that excludes from the outset
the use of a major class of grinders widely employed in the mining
industry, naming ball mills. The solubility of coffee in water, even in cold
water, does not allow the use of wet-processing equipment. This is
unfortunate, because wet-processing equipment is most appropriate when
a specific distribution spectrum is sought for, since it operates with
adjusted recycling of material. This only leaves the use of dry-processing
equipment, operating according to two commonly employed operating
principles: 'impact' grinding and 'gap' grinding.

5.2.1 Impact grinding

Impact grinding is accomplished by a series of blades, rotating at high speed, which exert a shock impulse on the particles encountered in their trajectory. This effect resembles grinding by compression, with the difference that, in the case under consideration, the reaction is given by the inertial force of the particle suddenly removed from its state of relative rest. This technique is utilized only in small-size grinders in the home, because of one major defect: it does not easily allow the regulation of particle size, since the ground product remains between the blades even after being ground and so undergoes further impacts, making powder fineness also dependent on the length of the operation.

5.2.2 Gap grinding

Gap grinding is based on the passage of the beans, usually by dropping, through a gap between moving tools, called 'cutting tools'. The geometry of the cutters causes a gradual reduction of the width of the gap during rotation, forcing the particles to come into contact with both cutters, which apply a force couple on each particle. According to the shape of the cutting surfaces, the force exerted may either be of a compressive or of a shear type. The most common styles of cutter pairs used in the coffee industry are (Figure 5.3):

■ **Roller cutters:** a couple of rifled cylinders with parallel axes, counter-rotating and juxtaposed so as almost to touch themselves at their generating lines. Beans are fed into the device by timed dropping, so as not to obstruct the gap between cutters. These tools are mainly used at industrial level.

Figure 5.3   Cutters: (left) roller cutters; (centre) conical cutters; (right) flat cutters

- **Conical cutters:** a male conical wheel rotating coaxially in a static truncated-cone-shape cavity. Again, feeding is performed by gravity, often by rejection, namely a column weight of the material to be ground. Excellent size versus throughput ratio.
- **Flat cutters:** a couple of disks with truncated-cone cavity, juxtaposed so as to almost touch their bases. One of them, usually the lower one, rotates coaxially. Feeding, by rejection or by forcing with worm screw system, is accomplished from the upper to the lower part. Ground coffee powder does not come out by gravity, but by the effect of the centrifugal force of rotation. These cutters are used both for industrial and professional grinding.

As already stated, the design of a grinding device must take into consideration the existence of two phases of comminution: the crushing of the bean (also called pre-breaking) and the grinding of the fragments to the desired fineness (also called finishing). These requirements can be met by two systems:

- A rather inexpensive system widely used in professional (coffee-bar usage) or home grinders fitted with conical or flat cutters, where cutters with unequal teeth are used. They are capable of forming gaps of different size, and of seizing whole beans as well as fragments.
- A system mostly reserved to industrial equipment because of its manufacturing complexity uses a series of grinding stages, each optimized for a specific particle size. In this configuration, each stage may also function as a regulator for the feeding of the following stage: in this case, rejection feeding is utilized only in the first stage, that is, in the stage that receives whole beans.

Obviously, a greater complexity of configuration is matched not only by higher equipment costs, but also by a greater difficulty in controlling stability. For this reason, industrial grinders are manufactured mainly according to two basic models:

- A grinder with one or two pairs of toothed rollers, in series, preceded by another pair of pre-breaking rollers, with capacities between 100 kg/h and several tons per hour.
- A grinder with a pair of flat cutters preceded by a conical pre-breaking stage, used for small to medium throughput (50–500 kg/h) of ground product.

5.2.3   Homogenization step

Besides the cutting tools, grinding equipment usually includes a homogenization and blending step for the ground product, consisting of rotating units with blades, worm screws, etc. Once the product has been ground, it is transported to the packaging section of the plant by means of conveyor belts, cups or pneumatic conveyors, sometimes under inert atmosphere.

5.3   METHODS FOR MEASURING GROUND PRODUCT FINENESS

The objective of grinding is to obtain a distribution of particles suitable for a specific purpose. This statement is demonstrated by a practical method applied by espresso bartenders for checking the fineness of ground coffee before introducing it into the espresso machine: the operator draws a single portion of coffee powder from the grinder, and uses it to prepare a cup of coffee. The quantity of liquid percolated in a given time is used as a measurement unit for regulating the primary variable of the grinder: the so-called 'notch', namely the cutters' position determining the shear gap.

This method has the clear advantage of allowing an immediate feedback, and an intuitive understanding of the cause–effect relationship. However, it may be prone to significant interference, because the quantity of beverage in a cup constitutes the final physical variable of the entire process, from grinding to extraction. As such, it may be influenced not only by the adjustment of the grinder, but also by other variables such as the ground coffee's waiting time before undergoing extraction, the compacting force exerted on the cake, and other parameters in adjusting the espresso machine (pressure, temperature, etc.; see also Chapter 7).

In order to relieve the evaluation of grinding effectiveness from any percolation tests, it is necessary to find a method that can be applied directly on the coffee powder. One such example is the simple, classical tactile examination performed by master grinding-operators for decades: it consists in assessing the coarseness of ground coffee by rubbing a pinch of powder between finger tips. However, such a subjective, individual analysis obviously pertains more to the domain of art than to that of science. Consequently, a major concern of the coffee industry is to devise an objective method for measuring ground coffee particle sizes.

5.3.1   Sifting methods

One of the earliest methods of analysis adopted (Allen, 1968) consists in shaking the ground product through a stack of graded-mesh sieves, with incrementally smaller-sized openings. The net weight retained on each sieve is recorded on linear, semilogarithmic or logarithmic charts (Prasher, 1987), showing a curve that characterizes size distribution. Various standardization proposals have been put forward for the shape of the sieve as well as for automated or manual shaking procedures (DIN, 1997). Among the many ingenious proposals, some consider sieving aided by vibrating chains, counter air-current as in DIN Norm 10765–1975, or solvent counterflow.

The sieving system has not met with much favour in the specific field of coffee grinding, specially when espresso is concerned, due to two basic considerations:

■ All methods have a poor repeatability, due to a substantial amount of oil coming up to the powder surface after grinding. The oil acts as an adhesive on the particles, forming random yet stable aggregates unable to pass through meshes. Moreover, particles passing through one sieve may commingle with other aggregates on the following sieves, thus distorting the objective results of the analysis. This drawback is more evident in espresso than in other brewing techniques, because of its darker roasting and finer grinding needs.

■ Sieving methods are long and laborious, and yield a characterization of the ground material which is often inadequate, since it is drawn from a too-limited series of sieves.

5.3.2   Imaging and sensing zone methods

Another intuitive approach to measuring ground coffee fineness consists in placing a powder sample on a microscope slide marked with a grid, and directly determining the size of a significant number of particles. This method is quite labour-intensive, even if eased nowadays by photographic microscopy techniques assisted by an automated specimen holder, and by computerized image analysis that provides histograms by particle size groups. As with all other micro-determination techniques, this measuring method also suffers from technical problems in sampling, and from the issue of representativeness of samples having the size of a few cubic millimetres, instead of tens or hundreds of grams used in sieving. A further difficulty lies in spreading out the particles in a thin layer, so as to allow their counting and measuring. Other methods (Svarovski, 1990) propose dispersing the powder in a solvent in order to determine physical characteristics such as electric conductivity (sensing zone counter), sedimentation rate, and density.

5.3.3   Laser diffractometry

Size distribution analyses based on optical principles appear much more rapid and promising. The most widely used method is laser diffractometry, suitable for ground coffee particles sized from 5 μm to 2000 μm. This technique uses the exact parallelism of coherent light emitted by laser sources in the visible or near-infrared. By means of very high-quality lens systems, it is possible to concentrate a beam of such light on an extremely small punctiform target (<0.1 mm) surrounded by concentric annular sensors. Particles passing through the beam produce a scattering, which disperses a little amount of light into a conical geometry. The smaller the particle, the greater the angle of the diffraction cone. The quantity of light striking each of the sensors displays a trend characterizing the particle sizes. Computer processing of this information produces a chart that associates size ranges with the concentration of particles contained in that range.

A difficulty set by this method lies in sample presentation: if performed dry, by pneumatic dispersion devices (Illy, 1994), it allows a short analysis time (a few minutes), which renders this method much more interesting than the more accurate but time-consuming wet presentation. A pneumatic dispersor is a tube to which a considerable quantity of coffee powder (up to 100 g) is continuously fed by vibrating conveyors. Compressed air is injected in various sites at different angles, exerting shear forces on particle clusters, breaking them up. The amount of feed and of air used must be calibrated to obtain the prescribed beam obscuration. Wet presenters require a much smaller quantity of powder (less than one gram), to be suspended in a convenient liquid (hexane, hot water) sometimes with the help of ultrasonication and surfactants. The suspension is then circulated through a shallow transparent cell placed across the beam's path. Typical analysis times range from a couple of minutes for dry presentation up to 20 minutes for wet presentation.

A picture of the size distribution of a powder sample is given by a histogram or, when many measure points are used, by a chart showing a grading curve (Figure 5.4).

The curve is plotted on a diagram in which the axis of the abscissa, usually on a logarithmic scale, carries the 'equivalent diameter' variable:

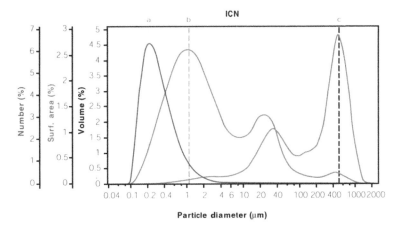

Figure 5.4 Particle size frequency distributions: (a) percentage at 0.205 mm (max. count); (b) percentage at 1.097 mm (max. surface); (c) percentage at 517.2 mm (max. volume)

this variable can be defined differently according to the technique adopted, but it is somehow linked with particle size. The percentages of particles belonging to each size group are plotted on the ordinate axis in differential (frequency chart) or integral form (cumulative curve). The percentage may refer to the number of particles, to their volume or mass or weight, or even to their surface. It is possible to pass from one chart form to the other by calculation, once the relationship between particle shape and density factors has been determined. The comparison of ground coffee batches obtained from different grinding equipments shows differences that characterize each ground coffee by its own grading curve, just as fingerprints identify a person.

Several companies sell turnkey computerized analysers, designed for general powders. They differ by bench length (that determines the maximum measurable particle diameter) and presentation accessories. An effort to adapt sample presentation devices to ground coffee is always needed, due to its sticky character and fouling tendency. Some brands worth mentioning are Coulter, Malvern, Fritsch and Sympatech.

5.4 PARAMETERS INFLUENCING GRINDING

The unit operation 'comminution' involves two types of input variables: those linked with raw material and those linked with grinding machinery.

5.4.1   Variability of the coffee blend

The first type of variable includes, chiefly, the variability in composition of the coffee to be ground. As explained in Chapter 4 on roasting, espresso needs a balance of flavours and aromas, obtained from a blend composed of different origins. Thus the grinder is confronted with beans that not only come from different botanical varieties and producing countries, but are also dried and processed by diverse methods. This often results in a lack of homogeneity in hardness: in addition, hardness is constantly susceptible to daily changes owing to dissimilarities in the batches, since it must be remembered that coffee is an agricultural commodity subject to the laws of climate and of natural evolution.

5.4.2   Roasting degree

Pyrolysis reactions forming volatile gases and aromas produce a considerable expansion of the bean, depending on final roasting temperature and rate of roasting (see 4.4). This expansion occurs at the expense of the elasticity of the cell walls, which become brittle and lose tenacity, changing from a nearly plastic behaviour in the green beans to an elastic behaviour, and eventually to a brittle failure in the dark roasted beans. If coffee is air-cooled after roasting, the high quantity of gas trapped in the cells may affect the grinding process stability. A way to induce degassing naturally, performed in industrial practice, is to allow the roasted beans to rest in a silo for a suitable number of hours.

5.4.3   Moisture of the roasted beans

As seen in 4.1, water quenching may result in a substantial residual moisture, which increases the bean's tenacity and hence the grinder's consumption of energy with a resulting over-heating of the coffee. Furthermore, contrary to air-cooling, quenching provokes cell-wall cracking from the sudden volume contractions resulting from the brutal temperature drop. This causes a larger degassing of the beans before grinding and modifies the resistance to comminution of the cell structure.

5.4.4   Cutters' distance

Machinery-related variables capable of affecting grinding results may be grouped in control variables and in disturbances. The major control variable, closely related to the 'notch' of the coffee-bar grinder, is the mechanical adjustment that allows regulation of the shear gap (or gaps) to obtain a finer powder. In flat or conical cutters this adjustment consists in drawing together coaxially the pair of cutting elements. The operation requires a good degree of precision and low working tolerance, and must be free from hysteresis; it is, however, easy to perform. In roller cutters, the two cutting cylinders are drawn together by translating the axis of the mobile cutter parallel to that of the fixed cutter: this delicate operation requires complex and costly mechanisms. In both cases, an important role is played by the mechanism controlling the separation between the cutters, which places the pair of rotating cutters almost into contact.

5.4.5 Other parameters

Equally important is the adjustment of the connected units, primarily the feeding and pre-breaking mechanism; such operation has a considerable effect on hourly throughput and on ground product quality, by contributing to the avoidance of flooding and overheating. Less important, and therefore less commonly performed, is the regulation of the cutters' rotation speed; such operation could either be performed by complex and delicate kinematic mechanisms or through more modern but equally expensive means, such as three-phase power supply frequency converters.

5.4.6 Disturbances

Disturbances are those hard-to-control variables causing equipment to perform irregularly, which industrial practice aims to minimize.

A major cause of loss of control in a grinder undoubtedly comes from energy dissipation during the application of mechanical power to break down the product. As in all mechanical operations, there is an inevitable element of friction, which adds to the heat generated by fracturing and causes differential expansion of metal parts. This is especially critical where tenths of a millimetre play an important role in controlling particle size distribution. Even though various systems for minimizing these harmful effects (thermostatation jacket with circulating liquid, injection of boiling-point liquid nitrogen between the cutters) have been proposed, the golden rule of grinding is still to wait for the machine to reach thermal stability before considering it 'operational', that is, before considering the ground product optimal.

As in all machinery, grinders are prone to wear and tear, and need proper maintenance. The parts that must be kept under control most carefully are of course the cutting tools, whose blades need periodical sharpening to recover their original performance. The result of poorly sharpened tools is an increase in adverse thermal effects, due to the higher compression-to-shear ratio applied to the coffee particles by blunt cutting edges. Worn cutters impart to the beverage a 'burnt', bitter taste by overheating the coffee.

Also worth mentioning are the disturbances caused by the electric power supply, or by foreign matter. Power supply disorders may affect rotation speed for a few seconds and destabilize a possibly already unstable equilibrium. Foreign matter that may be present in spite of the utmost attention and in spite of all the equipment installed for its elimination (pebble removers, magnets, etc.), may suddenly block the cutters and even damage them permanently.

5.5 PHYSICO-CHEMICAL MODIFICATIONS DUE TO GRINDING

The main influence of the grinding process on its feed, the roasted beans, is of course fragmentation. Its efficacy, often evaluated by the empirical method of rubbing a pinch of powder between the fingers to feel its coarseness, should be quantified by analysing the product by means of an objective measurement method, as seen in 5.3.

Grinding operations, besides the described effect on the physical structure, also impart to coffee some chemical modifications due to the temperature reached during grinding. It is quite common reading temperatures up to 80 °C on the ground coffee thermocouple, which probably correspond to real temperatures of over 100 °C. These levels are well below those reached in the roasting process, and the coffee beans are exposed to them for a shorter period than during roasting (only a few seconds, instead of minutes). Nevertheless, such temperatures speed up Maillard and oxidation reactions, with the detrimental effects on coffee taste that can be judged by comparative organoleptic tests.

Temperature acts also on a macroscopic phenomenon that is important for espresso beverage: the release of volatile aroma. When broken down by grinding, coffee cells release their content of pyrolytic gas formed at roasting. This gas is mostly composed of CO_2 and CO, accompanied by small amounts of hundreds of volatile chemical species (see 4.4). Although present only in low concentrations, volatiles are essential in bringing forth the typical coffee aroma. Most industrial grinding equipment is designed to handle the release of gas by conveying it externally through a filter. Techniques for aroma recovery have been proposed, such as cryogenic condensation, absorption in oil, or adsorption on porous material.

However, the unfractured cells still present in the larger-size particles (over 50 μm) keep their content of high pressure roasting gas, which will contribute to the formation of 'crema', the typical espresso foam (see 8.1.1). Inevitably, the gas will slowly escape during storage through micro-cracks, or through the natural porosity of the cell walls. This evolution is accelerated by high grinding temperature, which increases the pressure within the cell. While the pros and cons of degassing will be discussed in 6.1.2.1, it is certain that the worst method to degas ground coffee is by overheating it.

One last macroscopic effect induced by grinding consists in the appearance of coffee oil on the surface of the powder, resulting from the release of the lipidic content of so far undamaged cells. As already explained in 3.11.9 and 4.3.2.3, the major part of the coffee lipid fraction is composed of triglycerides of palmitic and linoleic acids, and of stearic and oleic acids in a lower percentage. This blend is very viscous at room temperature, but starts becoming fluid around 40 °C (Table 5.1). At higher temperatures, fluidity is such that oil can easily flow through micro cracks and cover the outer surface of the particles with a layer that regains its sticky, semi-solid state at room temperature. This phenomenon affects the cohesive behaviour of ground coffee particles when compacting (see 6.1.3.3), hence influencing the hydraulic resistance of the cake. Overheating on grinding may thus be responsible for erratic percolation behaviour when brewing espresso. It must be stressed therefore that temperature control in grinding equipment is one of the main requirements to obtain a powder fit for preparing consistently perfect espresso cups.

Table 5.1 Coffee oil viscosity

T (°C)	Viscosity (mPa s)
0	Solid
20	70.6
50	22.4
92	7.8

REFERENCES

Allen T. (1968) *Particle Size Measurement*. London: Chapman and Hall, pp. 165, 606.

DIN (1997) DIN-Taschenbuch 133: Partikelmesstechnik Deutsches Institut für Normung, Berlin.

Fayed M.E. and Otten L. (1984) *Handbook of Powder Science and Technology*. New York: Van Nostrand, pp. 562–606.

Illy E. (1994) System for controlling the grinding of roast coffee. US patent 5,350,123/27.

Petracco M. and Marega G. (1991) Coffee grinding dynamics: a new approach by computer simulation. *Proc. 14th ASIC Coll.*, pp. 319–330.

Pittia P. and Lerici C.R. (2001) Textural changes of coffee beans as affected by roasting conditions. *Lebensmitt. Wiss. Technol.* 34, 168–175.

Prasher C.L. (1987) *Crushing and Grinding Process Handbook*. Chichester: John Wiley & Sons, pp. 87–102.

Svarovski L. (1990) Characterization of powders. In M.J. Rhodes (eds), *Principles of Powder Technology*. Chichester: John Wiley & Sons, pp. 51–61.

CHAPTER 6

Storage and packaging

M.C. Nicoli and O. Savonitti

6.1 PHYSICAL AND CHEMICAL CHANGES OF ROASTED COFFEE DURING STORAGE

M.C. Nicoli

Roasted coffee is a shelf-stable product. In fact, thanks to the high temperatures attained in the roasting process and to its low water activity (a_w), no enzymatic and microbial spoilage occurs. However, during storage, coffee undergoes important chemical and physical changes, which greatly affect quality and acceptability of the brew. These events are responsible for staling of roasted coffee but they are also believed to be involved in the so-called ageing process. The border existing between ageing and staling is unclear, and the reactions which prevail in ageing with respect to staling are still unknown. Ageing is a short length of time, just after roasting, in which coffee is allowed to rest under proper technological conditions for improving its sensorial properties. Although many of the physical and chemical changes occurring in roasted coffee during storage are considered unavoidable, the rate at which they occur mostly depends on some environmental and processing variables such as oxygen and moisture availability, exposed surface as well as packaging conditions. The adoption of proper grinding and packaging conditions can greatly slow down staling reactions. However, they can be accelerated again, proceeding at a considerable rate, after the consumer opens the packaging. In fact, home storage conditions, mainly in terms of oxygen availability, relative humidity and temperature, are the key factors in determining the so-called secondary shelf life of the product (Full *et al.*, 2001).

Table 6.1 summarizes the most important physical and chemical events that take place in roasted coffee during storage.

Solubilization and adsorption represent the main mechanisms through which aroma compounds are retained into the bean structure. In fact,

Table 6.1 Physical and chemical changes occurring in roasted coffee during storage

Volatile solubilization	CO_2 release
Volatile adsorption	Oxidations
Volatile release	Oil migration

volatiles, as well as carbon dioxide, can exist entrapped in the pores, dissolved in coffee oils and moisture and probably adsorbed into active sites (Labuza *et al.*, 2001). This represents an important requisite for obtaining a high quality coffee cup. However, volatile compounds are easily released in the vapour phase due to diffusion mechanisms. The latter, together with oxidation reactions, are considered the main causes of coffee staling. The evolution of carbon dioxide, which is the most important non-aromatic volatile found in fresh roasted coffee, represents an additional problem during storage, mainly from a technological point of view. In fact, in the case of use of flexible pouches, it causes a rapid loss in the package integrity due to swelling and bursting. The amount of CO_2 formed depends on the degree of roasting and can be up to 10 ml/g (SPT), corresponding to 1–2% of roasted coffee weight (Sivetz and Desrosier, 1979). For this reason, different procedures such as degassing or pressurization, together with the adoption of suitable packaging solutions, have been developed in order to allow carbon dioxide to be released, minimizing aroma losses and oxidations (see 6.2).

Since roasted coffee is a dehydrated product, the major factors controlling its stability during storage are oxygen concentration (O_2), water activity (a_w) – particularly if roasting had been quenched by adding water – and temperature. Technological operations such as grinding, degassing and packaging may accelerate the above-mentioned changes due to the increase in the exposed surface, as well as in oxygen and moisture availability. In addition, since the pressure in the bean pores should be greater than atmospheric pressure, the external pressure applied to the coffee beans represents an additional variable to be considered for enhancing coffee stability during storage. In fact, as the external pressure becomes higher than the partial pressure of volatiles present in the beans, the degassing rate is reduced allowing a larger quantity of volatiles to be dissolved in the lipid fraction or adsorbed on the active sites (Clarke, 1987a; Labuza *et al.*, 2001).

6.1.1   The role of Maillard reaction products on roasted coffee stability during storage (see also 4.5)

Coffee staling takes its origin from roasting. Maillard reaction and pyrolysis, which are the main chemical events occurring in coffee during roasting, transform the beans into a very unstable and reactive system. In broad terms, the most macroscopic changes occurring in coffee during roasting and storage can be attributed to the formation of Maillard reaction products (MRPs) and carbon dioxide. The latter is a reaction product from both Maillard reaction and pyrolysis (Hodge, 1953; Yaylayan and Huyghues-Despointes, 1994). The formation of volatiles and CO_2 during roasting causes the expansion of the beans due to the internal build-up of gases which, along with the high temperatures, allows internal pores and pockets to be formed. However, the increased CO_2 formation in the advanced stages of the roasting process corresponds to a progressive reduction of the ability of the bean to entrap and retain CO_2 and other volatiles, as the bean loses its elasticity and surface cracks are formed under the push of the increased internal pressure (Massini et al., 1990). A comparison between the amount of CO_2 formed during roasting and retained into the bean during storage is shown in Figure 6.1.

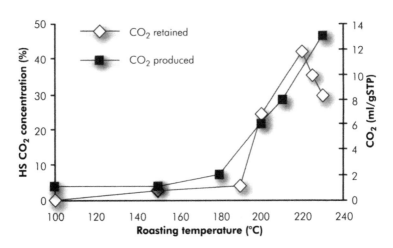

Figure 6.1   Comparison between amount of CO_2 produced during roasting and retained into the coffee beans (data elaborated from Barbera, 1967; Massini et al., 1990). Values of CO_2 retained are referred to the percentage of CO_2 present in the headspace in equilibrium with the sample after 5 minutes' storage after roasting

It can be observed that, as the bean temperature rises during roasting, increased amounts of CO_2 are formed. However, this corresponds to a progressive loss of the CO_2 retention capacity of the bean. Among the hundreds of compounds generally considered as MRPs, volatiles and melanoidins play a key role in determining the aroma and body of the coffee brew. Also, the development of the Maillard reaction during roasting is responsible for other important changes affecting coffee stability during storage. The most important effects exerted by MRPs in roasted coffee are summarized in Table 6.2.

Many of the low molecular weight MRPs are responsible for the formation of the typical coffee flavour. It is well known that the aroma profile of roasted coffee is given by several hundreds of compounds covering many different chemical classes. The development of the typical stale flavour of old coffee is the sum of two different actions: the partition of volatiles in the vapour phase and the development of off-flavours as the result of their oxidation. It has been observed that methanethiol, Strecker aldehydes and α-dicarbonyls are the most important low-boiling compounds responsible for the freshness of coffee aroma (Steinhardt and Holscher, 1991). These molecules, in addition to their high volatility, are particularly prone to oxidation. However, it must be pointed out that the aroma perception could also be affected by the ability of melanoidins to bind specific volatile molecules, as recently observed for some thiol compounds by Hofmann et al. (2001).

Most of the effects of MRPs on the shelf life of roasted coffee trace their origin to the ability of these compounds to act as both pro-oxidants and antioxidants. These contradictory functions may be explained by the complex and different chemical structures of MRPs. In broad terms, while melanoidins are generally recognized to be strong antioxidants, uncoloured compounds, such as those formed during the early and intermediate stages of Maillard reaction, may exert pro-oxidant activity, being

Table 6.2 Significance of MRPs in coffee during storage

Flavour	Shelf life
Volatile formation	Oxidation reactions:
	Degradation of aroma compounds
	Lipid oxidation slow-down
	Microbial growth inhibition
Volatile binding	

in most cases radicals (Namiki, 1990; Manzocco et al., 2001, 2002). During storage, radical reactions may occur leading to the formation of non-radical forms as well as to the generation of radical species chemically different from those initially present (Goodman et al., 2001).

Among melanoidins, recent findings suggest that the fraction of low molecular weight melanoidins (MW<1000–3000 Da) shows higher antioxidant efficiency than that of high molecular weight (MW up to 150 000) (Hofmann, 2000). These compounds seem to act both as primary and as secondary antioxidants since they have been found to behave as chain-breakers, reducing compounds and metal chelators (Homma et al., 1986; Nicoli et al., 1997). The antioxidant capacity of roasted coffee can greatly vary depending on the roasting degree. In Table 6.3 data of chain-breaking activity and redox potential with reference to light, medium and dark coffee samples are reported.

While the assessment of the chain-breaking activity allows estimation of the quenching rate of coffee compounds towards reference radicals, the redox potential gives indications on the effective oxidation/reduction efficiency of all the antioxidants present, including the 'slow' ones, which cannot be detected by the traditional chain-breaking kinetic assays. 'Slow' antioxidants are expected to play an important role in determining the shelf life of roasted coffee during storage (Nicoli et al., 2004). While slight changes in the chain-breaking activity can be observed from light to dark roasted coffees, indicating the existence of a balance between the anti-radical efficiency of naturally occurring antioxidants, which are degraded during roasting, and the heat-induced ones, the redox potential significantly decreases. This indicates that MRPs show higher reducing properties than natural antioxidants. In broad terms, this means that dark

Table 6.3 Chain-breaking activity and redox potential values of coffee samples having different roasting degree

Sample	Weight loss (%)	Chain-breaking activity ($-Abs^{-3} \, min^{-1} mg_{ss}^{-1}$)	Redox potential (mV)
Raw	0	0.320±0.13*	109
Light	14.5	0.273±0.03*	45
Medium	16.2	0.283±0.13*	−2
Dark	18.9	0.390±0.04	−35

*Data not significantly different.
The chain-breaking activity assay was performed using the DPPH test (Nicoli et al., 2001).

roasted coffee is expected to have a greater antioxidant reservoir potentially available during storage.

The presence of such reactive species in roasted coffee can explain the huge variety of redox and radical reactions occurring in coffee during storage and even, at an increased rate, after beverage extraction (Nicoli *et al.*, 2001; Anese and Nicoli, 2003). The oxidation of these reactive compounds has a double relation with roasted coffee shelf life: they are the main cause of aroma compounds degradation, but they also represent an efficient barrier against the oxidation of the lipid fraction. In fact, despite the very low a_w value, lipids in coffee show an extraordinary stability against oxidation thanks to the presence of lipid-soluble MR antioxidants (Whitfield, 1992). Figure 6.2 shows the influence of melanoidins on the stability of coffee lipids. It is interesting to note that samples previously freed from melanoidins showed lower oxidation stability as compared to that measured for lipids containing melanoidins. The antioxidant efficiency of melanoidins has been found to be strictly related to the intensity of the heat treatment used to allow their formation. The higher the intensity of heating, the higher their intrinsic antioxidant capacity (Anese *et al.*, 2000). For dark roasted coffee, the stability of the lipid fraction was observed to range from 4 to 10 months at 25 °C storage depending on oxygen availability and packaging conditions (Nicoli *et al.*, 1993).

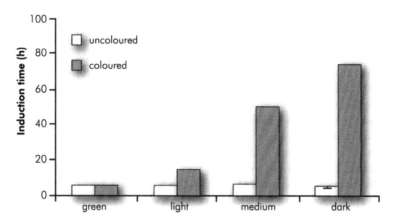

Figure 6.2 Influence of lipid-soluble melanoidins on oxidation stability of coffee lipids (from Anese *et al.*, 2000). The uncoloured samples were obtained by removing melanoidins with active carbon. Data are expressed as induction time (hours) prior to oxidation detected by Rancimat test

The ability of MRPs to act as antioxidants can in part explain their inhibiting action against the growth of some food-poisoning microorganisms. However, the extent of MRPs antimicrobial activity appears to be linked to their concentration and to the bacteria tested, indicating that different mechanisms other than radical scavenging activity may be involved, since diluted coffee shows a weaker antibacterial activity than concentrated coffees. In addition, the extent of the antimicrobial activity strictly depends on the species tested (Einarson, 1987; Sheikh-Zeinoddin *et al.*, 2000; Daglia *et al.*, 2000).

An increasing interest towards coffee MRPs is justified by the fact that, by virtue of their antioxidant activity, they are expected to exert a range of positive health effects on the human organism (O'Brian and Morissey, 1989; Faist and Erbesdobler, 1999). Thus, the preservation of the intrinsic antioxidant properties of roasted coffee during storage could be of a certain importance to obtain coffee products with high functional properties. Unfortunately, up to now no data are available about the influence of different environmental and processing conditions on the evolution of the overall antioxidant properties of roasted coffee during storage.

6.1.2   Staling kinetics

6.1.2.1   Degassing

Carbon dioxide formed during roasting is trapped in the cellular structure of the bean and is only released over a period of weeks following roasting, resulting in a 1.5–1.7% weight loss. The amount of gas released can be estimated at 6–10 litres per kilogram of beans depending on roasting degree, the higher figures being valid for dark roasted blends (Clarke, 1987b), recently confirmed by the results of Shimoni and Labuza (2000).

Degassing rate is inversely related to time from roasting. The massive degassing that takes place in the early hours after roasting slows down gradually, and it may take months for all the CO_2 to be released from the beans. The process is slow because much of the CO_2 is bound to the bean structure. From a physico-chemical point of view, carbon dioxide in the coffee bean can be solubilized partly in the lipid phase, partly in the water (moisture). Some will be adsorbed on the apolar sites, and, finally, a part will be blocked within the pores, which have collapsed because of the phase changes in the structure of the bean due to the roasting process.

The amount of CO_2 adsorbed in the lipid and aqueous fractions can be roughly estimated at 15×10^{-5} mg of CO_2 per gram of fresh roasted coffee in the lipid phase, and 2.13×10^{-5} mg of CO_2 per g of fresh roasted and ground coffee in water, considering coffee in air at normal CO_2 pressure (Shimoni and Labuza, 2000; Labuza *et al.*, 2001).

The driving force at the basis of carbon dioxide and volatile release from roasted coffee is given by a diffusion flow due both to concentration and pressure gradients. The release begins at a considerable rate just after the end of roasting; then it gradually slows down as the CO_2 and volatile concentration in the headspace increases, reducing the mass transfer driving force.

Figure 6.3 shows an example of CO_2 and volatiles released from roasted coffee beans and roasted ground coffee.

Recent studies carried out by Labuza *et al.* (2001) suggest that CO_2 release from ground coffee is controlled by two different mechanisms: at the beginning the diffusion is regulated by pressure gradients between the outside and the inside of the coffee particles, then by molecular diffusion. The time at which the changeover in mechanism occurs in STP is about 120 min. The few data available about volatile partition kinetics from roasted coffee packed in air indicate that the rates of CO_2 and volatiles release are of the same magnitude. The volatiles present in the headspace at equilibrium with the product after three days' storage were 50%, 35% and 18% at 4, 24 and 40 °C respectively. As regards CO_2 evolution, the residual gas percentage in the headspace after the same storage time and temperature was 63%, 37% and 18% (Nicoli *et al.*, 1993).

As mentioned above, the rate of volatile compounds and gas release from coffee is strongly affected by temperature. Additional factors are pressure, particle size distribution, and during the degassing process, packing, bed depth and entrainment gas composition.

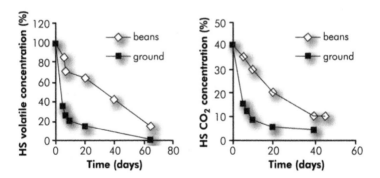

Figure 6.3 Changes in headspace volatile and CO_2 concentration, expressed as percentage, in (left) roasted coffee beans and (right) roasted ground coffee stored at 25 °C in air (Nicoli *et al.*, 1993)

6.1.2.2   Effect of temperature

The temperature dependence of CO_2 and volatile release from roasted coffee can be well described by the Arrhenius equation. An example of an Arrhenius plot, referring to volatile release from roasted coffee beans, is shown in Figure 6.4. The corresponding activation energy was 28.1 kJ/mol, which gives a temperature sensitivity, expressed in terms of Q_{10}, of around 1.5. It means that, in the experimental conditions adopted, for any 10 °C increase, the rate of volatile release increased 1.5-fold. This result agrees with data on the temperature dependence of diffusion coefficients for CO_2, recently reported by Labuza et al. (2001). Moreover, the Q_{10} of CO_2 diffusion coefficients calculated for ground coffee were shown to be approximately twice those of the beans.

The importance of the temperature variations due to day–night fluctuations and to transport is difficult to assess. It may be assumed that, because of temperature changes, competing degradation reactions occur at different times, so that the resulting deterioration is likely to occur faster than at constant temperature. Besides temperature, additional factors affecting staling reaction rates of coffee during storage are a_w and O_2 availability.

6.1.2.3   Effect of moisture

Water in roasted coffee may have different origins: despite the strong dehydration occurring during the first stage of the roasting process, new

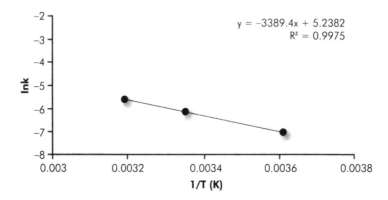

Figure 6.4 Arrhenius plot of volatile release from roasted coffee beans. Data are expressed as ln volatile release rate vs. temperature^{-1} (K). (Elaborated from Nicoli et al., 1993)

water molecules are formed as a consequence of the Maillard reaction. Additional water may come by the use of water quenching instead of air-cooling. Furthermore, the eventual sorption of ambient moisture from roasted coffee during storage represents an additional cause of moisture and hence of a_w increase. Moisture sorption isotherms at 22 °C of roasted coffee beans and ground coffee are shown in Figure 6.5. Due to the higher exposed surface, which affects concentration and availability of polar active sites, ground coffee shows a considerable higher capacity to bind water molecules compared to beans.

In addition, depending on a_w, the bean structure can be brittle or elastic as shown in Figure 6.6.

It is well known that at normal storage temperatures the bean structure is brittle and easy to grind, as shown by the low fracture force measured for air-cooled roasted coffee samples with a_w values ranging at 0.1–0.2. This is due to the fact that roasted coffee beans are greatly below the glass transition temperature, which was recently observed to range from 130 to 170 °C (Karmas and Karel, 1994; Eggers and Pietsch, 2001). However, increasing a_w, the fracture force was shown to double up to a_w 0.85. This is due to the plasticizing effect of water responsible for a decrease in the glass transition temperature (Tg), making the coffee structure rubbery and more elastic. For a_w values higher than 0.85, solubilization of coffee compounds and a sharp decrease in the fracture force are detected.

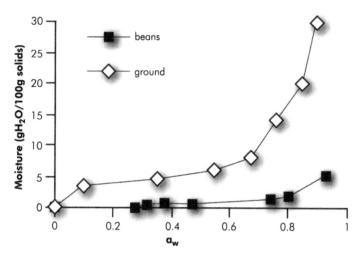

Figure 6.5 Moisture sorption isotherm at 22 °C of roasted coffee beans and ground coffee. (Data from Labuza et al., 2001; Pittia, 2002)

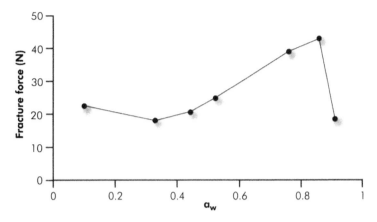

Figure 6.6 Changes in fracture force of dark roasted coffee beans as a function of water activity (Pittia, 2002)

6.1.2.4 Oxygen uptake

As mentioned above, oxidation reactions represent an additional cause of quality loss of roasted coffee during storage. In general terms the rate of all the oxidation reactions occurring in coffee, including those of aroma compounds, increases with increasing oxygen pressure, temperature and a_w. A study of the influence of these three variables on the overall oxidation rate of ground roasted coffee was recently reported by Cardelli and Labuza (2001). The oxygen uptake and sensorial evaluations were used as indicators of product acceptability. By matching the map of rate of food deterioration reactions as a function of water activity with the water sorption isotherm of roasted and ground coffee at room temperature it could be possible to understand what type of chemical reactions are involved in coffee deterioration. In coffee, moisture content ranges from 2% to 5%, which corresponds to 0.2–0.4 water activity. Using these data, it is evident that the oxidation of the coffee lipids is the prevalent reaction.

The influence of a_w and temperature on the shelf life of roasted ground coffee, calculated on the basis of the oxygen uptake index, is illustrated in Figure 6.7. On the basis of these results it can be stated that temperature has a smaller effect than moisture on the shelf life of the product.

The increase in the rate of degradation can be quantified, as a function of oxygen, by introducing a QO_2 index, which represents the relative variation of the rate of degradation caused by an increase in oxygen content of 1%. According to Cardelli and Labuza (2001), QO_2 is 10.5,

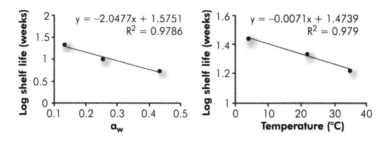

Figure 6.7   Water activity and temperature dependence of the shelf life of ground and roasted coffee (Labuza *et al.*, 2001)

when oxygen concentration is low (0.1–1.1%). It is 1.1 when oxygen increases up to 5%. Further increase of oxygen, up to 21%, does not affect the index, which remains 1.1. Summarizing, it can be stated that the rate of oxygen uptake increases 10-fold, from 0.1 to 1.1%; above 1.1% of oxygen concentration, for each 1% increase of oxygen there is a consequent increase for the rate of degradation of 10%.

6.1.3   Other physico-chemical changes

O. Savonitti

6.1.3.1   *Volatile compounds*

The aroma complex formed during roasting consists of several hundred compounds covering many different chemical classes. Some of them are stable and undergo no change during storage other than loss due to evaporation. Apparently, such compounds do not contribute much to the quality of the perceived aroma. Aroma-impact components are mainly unstable substances particularly prone to oxidation; some, aldehydes and ketones, are present in intermediate states of oxidation, whilst many more, alcohols, phenols, thiols, etc., are in a reduced form. There is then a significant risk of aroma changes occurring as a result of oxidation, particularly of compounds such as the thiols, which have very low threshold values.

The mechanisms responsible for the deterioration of aroma constituents have not yet been fully elucidated; immediately after roasting it may be simply due to evaporation and physical loss of the most volatile compounds; conversion to other volatile or non-volatile compounds, retained in the coffee matrix, also takes place; intermolecular reactions are also likely to occur – for instance, α-dicarbonyls and aldehydes can bind to compounds containing free amino groups. These reactions are apparently influenced by the presence of atmospheric oxygen, suggesting an oxidative route to degradation. The change in aroma profile cannot be simply explained by the loss of the compounds affected, as the reaction products themselves may make a sensory contribution. In the case of thiols, for instance, the corresponding condensation products, the disulphides, have a major sensory impact, different from that of the fresh product. Afterwards, oxidation reactions become predominant, as indicated by the correlation between deterioration and oxygen uptake; and oxidation of the aroma complex will lead to a further dramatic shift in aroma quality.

Oxidation, responsible for the typical stale flavour of old coffee, is the sum of two actions: the loss of pleasant aroma components accompanied by the formation of off-flavours. A study has concentrated on identifying the compounds responsible for the freshness of coffee aroma, defined as the fine and pleasant smell arising from freshly roasted beans (Steinhart and Holscher, 1991). The conclusion was that low-boiling compounds – such as methanethiol, Strecker aldehydes and α-dicarbonyls – were the most important; in particular, loss of methanethiol was considered as the most important indicator. All of these compounds are volatile and extremely reactive species, therefore, easily oxidized. Figure 6.8 shows that their disappearance, particularly noticeable among those with the lowest boiling point, occurs mainly over the first three weeks of storage, and is principally due to evaporation.

Successive research during the 1990s led to a set of 28 compounds in roasted and ground coffee as the most odour-active components associated to fresh coffee aroma (Grosch, 1999, 2001; Czerny and Schieberle, 2001).

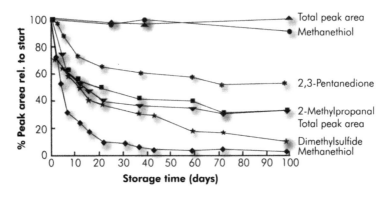

Figure 6.8 Loss of volatile compounds during storage

One of these, namely methanethiol, is frequently used in many freshness indices for coffee samples stored in different conditions (Sanz et al., 2001). Many of these compounds can be used, in a definite linear combination, as indicators of the change of freshness for a roasted and ground coffee exposed to air at different temperatures (Cappuccio et al., 2001).

The loss of volatile substances from coffee is not only related to their volatility, but also to the way they are trapped in the cells. In fact, two additional mechanisms of aroma retention play an important role. The cell wall consists of polar polysaccharides and melanoidins, which can adsorb and retain a wide range of volatile substances, particularly polar ones. Oil, on the contrary, preferentially retains lipophilic substances, such as alkylpyrazines, etc. In order to evaporate, volatile particles 'trapped' in the cell structure of the beans must first migrate across the cell walls. In ground coffee this is made easy by the small particle size, so that aroma loss occurs at a higher rate in ground coffee than in coffee beans and is inversely proportional to particle size. Hence, finely ground espresso blends lose their aroma just a few hours after being exposed to air, and start smelling stale due to oxidation, if not stored in packages. However, even in this case coffee aroma can undergo modifications through time, if stored in packages somehow permeable to oxygen. Coffee staling is associated with a degradation of some particular odorants, like odour-active aldehydes, α-dicarbonyls and thiols (2-furfurylthiol has the main impact), rather than to the creation of off-odorants. This is due again to the presence of oxygen, because of the non complete impermeability of the package (Czerny and Schieberle, 2001).

6.1.3.2 Non-volatile compounds

Production of carbon dioxide during roasting forms an effective barrier, which excludes most of the atmospheric oxygen from the cellular structure so delaying the onset of oxidation. As degassing goes on, lipid oxidation can no longer be stopped by the antioxidant activity of the Maillard reaction products formed during roasting (Homma, 2001; Steinhart et al., 2001), and in a few months, at the end of degassing, the beans are rancid.

Ground coffee is particularly sensitive to oxidation because grinding removes the CO_2 barrier almost completely, while at the same time the oil–air contact surface dramatically increases, promoting oxygen uptake.

Lipids, an important fraction in roasted coffee, are easily oxidized during storage (Huynh-ba et al., 2001). The major unsaturated fatty acid in coffee is linoleic acid ($C_{18:2}$), which contains two double bonds and is, therefore, rather susceptible to oxidation. Figure 6.9 shows the reduction in linoleic acid content of rancid beans with respect to fresh beans; while the major saturated fatty acid, stearic acid ($C_{16:0}$), remains practically unchanged (Fourny et al., 1982).

An important contribution to oxidation off-flavours is the formation of volatile aldehydes, such as *trans*-2-nonenal, the main breakdown product of lipid oxidation, which has an odour threshold level of 0.08 µg/l in water (Flament, 2002).

6.1.3.3 Oil migration

Oil migration starts during roasting and goes on during degassing because carbon dioxide tends to push oil outwards through the cell pores. The increase in oil viscosity at lower temperatures slows down the process.

Oil migration to the surface of the beans, where the risk of oxidation is maximal, is particularly important in fine ground dark espresso blends, since dark roasting leads both to fast degassing and to increased porosity, from disruption of cell walls. A further problem linked with oil exudation is the increase in stickiness of the particles, which tend to aggregate into lumps, making brewing irregular. The aggregation of particles on storage is worsened by the absorption of moisture, as discussed in 6.1.2.3.

6.1.3.4 Effect of light

Light plays a catalytic role in many chemical reactions; in the case of espresso (arabica) blends, particularly rich in unsaturated fatty acids, light

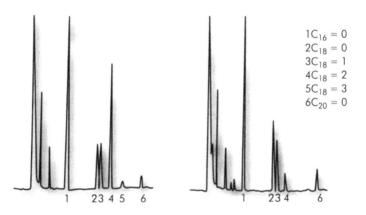

$$1C_{16} = 0$$
$$2C_{18} = 0$$
$$3C_{18} = 1$$
$$4C_{18} = 2$$
$$5C_{18} = 3$$
$$6C_{20} = 0$$

Figure 6.9 Fatty acid content of (left) fresh and (right) rancid beans

catalyses the prime trigger of their auto-oxidation reaction, i.e. the formation of H·, R· (alkyl) and ROO· (peroxide) free radicals which then cause the reaction to propagate.

6.2 PACKAGING OF ROASTED COFFEE

O. Savonitti

All the potential causes of deterioration highlight the importance of packaging. Indeed, it may be stated that, because of its instability, roasted coffee should either be consumed straight away or packaged in a water- and oxygen-tight container. Special packaging materials and techniques, with careful process monitoring, are required not to jeopardize the final quality by poor manufacturing practices.

6.2.1 Packaging materials

In order to meet total quality requirements (see Chapter 1) the package should:

- act as a barrier against water and moisture;
- act as a barrier against atmospheric oxygen;
- preserve coffee aroma and keep out foreign odours;
- be grease-proof;
- be light-proof;
- allow the carbon dioxide released during degassing to escape;
- be chemically inert;
- be hygienically safe and suitable for foodstuffs;
- be long-lasting;
- be sturdy and withstand pressure variations;
- be cheap;
- be practical;
- be environmentally friendly;
- be consumer-friendly.

Most of these requirements are aimed at preventing spoilage; some are exclusively intended to add value to consumers. Packaging materials should be grease-proof to prevent air oxidation of coffee oil leaking through the package, which would then spoil the beans. Chemical inertia of the material is also important, and metals or other substances, which could act as oxidation catalysts, should be avoided. Table 6.4 lists the materials meeting most of these packaging requirements.

Table 6.4   Materials commonly used for packaging roast coffee

Material	Advantages	Disadvantages
Tinplate	Total barrier Resistant to pressurse Satisfactory ecobalance* Recyclable Sturdy	Expensive Its rigidity prevents an optimal use of space
Aluminium	Total barrier Resistant to pressure Recyclable	Expensive Its rigidity prevents an optimal use of space Poor ecobalance Difficult to sort from other wastes
Glass	Total barrier Good ecobalance Recyclable Resistant to pressure	Fragility Heavy material Expensive Its rigidity prevents an optimal use of space Difficult to sort from other wastes
Combined flexible multi-ply polymer aluminium	Inexpensive Simple manufacturing technology Large flexibility and optimal use of space Satisfactory ecobalance	Non-tight barrier Resistant only to negative pressures Poor strength
Combined multi-ply aluminium cardboard	Inexpensive Total barrier Satisfactory ecobalance	Difficult to sort from other wastes Non resistant to pressure Poor strength Complex manufacturing technology
Flexible combined multi-ply polymer	Inexpensive Simple manufacturing technology Satisfactory CO_2 permeability No need to sort for other wastes	

*Ecobalance is defined as the total pollution resulting from energy consumption, and that resulting from the product, during manufacture, use and disposal.

Commonly used materials are the inexpensive flexible polymer-aluminium multi-ply, ensuring an efficient barrier and an optimal use of space all along the life cycle, and tinplate, particularly useful for positive pressure-resistant packages. The aluminium contained in multi-ply materials constitutes the only completely effective barrier; the inner layer is a film of waterproof weldable material; the centre layer is an airtight aluminium film and the outer layer consists of a material that makes the structure more rigid and resistant to mechanical stress.

None of the materials listed in Table 6.4 ensures the level of permeability to carbon dioxide required to keep a stable internal pressure during degassing; other means must be applied.

Blow-moulded rigid plastic materials (HD-PE or PET) have lately been introduced for roast and ground coffee packaging: one-way valves on easy-peel-off membranes or applied directly on the packaging allow degassing and avoid deformations. This type of container is light, resistant to breakage, can be easily piled up, and recycled. Production technology is well known and the form can be chosen at will. However, the material does not provide a perfect barrier to gases.

6.2.2 Packaging techniques

The techniques used for packaging roast coffee are listed in order of increasing product protection:

- air packaging
- vacuum packaging
- inert gas packaging
- pressurization
- combined use of one of the previous techniques with active packaging.

6.2.2.1 Air packaging

Air packaging consists in simply filling and hermetically sealing the package; coffee is protected against humidity, off-flavours and light, but the presence of air inside the package means high oxygen levels and consequently shortened shelf life. This technique can only be used with degassed coffee to prevent swelling and possible explosion of the airtight package. If the coffee, particularly coffee beans, is not degassed, swelling can be prevented by fitting a one-way safety valve on the package allowing carbon dioxide to escape without letting air in. As CO_2 is heavier than air, it tends to stratify at the bottom expelling most of the oxygen gas; the flushing effect reduces residual oxygen and increases product shelf life. Air packaging using a one-way safety valve is an acceptable technique for air-cooled coffee beans since they contain large quantities of gas; its main disadvantage is that most of the aroma escapes together with CO_2, so dulling cup taste.

6.2.2.2 Vacuum packaging

Vacuum packaging meets two different objectives: air extraction with lowering of oxygen level, and use of flexible materials. The technique, which can also be used with rigid materials such as tinplate, is commonly applied to flexible materials to make the coffee 'bricks' sold in supermarkets. The box shape is obtained by an intimate contact between the material and the coffee, which must be completely degassed to avoid a decrease in compactness of the package provoked by a release of carbon dioxide making it floppy or swollen. This is why bricks are generally used only for packaging water-cooled ground coffee. Disadvantages of vacuum packaging in flexible containers are linked with water-cooling, which shortens product shelf life and may decrease yield. On the other hand, in rigid containers the internal vacuum widens the difference between the partial pressure of volatile aroma and ambient, so that more aroma volatilizes to saturate the headspace, dulling the cup. Furthermore, even if rigid containers can be strengthened by ribbing to prevent collapse, the vacuum level still remains less than in flexible bags.

6.2.2.3 Inert gas packaging

In inert gas packaging the air inside the container is replaced by inert gas either through the compensated vacuum technique or by flushing the inside of the package with inert gas. In the former case, first a vacuum is created in the package, then enough inert gas to balance the internal and external pressures is admitted. In the latter case, a drop of liquefied inert gas is placed on the bottom of the package, which evaporates pushing the air out. This process generally uses nitrogen or carbon dioxide which, although not an inert gas, behaves as such in a moisture-free environment and, moreover, is naturally present in roast coffee. The use of an inert gas for coffee packaging increases shelf life three-fold (see Table 6.5) with respect to vacuum packaging. Even with the same type of package – and therefore with the same permeability to oxygen and water vapour – the type of conditioning process will impact on the product's shelf life, and consequently on the cup (Alves et al., 2001).

The pressure in the package is in equilibrium with the atmosphere at the moment of sealing. Just like in air packaging, to prevent internal pressure rising due to degassing, coffee must be either packed after degassing or the package fitted with a one-way valve.

From a legal standpoint the gas added to the package is considered a processing aid rather than an additive, since it dissipates on opening.

Table 6.5 Barrier properties of packaging materials for roasted coffee

Material	Permeability*	Barrier
Tinplate, glass aluminium		Complete
Three-ply with aluminium	<0.5	Very high
Three-ply with metal	0.5–3.0	High
Two-ply with metal	3.0–30	Moderate
Two-ply	30–150	Low
Single film, treated paper	>150	Very low

*Values expressed in:
cm^3/m^2 24 h bar at 23 °C and 0% R.H. for gases: gas volume (cm^3 of gas) that passes through 1 m^2 of surface in one day (24 hours), with one bar of pressure difference between the two layers at 23 °C and 0% of relative humidity g/cm^2 24 h bar at 38 °C and 90% R.H. for water values gas mass (grams of water vapour) that passes through 1 m^2 of surface in one day (24 hours), with one bar of pressure difference between the two layers at 38 °C and 90% of relative humidity.

6.2.2.4 Pressurization

Pressurization is the same as inert gas packaging except that the internal pressure is higher than atmospheric pressure. If coffee is immediately packaged after air-cooling, the pressure normally rises due to degassing. The packaging technique is the same as in compensated vacuum packaging, but in order to withstand pressure containers must be made of rigid materials, normally tinplate or aluminium; they must also be equipped with a safety valve opening when pressure increases by 0.5 atmospheres due to the large amount of gas released (see 6.1.2.1 and Figure 6.1).

Pressurization has an 'ageing' effect with quality improvement after 10–15 days. Indeed, an espresso cup prepared from an aged product has an improved body and aroma. The ageing may be explained by a binding of aroma constituents to the oil trapped within the cell structure. During degassing, pressure rises in the container and the reduced pressure gradient between the packaging environment and the bean structure produces two combined effects underlying the ageing mechanism: first, a decrease in degassing rate reduces oil migration so that more of it remains in the beans (Clarke, 1987b, p. 204); second, the pressure in the package becomes higher than the partial pressure of most volatile compounds present in the cells, allowing for a larger quantity to be dissolved in the lipid phase or to bind to melanoidins.

Pressurization, by creating high pressure within the bean, also has the effect of spreading the oil on the cell walls to form an effective barrier against oxygen; this effect is clearly visible in Figure 6.10, which shows the cross-sections of, on the left, a non-pressurized roast bean, and, on the right, a pressurized one; caramelized material, containing lipids (dissolving volatile aroma), is shown in orange-yellow, while cell walls are reddish. (PIC IC B&W)

Due to the difference between internal bean pressure (which at the end of the degassing process reaches equilibrium with the pressure in the container) and atmospheric pressure, when the package is opened, there is still a residual amount of CO_2 in the beans to be released. Moreover, when oxygen enters the beans at the end of the degassing phase, oxidation is reduced because the aroma constituents are dissolved in the compacted lipid aggregates, so that, even once the package is open, a pressurized blend retains its fragrance longer than a blend packaged by other techniques.

6.2.2.5   Active packaging

The need to stress the performance of existing passive packages and related packaging techniques on one hand, and the need to guarantee an adequate product freshness on the other has brought the concept of 'active packaging' into consideration for coffee as well. Rooney (1995) defined active packaging as 'packaging that performs a role other than an inert barrier to the outside environment'. A more recent definition of active packaging is 'packaging that changes the condition of the packaged food to extend shelf life or improve food safety or sensory properties, while maintaining the quality of the packaged food'. This second definition has

Figure 6.10 Cross-section of cells from (left) non-pressurized and (right) pressurized beans

been chosen for the European FAIR project Ct.98-4170. (De Kruijf *et al.*, 2002).

Active packaging systems can be classified into active scavenging systems (absorbers) and active-releasing systems (emitters). In the specific case of active coffee packaging, it includes oxygen and carbon dioxide scavengers. The oxygen scavenger has the precise task to eliminate residual oxygen present within the headspace of the packaging in the case of high barrier packages and conditioning in inert gas, or to eliminate the oxygen present, in the case of air packaging. In both cases, the elimination of oxygen contributes to extending the stability of the product during the storing phase. Moreover, independently from the conditioning used, in the case of permeable packaging, the oxygen absorber has the task of capturing oxygen entering through the walls of the package until it becomes saturated.

In general, the existing oxygen scavenger technologies are based on one or more of the following principles: (1) iron powder oxidation, (2) ascorbic acid oxidation, (3) photosensitive dye oxidation, (4) enzymatic oxidation, (5) unsaturated fatty acids and (6) yeast immobilized on solid material. They may be used as sachets inserted into the package, or be bonded by a label to the wall of the package. They can be used as linings in the closure of the package or can be dissolved, dispersed or immobilized in the polymer of the package material. The most commonly applied scavengers in coffee packaging are based on iron powder oxidation; their advantage is that they are in a position to reduce the level of oxygen below 0.01%, a value much lower than 0.3–3%, the residual values typically reached when applying a modified atmosphere packaging technique.

Carbon dioxide scavengers have the task of removing a proportion of the carbon dioxide released from the coffee in order to prevent the package from bursting when the coffee has not been previously degassed. They can represent an alternative use to the one-way valves or to materials permeable to carbon dioxide (which are in general also very permeable to oxygen and water vapour) (Vermeiren *et al.*, 1999).

A combined solution, such as dual action oxygen and carbon dioxide scavenger systems in the form of sachets or labels, is common in canned and foil-pouched coffees in Japan and the USA. This dual action system typically contains iron powder as oxygen scavenger and calcium hydroxide that scavenges carbon dioxide in the presence of high moisture levels (http://www.atalink.co.uk/pira/html/menu-packaging.htm 'A fresh approach', Brian Day Campden and Chorleywood Food Research Association). Some oxygen scavengers are subject to competition between oxygen and carbon dioxide; for this reason the dual action system contains alkaline hydroxide specific for carbon dioxide (Matsushima *et al.*, 1995).

6.2.3 Control parameters

The packaging process starts from the selection of the packaging materials and techniques, through conditioning, when coffee is filled into the packs and sealed, to the final phase when bags and cans are stored and taken to the place of consumption. Each step of this long chain must be carefully monitored to prevent or at least correct all possible defects so that the quality (see 1.3) of the ready-to-use coffee blend complies with regulations and is up to consumers' expectations. Various parameters are monitored throughout the different phases of the packaging process.

6.2.3.1 Packaging material

The packaging material must be adapted to the packaging technique and to the intended shelf life. As mentioned in 6.2.1, the most important packaging requirements concern the barrier properties of the chosen material, i.e., its permeability to gas and moisture. Such barrier properties can be classified as shown above in Table 6.5.

Glass and metals are air- and watertight; the choice of packaging materials based on their barrier properties, therefore, only applies to multi-ply materials, which can be prepared from different polymers. Their permeability is a function of the type of polymer, of possible surface treatments on the polymer, their thickness and the combination chosen.

In addition to the packaging material, particularly in the case of tight containers, the packaging shape must also be decided. The main parameters to be considered in this choice are volume of headspace, which should be as small as possible to limit the amount of aroma volatilized to reach saturation equilibrium, and the ratio of volume to surface area, which should also be as low as possible, particularly when incomplete barrier materials are used, since permeability depends on surface area.

6.2.3.2 Choice of the packaging technique

As seen in 6.1.1, oxygen consumption is the main cause of deterioration of roast coffee, which should therefore be protected from entering into contact with oxygen as soon as roasted. The lower the level of residual oxygen in the package, the longer the shelf life of the product. But the lower the amount of residual oxygen the more effective the package should be as a barrier to keep oxygen at the same level throughout the shelf life of the product. The best techniques to improve shelf life are, in decreasing order, pressurization, inert gas and vacuum packaging; the technique must be chosen as a function of the desired shelf life.

The choice of the most suitable material is also influenced by the degree to which it can withstand positive internal pressures, when coffee is pressurized, or negative ones, when it is vacuum packaged. Table 6.6 shows the implications of these two parameters on shelf life and packaging requirements.

6.2.3.3 Conditioning

When coffee is air-packed no conditioning is needed; packages are just filled up and sealed, the only parameter to be monitored being air tightness, as the package needs only to protect coffee from moisture pick-up.

In vacuum packaging, conditioning consists in creating a 300–500 mbar vacuum inside the pack before sealing it. The vacuum level must then be monitored, particularly in rigid cans, which might otherwise collapse. If the vacuum is replaced by inert gas, the applied gas pressure and/or the internal pressure of the package must be checked. Pressure levels near atmospheric pressure are normally applied, while valveless packages are conditioned under a slight vacuum.

In pressurization, conditioning consists in exactly the same process, but the cans are pressure-resistant and coffee is not degassed. In this case, in

Table 6.6 Packaging parameters

Technique	Residual O_2(%)	Shelf life (mth)	Absolute P_{int} (Atm)	Material
In air:				
tight	16–18	1	nRT/V	Rigid
with valve	10–12	3	1.01*	Indifferent
Under vacuum	4–6	4–6	0.3	Better flexible
Under inert atmosphere	1–2	6–8	1.01*	Indifferent
Under pressure	<1	>18	Up to 2.2	Rigid

*Pressure at which the valve opens.

addition to residual oxygen, the pressure in the package should also be monitored once degassing has made the package swell.

The airtight seal, already important in air packaging, becomes critical with all the other types of packages because any leak in the packaging material would allow oxygen in, drastically reducing shelf life.

Pressure is measured by standard pressure gauges; the airtight seal is checked by measuring residual oxygen and/or, in the case of positive internal pressures, by plunging the package in water and checking for leaks. Various sensor methods are available to measure residual oxygen levels.

6.2.3.4 Logistics

In logistics, pressure, temperature and time are the key parameters.

The pressure level must be monitored to avoid the risk of swelling or even explosion from a build-up of pressure in valveless packs. This is why air transportation of coffee should be avoided, unless pressurized compartments are available.

The spoilage rate of roasted coffee increases seven-fold with every 10 °C rise in temperature. Coffee should therefore be stored at low temperature. Furthermore, low temperatures slow down evaporation of volatile compounds, so that coffee kept at low temperature retains more aroma than at normal temperature, as recently confirmed on roasted and ground coffee in air by Cappuccio et al. (2001). Cooled coffee beans should, however, be left at room temperature for a few hours before grinding to trap the volatile aroma present in the cells in the oil and melanoidins; a large aroma loss at grinding might otherwise dull the flavour of the cup. In pressurized coffee the binding of aroma by oil is hindered if the oil becomes too viscous, as is the case at temperatures approaching 0 °C; pressurized coffee should therefore be kept at room temperature for a few days after packaging. Temperature fluctuations should be avoided as they accelerate spoilage.

The last parameter to be monitored is time, i.e. the total storage period, which should always be kept shorter than the actual shelf life of the product. On average, coffee is consumed within three months from production which makes time a critical factor only for the least sophisticated packaging techniques.

6.2.3.5 Testing of finished products

Although lipid oxidation plays a major role in coffee deterioration, the peroxide test, commonly performed to assess rancidity in foods, is not sensitive enough for coffee. Peroxides reach significant levels only in the very last stages of spoilage, when coffee has already become undrinkable. Linoleic acid content is another parameter of little use to analysts, as it requires reference values measured on the same blend when fresh. Electron spin resonance (ESR) measurements have also been carried out to detect the presence of free radicals; the intensity of the ESR signal has been found to be inversely related to the level of perceived acidity, which rises with deterioration; however, even this technique is not sensitive enough.

The concentration of specific volatile compounds in the headspace of the package or of the cup, measured by gas chromatography, has been used to determine the degree of spoilage. For instance, the ratio of 2-methylfuran to butanone, the 'aroma index' M/B, drops from 3.2 in fresh coffee to 2.3 in spoilt coffee. However, such indices depend on blend composition and on roasting technique, normally unknown to the analyst, so that these parameters can only be relied on to a very limited extent.

None of the test methods available can yet replace sensory evaluation, and tasting remains the most reliable of all test methods.

REFERENCES

Alves R.M.V., Mori E.E., Milanez C.R. and Padula M. (2001) Roasted and ground coffee in nitrogen gas flushing packages. *Proc. 19th ASIC Coll.*, CD-ROM.

Anese M. and Nicoli M.C. (2003) Antioxidant properties of ready-to-drink coffee brews. *J. Agric. Food Chem.* 51, 942–946.

Anese M., De Pilli T., Massini R. and Lerici C.R. (2000) Oxidative stability of the lipid fraction in roasted coffee. *Ital. J. Food Sci.* 12, 457–463.

Anese M., Manzocco L. and Maltini E. (2001) Determination of the glass transition temperatures of solution 'A' and HMW melanoidins and estimation of viscosities by WLF equation: a preliminary study. In J. Ames (ed.), *Melanoidins in Food and Health Volume 2*, Proceedings of COST Action 919 workshop held at the Technical University of Berlin, 7–8 April 2000 and the Institute of Chemical Technology, Prague 22–23 September 2000, pp. 137–141; EUR 19684 EN, Belgium.

Barbera C.E. (1967) Gas-volumetric method for the determination of the internal non odorous atmosphere of coffee beans. *Proc. 3rd ASIC Coll.*, pp. 436–442.

Cappuccio R., Full G., Lonzaric V. and Savonitti O. (2001) Staling of roasted and ground cofffee at different temperatures: combining sensory and GC analysis. *Proc. 19th ASIC Coll.*, CD-ROM.

Cardelli C. and Labuza T.P. (2001) Application of Weibull hazard analysis to the determination of the shelf life of roasted and ground coffee. *Lebensm. Wiss. Technol.* 34, 273–278.

Clarke R.J. (1987a) Packing of roast and instant coffee. In R.J. Clarke and R. Macrae (eds), *Coffee: Volume 2 – Technology*. London: Elsevier Applied Science, pp. 201–215.

Clarke R.J. (1987b) Roasting and grinding. In R.J. Clarke and R. Macrae (eds), *Coffee: Volume 2 – Technology*. London: Elsevier Applied Science, pp. 73–197.

Clarke R.J. and Macrae R. (eds) *Coffee: Volume 2 – Technology*. London: Elsevier Applied Science.

Czerny M. and Schieberle P. (2001) Changes in roasted coffee aroma during storage – influence of the packaging. *Proc. 19th ASIC Coll.*, CD-ROM.

Daglia M., Papetti A., Gregotti C., Bertè F. and Gazzani G. (2000) In vitro antioxidant and ex vivo protective activities of green and roasted coffee. *J. Agric. Food Chem.* 48, 1449–1454.

De Krujf N., Van Beest M., Rijk R. J., Sipilainen-Malm T., Paseiro Losada P. and De Meulenaer B. (2002) Active and intelligent packaging: applications and regulatory aspects. *Food Addit. Contam.* 19, 144–162.

Eggers R. and Pietsch A. (2001) Technology I: Roasting. In R.J. Clarke and O.G. Vitzhum (ed.), *Coffee: Recent Developments*. Oxford: Blackwell Science, pp. 90–107.

Einarson H. (1987) The effect of time, temperature, pH and reactants on the formation of antibacterial compounds in the Maillard reaction. *Lebensm. Wiss. Technol.* 20, 51–55.

Faist V. and Erbesdobler H.F. (1999) In vivo effects of melanoidins. In J. Ames (ed.), *Melanoidins in Food and Health Volume 1*. Proceedings of COST Action 919. Workshop held at the University of Reading, UK, 2–4 December 1999, pp. 79–88; EUR 19684 EN Belgium.

Flament I. (2002) *Coffee Flavor Chemistry*. New York: Wiley & Sons, p. 118.

Fourny G., Cross E. and Vincent J.C. (1982) Etude préliminaire de l'oxidation de l'huile de café. *Proc. 10th ASIC Coll.*, pp. 235–246.

Full G., Savonitti O. and Cappuccio R. (2001) Staling of roasted and ground coffee at different temperatures: combining sensory and GC analysis. *Proc. 19th ASIC Coll.*, CD-ROM.

Goodman B.A., Pascual E.C. and Yeretzian C. (2001) Free radicals and other paramagnetic ions in soluble coffee. *Proc. 19th ASIC Coll.*, CD-ROM.

Grosch W. (1999) Key odorants of roasted coffee: evaluation, release, formation. *Proc. 18th ASIC Coll.*, pp. 17–26.

Grosch W. (2001) Chemistry III. Volatile compounds. In R.J. Clarke and O.G. Vitzhum (eds), *Coffee: Recent Developments*. Oxford: Blackwell Science, pp. 68–90.

Hodge J.E. (1953) Chemistry of browning reaction in model systems. *J. Agric. Food Chem.* 1, 928–943.

Hofmann T. (2000) Isolation, separation and structure determination of melanoidins. In J. Ames (ed.), *Melanoidins in Food and Health – Volume 1*, Proceedings of COST Action 919, Workshop held at the University of Reading, UK, 2–4 December 1999, pp. 31–43; EUR 19684 EN.

Hofmann T., Czerny M., Calligaris S. and Schieberle P. (2001) Model studies on the influence of coffee melanoidins on flavour volatiles of coffee beverages. *J. Agric. Food Chem.* 49, 2382–2386.

Homma S. (2001) In R.J. Clarke and O.G. Vitzthum (eds), *Coffee: Recent Developments*. Oxford: Blackwell Science, pp. 50–68.

Homma S., Aida K. and Fujumaki M. (1986) Chelation of metals with brown pigments in coffee. In M. Fujimaki, H. Kato and M. Namiki (eds), *Amino-Carbonyl Reactions in Food and Biological Systems*. Amsterdam: Elsevier, pp. 165–172.

Huynh-Ba T., Courtet-Compondu M.C., Fumeaux R. and Pollien Ph. (2001) Early lipid oxidation in roasted and ground coffee. *Proc. 19th ASIC Coll.*, CD-ROM.

Karmas R. and Karel M. (1994) The effect of glass transition on Maillard browning in food models. In T.P. Labuza, G.A. Reineccius, V. Monnier, J. O'Brien and J. Baynes (eds), *Maillard Reaction in Chemistry, Food and Health*. Cambridge: The Royal Society of Chemistry, pp. 164–169.

Labuza T.P., Cardelli C., Andersen B. and Shimoni E. (2001) Physical chemistry of carbon dioxide equilibrium and diffusion in tempering and effect on shelf life of fresh roasted ground coffee. *Proc. 19th ASIC Coll.*, CD-ROM.

Manzocco L., Calligaris S., Mastrocola D., Nicoli M.C. and Lerici C.R. (2001) Review of non-enzymatic browning and antioxidant capacity of processed foods. *Trends Food Sci. Technol.* 11, 340–346.

Manzocco L., Calligaris S. and Nicoli M.C. (2002) Assessment of pro-oxidant activity of foods by kinetic analysis of crocin bleaching. *J. Agric. Food Chem.* 50, 10.

Massini R., Nicoli M.C., Cassarà A. and Lerici C.R. (1990) Study on physico- and physico-chemical changes in coffee beans during storage. Note 1. *Ital. J. Food Sci*, 2, 123–130.

Matsushima T., Oguro N. and Ichiyanagi S. (1995) Stability improvement of roasted and ground coffee by oxygen absorbent. *Proc. 16th ASIC Coll.*, pp. 426–434.

Namiki M. (1990) Antioxidants/antimutagens in food. *Crit. Rev. Food Sci. Nutr.* 29, 273–300.

Nicoli M.C., Anese M., Manzocco L. and Lerici C.R. (1997) Antioxidant properties of coffee brews in relation to the roasting coffee. *Lebensm. Wiss Technol.* 30, 292–297.

Nicoli M.C., Anese M. and Calligaris S. (2001) Antioxidant properties of ready-to drink coffee beverages. *Proc. 19th ASIC Coll.*, CD-ROM.

Nicoli M.C., Innocente N., Pittia P. and Lerici C.R. (1993) Staling of roasted coffee: volatile release and oxidation reactions during storage. *Proc. 15th ASIC Coll.*, pp. 557–566.

Nicoli M.C., Toniolo R. Anese M. (2004) Relationship between redox potential and chain-breaking activity of model systems and foods. *Food Chem.* 88, 79–83.

O'Brian J. and Morissey P.A. (1989) Nutritional and toxicological aspects of Maillard browning reaction in food. *Crit. Rev. Food Sci. Nutr.* 28 (3), 221–248.

Pittia P. (2002) Personal communication.

Radtke R. (1979) Zur kenntnis des sauerstoffverbrauchs von rostkaffee und seiner auswirkung auf die sensoriesch ermitelte qualitata des kaffeegetranks. *Chem. Mikrobiol. Technol. Lebensm.* 6, 36–42.

Rooney M.L. (1995) Overview of active food packaging. In M.L. Rooney (ed.), *Active Food Packaging*. New York: Blackie Academic and Professional/Chapman and Hall, pp. 1–37.

Sanz C., Pascual L., Zapelena M.J. and Cid M.C. (2001) A new 'aroma Index' to determine the aroma quality of a blend of roasted coffee beans. *Proc. 19th ASIC Coll.*, CD-ROM.

Sheikh-Zeinoddin M., Perehinec T.M., Hill S.E. and Rees C.E.D. (2000) Maillard reaction causes suppression of virulence gene expression in *Listeria monocytogenes. Int. J. Food Micr.* 61 (1), 41–49.

Shimoni E. and Labuza T.P. (2000) Degassing kinetics and sorption equilibrium of carbon dioxide in fresh roasted and ground coffee. *J. Food Process Eng.* 23, 419–436.

Sivetz M. and Desrosier N. (1972) *Coffee Technology.* Westport, CT: AVI Publishing.

Steinhardt H. and Holscher W. (1991) Storage related changes of low-boiling volatiles in whole beans. *Proc. 14th ASIC Coll.*, pp. 156–174.

Steinhart G., Luger A. and Piost J. (2001) Antioxidative effect of coffee melanoidins. *Proc. 19th ASIC Coll.*, CD-ROM.

Vermeiren L., Devlieghere F., Van Beest M., de Krujf N. and Debevere J. (1999) Developments in the active packaging of foods. *Trends in Food Sci. Technol.* 10, 77–86.

Whitfield F.B. (1992) Volatiles from interactions of Maillard reactions and lipids. *Crit. Rev. Food Sci. Nutr.* 31(1/2), 1–58.

Yaylayan V.A. and Huyghues-Despointes A. (1994) Chemistry of Amadori rearrangement products: analysis, synthesis, kinetics, reactions and spectroscopic properties. *Crit. Rev. Food Sci. Nutr.* 34 (4), 321–369.

CHAPTER 7

Percolation

M. Petracco

For all types of coffee beverages, brewing is intrinsic to the relevant definition (Petracco, 2001). This holds especially in the case of espresso, where, among all subsequent technological processes through which coffee must go before finally getting into the cup, the last one – percolation – is the stage crucially contributing to the beverage's qualities.

On the one hand, it is possible to sort, roast, grind and package coffee following all the established rules for an espresso coffee, as described in the preceding chapters, and finally prepare a cup by infusion or drip filtering. The beverage so obtained certainly cannot be called an espresso. On the other hand, the inverse operation could be reasonably well performed: it could be possible to prepare an espresso cup (even if not the best one) by using a low quality raw material, which has been inadequately roasted and improperly ground, provided that it is transferred into the cup by an extraction process following the espresso percolation process, as will be discussed in 7.4 below. With the importance of extraction as an integral part of the definition of espresso ascertained, we have to face the necessity of giving a scientific definition of percolation, and establishing the conditions for a correct espresso percolation.

7.1 CONCEPTUAL DEFINITIONS

7.1.1 Definition of percolation

Percolation takes place each time a fluid flows through a porous medium. In hydrodynamics, this concept materializes in many aspects of our everyday life: rainwater percolates through sand in the soil; the engine lubricant percolates through the oil filter. While such phenomena are cited in common language as filtration, this term is used improperly. In fact, filtration strictly refers to the process of separation undergone by a suspension (solid particles dispersed in a liquid) flowing through a filtering baffle: the baffle may consist of a bed of packed particles, a fibrous tangle or simply a perforated plate. Unlike other forms of liquid/solid separation (sedimentation, centrifugation, electro-deposition, etc.), filtration cannot take place without percolation; conversely, a wide range of percolation phenomena, where there is removal of mass from the solid bed instead of accumulation, bear no relationship to filtration but rather to extraction.

The term percolation is, in a more general meaning, used also in the language of topology, the branch of geometry that studies the characteristics of forms. Under this meaning, an abstract property of space can spread from one area to another according to fixed mathematical laws (Stauffer, 1985). A classical example is Bernoulli's percolation through a lattice randomly composed of conducting or insulating segments. If at least one uninterrupted series of conducting segments stretches from one end of the lattice to the other, electric current may flow through the lattice, which is thus called 'percolating' (Mandelbrot, 1977). This reference is pertinent to our case, as the application of topological, or more generally, mathematical methods may contribute to the understanding and modelling of hydrodynamic phenomena and, in particular, of espresso coffee brewing.

7.1.2   Definition of coffee cake

Back to hydrodynamics, it is useful to explain what is meant by 'porous medium'. It is an aliquot of space, possibly bounded by walls impermeable to fluids, and partially filled with a solid material leaving empty spaces, filled in their turn with a pre-existing fluid that can be displaced by the percolating fluid (Bear and Verrujt, 1987). A simple example: a bucket full of gravel has interstices between pebbles, which are filled up by air. By pouring water into the bucket, the liquid takes up the space previously held by air and fills up the interstitial spaces. The measure of porosity is called 'fraction void', ε, which is defined (Kay, 1963) as

$$\varepsilon = \frac{\text{empty spaces volume}}{\text{total volume of the porous body}}$$

A porous medium may be composed of a connected structure of interstices, where the spaces are intercommunicating. It can therefore be filled up by a fluid, a phenomenon called invasion. Conversely, the interstices may form a non-connected structure (e.g. insulating foam, made up of many gas cells or bubbles entrained in a solid polymer matrix). The first case concerns us most, even if ground coffee does not have the same porosity structure of other typical percolating media with connected pores, such as tangles of fibres (the most common example is felt) or sponges (where solid fibres have grown together to form a single body). Actually, ground coffee behaves as a bed of particles, namely an aggregate of solid granules in mutual contact under the force of their own weight, which do not penetrate each other owing to their solid state. Such a bed may be unstable and prone to rearrangement by vibration, or remain stable under gentle vibrations but get reshuffled under stronger forces (e.g. overturning). Particles may also stick together, forming a mass having its own shape and capable of being handled as one whole solid (e.g. sintered metals, made up of metal pellets fused together by mutual friction).

Apart from an ideal case, all particles are formed by materials that are to a certain extent elastic or plastic; therefore, particles are subject to the force of their own weight and to other external forces causing reversible or irreversible deformations. Particles form beds that can be compacted by compression, with reduction of the total apparent volume of the bed. Owing to this compaction and to a certain degree of adhesion of the external particle layer, beds can take up a shape stable enough to withstand handling (with due care in the case of coffee).

This last description provides an adequate definition of a ground coffee bed, also known as a 'cake', ready to be extracted in the espresso machine. By contrast, ground coffee used in other brewing methods, as for instance drip filter coffee, lets percolation occur through particle beds that are much less tightly aggregated than espresso cakes.

7.2 PHYSICAL AND CHEMICAL CHARACTERIZATION OF THE PERCOLATION PROCESS

7.2.1 Physical description

From a physical point of view, preparing a cup of espresso coffee means percolating a definite quantity of liquid through a compact bed of ground coffee particles, the cake. The latter is usually contained in an impermeable metal holder, generally made of stainless steel for its inalterability and easy cleaning. The holder is popularly called a 'filter', in spite of the inappropriateness of this term as pointed out above: it is approximately cylindrical in shape, with a perforated bottom. Percolation takes place when liquid penetrates the interstices between the particles at a certain spatial velocity, displacing the pre-existing air. The concept of spatial velocity, borrowed from chemical reactor theory, refers to the flow of liquid per unit section of the bed. When dealing with espresso, the liquid involved in this process is water, heated to a temperature just above 90 °C by means of direct or indirect heating.

The driving force that generates the flow is produced by the pressure drop undergone by the hot water that enters the cake with positive hydrostatic pressure. Pressure, a well-known notion of general physics, defines a force exerted perpendicularly over a unit of area. In the domain of fluid dynamics, pressure refers to an intensive-type property, defined instant by instant in any point of space taken up by a fluid; it can be represented by means of a scalar field. The existence of a pressure field within a fluid yields a potential energy (Bernoulli's piezometric energy) that can be easily transformed into kinetic energy, thus giving speed to elementary masses of fluid.

Pressure energy is partly dissipated as heat by the friction generated when water flows through the particles, and partly gradually transformed into kinetic energy of the beverage which flows into the cup, according to the general equation (Bullo and Illy, 1963):

$$P_e = P_r + P_c + P_a$$

where:

P_e = driving pressure (water pressure at the filter inlet)
P_r = pressure drop from bed resistance
P_c = kinetic pressure drop
P_a = atmospheric pressure

The average water velocity must remain constant through the cake in order to provide a continuous jet, while pressure decreases along the percolation axis until it reaches atmospheric level. At this point, the 'liquor' obtained by percolation flows into the cup underneath by gravitation.

It is beyond dispute that every percolation method requires some pressure (at least several millimetres of water column height) to overcome the head loss required for obtaining a flow through ground coffee beds; nevertheless, a higher relative pressure (one order of magnitude, or two) is needed when dealing with finer grinds. Even higher pressure, up to 10 atm (corresponding to 103 327 mmH$_2$O: four orders of magnitude higher), must be used when trying to pass water through a compacted cake of finely ground coffee, as happens in espresso brewing. The energy expended during this kind of operation produces interesting effects, like driving of micron-size solid particles and oil droplets into the cup. This may influence the beverage's properties dramatically.

Hydrostatic pressure is generally generated by a device constituted of a rotary or reciprocating pump: some set-ups nevertheless utilize compressed air or steam pressure to pressurize a water tank, or exploit the centrifugal force of a rotary system to push water through the cake. Even a head formed by a water column (tens of metres high, anyway!) might serve the task. A description of technological devices that have been developed in nearly a century, since the birth of espresso coffee, to produce this characteristic beverage will be discussed in 7.4 below. However, it must be stressed that an accurate design and a precise mechanical manufacture of espresso machines are most essential because these machines, even the more economical models for home usage, are far from being simple (see later, Figure 7.6).

7.2.2 Physico-chemical description

The intimate contact of water with roasted coffee is the cardinal requirement for producing a coffee beverage. From a chemical point of view, percolation induces two processes: the extraction of the water-soluble substances and the emulsification of the insoluble coffee oils. Both phenomena together are responsible for the considerable differences between pure water and the espresso brew (Table 7.1).

Solid-liquid extraction produces the removal of a soluble fraction – in the form of a solution – from an insoluble permeable solid phase where it belongs. Two steps are always involved in extraction: contact of solid and solvent to effect mass transfer of solubles to the solvent, and separation of

Table 7.1 Espresso brew contrasted with pure water

Colour	Brown-black, due to partially defined pigments (melanoidins)
Refractive index	Increased by the presence of solutes
Surface tension	Reduced by surface-active substances present
pH	Lowered by organic acids and by phosphoric acid
Electric conductivity	Higher, owing to the ions presents
Viscosity	Higher, mainly due to the oils in emulsion
Density	Higher, due to the high solute content

the resulting solution from the residual solid, this latter step is usually done by filtration. The mechanism of the first step is favoured by increased surface per unit volume of solids to be extracted, and decreased radial distances that must be traversed within the solids, both of which are favoured by decreased particle size as it occurs in coffee ground for espresso.

The main dependent variable that can be objectively measured (along with the obviously all-important sensory ones, which are largely based on subjective evaluation) is brewing yield, namely the ratio between the mass of the coffee material that passes into the cup and the total coffee material used (the balance to be disposed as spent grounds). Brewing temperature exerts a preponderant role on yield variability, which may range between 20% and 30%, and darker roasting procures higher yields. It is worth mentioning that yield is a different concept from beverage concentration – or strength, measured in grams of extracted matter per litre of beverage – and can be set independently, if a different brewing formula (coffee/water ratio) is chosen.

The general principle ruling extraction is Fick's law of diffusion, which gives the quantity s of solute diffusing from a solid particle surrounded by a liquid. It can be written in the following form (Barbanti and Nicoli, 1996):

$$s = k \ T/\eta \ A/x \ (C-c)\theta$$

where:

k = constant, depending on molecular factors
T = absolute temperature
η = viscosity of the liquid, f (T)
A = layer cross section around the particle
x = layer depth
C = solute concentration in the solid
c = solute concentration in the liquid
θ = contact time

In espresso percolation, the short extraction time does not allow equilibrium to be reached. This calls for consideration of reaction kinetics too, especially the branch dealing with the diffusion of water entering, and of molecules of various solutes leaving, a cellular structure. Optical and electronic microscopic examinations of the cell structure of coffee show the presence of soluble substances both outside and inside the cell wall already in the green bean (Dentan, 1977), and no evidence of pores or channels running through the cell wall has yet been produced.

Therefore, it may be assumed that percolation provokes an extraction mainly by washing out the outer surface of the particle, rather than by diffusing from the interior of the particle to the solution.

A naive way of thinking could suggest that the more one extracts from one's purchased material, the better: nothing is less true, quality-wise. Since the many chemical species identified in roasted coffee (more than 1800 so far) exhibit different extraction rates (Lee *et al.*, 1992), it is logical that extraction yield relates to sensory quality too, leading to the two celebrated converse brewing errors: under- and over-extraction. The former is due to lower yields (deriving from whatsoever reason, like low water temperature, short contact time or too coarse grind), when the most soluble substances – the acidic and sweet ones – are predominantly driven into the cup, producing the so-called under-extracted beverage. The latter happens when higher yields are obtained, by forcing more substances – the bitter and astringent ones – into the cup. It is therefore evident that brewing yield should be adjusted according to personal taste, in agreement with the nature of the beans used (origin, blend, roasting).

Only few authors have tackled so far the study of coffee extraction kinetics (Voilley and Clo, 1984; Nicoli *et al.*, 1987; Cammenga *et al.*, 1997). The only substance that has been studied to a considerable extent is caffeine, the constituent producing the most remarkable physiological effect (Spiro and Selwood, 1984; Spiro and Page, 1984; Spiro and Hunter, 1985; Zanoni *et al.*, 1991; Spiro, 1993). Caffeine exhibits high solubility in hot water (Macrae, 1985; Cammenga and Eligehausen, 1993); nevertheless, its quantitative extraction may be obtained only by coffee preparation methods that permit a long contact period between water and ground coffee (Peters, 1991). In espresso percolation, caffeine extraction yield is usually within the range 75–85% (Petracco, 1989); this relatively low figure may be explained by the fact that caffeine is extracted from within the cell, by diffusion, and not by simple washout kinetics.

The other important chemical phenomenon occurring during coffee percolation is the emulsification in water of non-soluble matter, mostly composed of lipids, present in coffee in quantities ranging from 8 to 16% dry basis (see 4.3.2.3), that are mainly contained inside the cells of roasted coffee. By means of the energy transferred to the bed by the water pressure drop, the lipid fraction in broken cells may easily form an emulsion. Optical microscopy of emulsified lipids in an espresso brew (see Figure 8.2) (Heathcock, 1988) shows that they are present as small droplets varying from a minimum of $0.5\,\mu m$ to a maximum of $10\,\mu m$ in diameter.

As will be seen in 8.1.2, the stable emulsion formed during espresso percolation does not coalesce spontaneously, even after weeks of rest. None of the artifices commonly used to break up emulsions (filtration, dilution, addition of electrolytes, etc.) overcomes the reluctance of espresso coffee to be separated into two distinct liquid phases. The small diameter of droplets is a factor that helps in precluding coalescence; moreover, the density of the disperse phase is not very different from that of the solution – this prevents droplets from buoying up and concentrating on the free surface. The key factor of emulsion's stability, whose importance in the sensory context will be discussed in 9.5, is constituted by a layer of surface-active substances surrounding each droplet and thus generating the chemical stability (Navarini et al., 2004).

7.3 MODELLING OF THE PERCOLATION PROCESS

The first attempt to set up a mathematical model for the espresso brewing process dates back to the early 1960s (Bullo and Illy, 1963), when the key role of the pressure drop due to the hydraulic resistance of the ground coffee cake was first recognized. As a starting approximation, it can be assumed that the percolation process can be hydraulically defined by an equation associating five variables, sufficient to characterize the process in its macroscopic physical aspect:

$$V = \int_0^t \frac{\Delta p(t)}{R[t, T(t)]} \, dt$$

where:

    V = volume of beverage in the cup
    T = temperature inside the coffee cake
    t = percolation time
    Δp = pressure difference above and under the cake
    R = hydraulic resistance of the cake.

This choice of variables is not the only possible one: for instance, the volume of beverage in the cup, V, might be replaced by its derivative in relation to time, the instantaneous flow Q. Moreover, the hydraulic resistance R depends on sub-variables as the dose and particle size of the ground coffee, as well as the compacting of the bed. Anyway, five variables are sufficient to characterize the process in its macroscopic physical aspect, in agreement with the traditional rule of the *barista* (the espresso bartender):

To prepare an espresso cup correctly, it is necessary to set the right temperature and pressure, then to adjust the hydraulic resistance (by grinding and compacting), until the right volume of beverage is obtained at the right time.

To model the phenomenon theoretically, that is to forecast the value of a dependent variable once the values of the independent ones have been set, a semi-empirical approach uses flow, pressure and temperature continual sensors to obtain flow curves as a function of time (Petracco and Suggi Liverani, 1993). Classical hydraulics would apply Darcy's law – 'The loss of pressure in a pipe is proportional to the flow rate of the liquid through it' – or:

$$\Delta p = RQ$$

where:

$Q = $ flow
$R = $ hydraulic resistance of the system

But two observations that lie in clear contradiction with this equation emerge from the experimental curves (Figure 7.1):

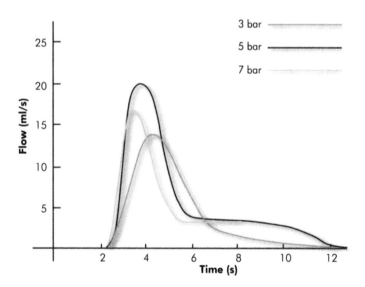

Figure 7.1   Pressure dependence of flow during percolation

- flow is not constant for a constant pressure drop, but after an initial transient peak it decreases in time until it reaches an apparently asymptotic value (dependent on temperature);
- the mean flow is not proportional to the pressure applied, but increases with it up to a certain value, then remains constant or even decreases.

The anomalous behaviour of the pressure is not linked to the extraction reactions, started by high water temperatures not far from 100 °C. It is actually feasible to percolate ground coffee with as cold water as possible (in practice not lower than 4 °C to prevent freezing) so as to minimize any influence of solubilization on the physical phenomenon. Data show that the asymptotic non-monotonic behaviour persists also under such conditions, even if higher flows are needed (Figure 7.2).

However, the flow's quasi-exponential decay suggests that the bed of ground coffee particles should present a time-dependent geometry (Baldini, 1992). An indication that some form of modification does occur inside the bed derives from electron microscopy of the bed structure (Figure 7.3), which is made up of coarse particles containing fragments of cell walls, called fines. The latter are assumed to migrate to and eventually to concentrate at the bottom of the bed. Indirect confirmation comes from

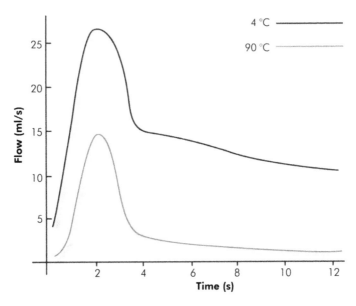

Figure 7.2   Temperature dependence of flow during percolation

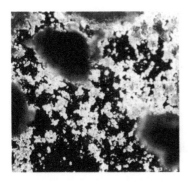 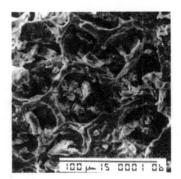

Figure 7.3   Microstructure of ground coffee particles

tests where the direction of percolation was reversed, using an extraction chamber that could be turned upside-down.

The direct flow behaves normally, remaining unchanged even when the pump is stopped for a brief moment. Surprisingly, flow increases when the percolation chamber is inverted (Figure 7.4). This effect can be explained by assuming that the finest particles, which concentrate in a lower section of the cake causing an increase in the hydraulic resistance, now counter-migrate, with an increase in hydraulic conductivity in the opposite direction, until the system reaches a new steady state.

Attempts have been made (Baldini and Petracco, 1993) to formulate a law providing an explanation for this experimental behaviour, but finding

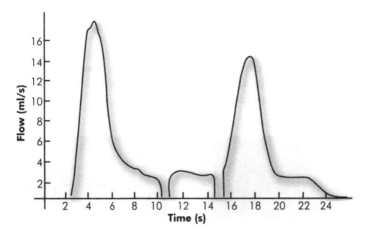

Figure 7.4   Direct/inverse discharge curve

an exact analytical solution to the consequent system of differential equations still remains a difficult, if not impossible, task. Computer-aided calculation of the numerical solution to these equations becomes feasible only by fixing the boundary conditions; the coefficients, however, must always be determined experimentally. The theory has yet to be expanded to include not only the physical, but also the chemical aspect of percolation and possibly to allow also forecasting sensory features like foam, body and flavour.

Another method, alternative to differential equations, is offered by 'molecular ontology', a branch of 'naive physics' that studies physical phenomena starting from experimental and qualitative observation via a heuristic approach (Bandini and Cattaneo, 1988). A model is constructed by breaking down a physical phenomenon into its constituent elements (in the present case, coffee particles and elemental volumes of water), and by describing their local interactions, whose synergism builds up the system's overall behaviour.

Simulations of espresso percolation obtained by powerful computers (Figure 7.5) demonstrate that particles with a given possibility of movement tend to migrate and to accumulate in a specific critical section simulating a metallic filter (Bandini et al., 1997). This technique shows micro-vortexes and allows the trajectories of the liquid and the particles to be followed (Cappuccio and Suggi Liverani, 1999).

7.4   THE ESPRESSO MACHINE

As can be seen from what has been said so far, preparing an espresso coffee requires equipment able to:

■   keep a portion of ground coffee in the form of a compact cake inside a container that allows the brew to drip out while holding back the particles (spent grounds);

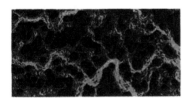 

Figure 7.5   Stages of cake percolation simulation

■ impart to the water a temperature close to 90 °C and a pressure customarily fixed at 9 relative atmospheres.

Such equipment is called an 'espresso coffee machine', often shortened into 'espresso machine' or, in Italy, 'coffee machine'. Coffee machines are available on the market in many models, ranging from compact light ones for occasional home usage to the professional computerized units. The main parts of the machine are the pump, the heat exchanger and the extraction chamber, as sketched in Figure 7.6.

The origin of espresso coffee machines dates back to the beginning of the 1900s, when the pioneer Bezzera patented a machine in which pressure was applied to water by simple boiling in an autoclave. The drawback of that design is that the attainment of the desired pressure levels (around 1.5 atm) required a very high temperature, well above 100 °C. These conditions cause the extraction of substances normally insoluble, imparting an unpalatable bitter flavour to the brew. With the knowledge of the espresso technique available now, a brew obtained with such a machine could no longer be called an espresso, due to the much lower pressure applied; this principle is still widely employed today in Italian home coffee brewing by the commonly used inexpensive machine called a 'Moka' (Petracco, 2001).

This drawback can be avoided by separating the generation of pressure from that of temperature. After the compressed-air machine, whose technology was probably ahead of its time (Illy, 1935), a lever-system was developed in 1945 by Achille Gaggia. It consisted in a spring which, wound up by the muscular force of the *barista*, released energy – via a

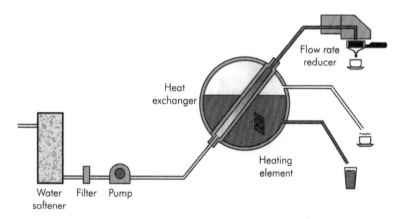

Figure 7.6 Schematic flow diagram of an espresso machine

piston – to the water contained in a pressurization chamber. This system is still sometimes favoured in traditionalist coffee houses; its performance depends on the operator's manual ability in properly dosing and applying force and time, rendering espresso preparation an 'art' with uncertain results. The system has the advantage of allowing the manufacture of machines that do not require electric energy, as heating can be accomplished by a gas burner, and no pump is needed to pressurize the water.

7.4.1   The pump

The spreading of electricity allowed the automation of espresso machines, mainly through the introduction of a pump as the source of pressure. Professional pumping devices consist in an impeller with fins, driven by an electric motor, and in an adjustable bypass valve. The purpose of the latter is to withdraw part of the water delivered by the rotor at the discharge outlet of the pump, and to recycle it to the suction inlet: this power dissipation is legitimate because it enables the pump head (back pressure) to remain constant, independent from the variability of the hydraulic resistance (coffee cake) downstream. Also, the repeatability of percolation has considerably benefited from the use of such pumps, because an electric motor is far less prone to imprecise behaviour than muscular force.

Machines for home usage, where neither the encumbrance nor the cost of a centrifugal pump is justified, utilize preferably a vibration pump. This is a volumetric pump where a small piston is set in a vibrating reciprocating motion by an electromagnet coupled with a return spring. These pumps exhibit a very high head (up to 30 atm), heavily depending on the hydraulic resistance downstream. As a consequence, the quality of the beverage in the cup is strongly influenced by the amount of ground coffee used, and by the degree of grinding and of compacting. A less common alternative, used to generate water pressure in commercial domestic machines, is centrifugal force.

7.4.2   The heat exchanger

The invention of waterproof electric resistors, which permitted the manufacture of direct heat exchangers, was another major innovation. In modern espresso machines (see Figure 7.6), the heating of water is based on a low-pressure boiler of a capacity of 3–10 litres, heated by a waterproof electric resistor immersed in it (seldom by a gas burner). This reservoir supplies hot water for tea and other infusions, and saturated steam for *cappuccino* steamed milk. In addition, it works as heat storage to warm up the espresso percolation water, which runs through a completely different circuit. This water comes directly from the tap and is driven by the pump through a coil immersed inside the boiler, where it reaches dynamically the desired temperature (many degrees lower than the temperature in the boiler itself). This system, particularly cumbersome and expensive, is applied in all professional machines and only in some of the classiest home-use machines, where it permits an ample production of steam for *cappuccino* preparation.

In smaller machines, the boiler is replaced by a direct heat exchanger, consisting of a metal block with an enclosed electrical resistance fed by a bimetallic thermostat or, in more modern models, by an electronic power supply driven by a thermocouple sensor. Cold water is forced by the pump to pass through a sinuous circuit inside the block, and then passes directly through the coffee cake beneath. In professional machines, conversely, temperature is regulated by varying the boiler pressure (see also 7.5.6).

7.4.3   The extraction chamber

The extraction chamber consists in an upper block, on which is inserted (like a bayonet) a filter-holder cup. The ground coffee portion is poured and compacted into this filter. Since the block protrudes from the body of the machine, it is recommended to heat it by means of a small calibrated flow of boiler water passing through a built-in tube. When the pump is switched off, water is no longer able to flow through the cake and remains under pressure within the chamber, due to the increased bed resistance caused by hydration of the spent coffee grounds. If not overcome, this residual pressure would cause troublesome splashes when removing the filter-holder from its block. Therefore, the hydraulic circuit of all espresso machines is equipped with some kind of device that allows depressurization of the extraction chamber at the end of percolation, and this is usually achieved by calibrated check valves, or servo-controlled valves synchronized with the pump power switch.

7.4.4   Water pre-treatment

Another all-important piece of equipment is the pre-treatment unit for the water, as the latter usually comes directly from the mains water supply system with substantial calcium and magnesium content. Any system of heating hard water must forestall the formation of alkaline-earth carbonate deposits on heated surfaces, and the consequent harmful decline in heat exchange efficiency (Cammenga and Zielasko, 1997). An operation most commonly implemented to prevent malfunctioning in professional machines is water softening. It consists in passing the water through a bed of ion-exchange resins, which capture calcium ions present in the water and replace them with sodium ions, whose salts are completely soluble. Water softeners need regular regeneration by washing with sodium chloride: this is a vital operation for ensuring efficient water conditioning and a long service life of the espresso machine. Finally, the effect of water acidity and hardness on espresso cake may affect extraction (Fond, 1995). More details about water influence on espresso will be found in 7.5.5 below.

7.5 PARAMETERS INFLUENCING PERCOLATION

The three main features that lead to the definition of espresso are:

■ 'on the spur of the moment': the brew must be prepared just before serving it;
■ 'in a short time': brewing must be fast;
■ 'under pressure': mechanical energy must be spent within the coffee cake.

Whereas the feature 'on the spur of the moment' (or, more colourfully, 'extemporaneously') refers not so much to the coffee brewing method as to the 'espresso lifestyle', the other two features are strictly related to percolation.

These general features are necessary for the identification of an espresso. Nevertheless, they are insufficient to establish whether that cup has been prepared correctly, that is, whether it falls within the interval of the n-dimensional space of the fundamental preparation variables that allows the consumer to expect a sensory experience that is up to the well-established fame of espresso. In this interval, the ideal espresso is represented by a point, and probably not always by the same point nor by one single point. Reaching this point is an art, and is so related to the subjective sphere of the individual that it cannot be completely clarified in a technical book. The present work can, however, strive to fix the boundaries of the n-dimensional interval. They could be imagined as the limits of an 'excellent espresso', with each one being a necessary, but not a sufficient condition (Swartz, 1997).

The first difficulty encountered is establishing the number of dimensions needed to describe an 'excellent espresso'. Certainly, they are more than those five variables mentioned in 7.3, because each of the latter is the aggregation of several sub-variables. Moreover, some variables have not as yet been mentioned, inasmuch as, if we wished to tackle every aspect of coffee preparation, a myriad of variables should be considered (Illy, 2002). For practical reasons, we shall limit ourselves to discussing the following ten:

- ground coffee portion
- particle size distribution
- cake porosity
- cake shape
- water quality
- temperature
- pressure
- percolation time
- cake moistening
- machine cleanliness

The first five variables refer to the raw materials: four of these are related to the product to be extracted (the ground coffee cake) and can be seen as factors influencing its hydraulic resistance; an additional one is dedicated to the second main raw material of espresso coffee, water, which is often neglected because of its inexpensiveness. The last five variables refer to the machine. Three of them, temperature, pressure and percolation time, are, together with hydraulic resistance, the primary independent variables of the percolation equation (see 7.3). The last two, cake moistening methods and machine cleanliness, are also important, yet neglected, factors.

7.5.1 Ground coffee portion

A 'portion' is the weight of roast and ground coffee required for preparing one cup. The acceptable extreme range of values for a portion may be from a minimum of 5 g to a maximum of 8 g. Minimum values are sometimes used in coffee houses that wish to 'save' on coffee: these values are only barely tolerable when dark roasted coffee is used, for it has a higher content of soluble substances. Maximum values are often encountered in home-brewed coffee, as the host inclines to overdose the serving in order to offer the guest the best possible coffee cup. This practice is risky, however, because, as will be seen in 7.5.4, an excessive amount of ground coffee does not permit sufficient expansion during cake wetting. This causes over-compacting (deriving from the imbibition forces), which disturbs percolation and may cause the deposit of solids into the cup.

In professional practice, a double portion is generally used because coffee-bar espresso machines are designed to meet consumer demand in peak hours. This need is fulfilled in two ways: the simultaneous use of more than one percolation head (up to five, or most often three per machine) or by preparing two cups per head at the same time. The social role of an espresso, particularly in Italy, is such that a person wishing to enjoy a cup of espresso hardly ever enters a coffee house alone, therefore espresso machines are usually optimized for a double portion, equal to 13 g of ground coffee. Accordingly, the optimal portion used for one cup (namely the single serving asked for by an individual customer) is of 6.5 g in the case of arabica blends, something less for robusta blends because of higher extraction yield. Also roasting degree influences yield, up to a point: therefore darker roasts allow for slightly reduced dosages.

Unlike in quality-control laboratories, where the portion is determined by weighing on analytical balances (with a precision of 0.1 g), in espresso bars it is measured volumetrically using the so-called 'grinder-doser'. It is an additional appliance fitted on the professional grinding machine (with a capacity of at least 2 kg of ground coffee per hour), consisting in a device with rotary paddles into which ground coffee powder drops from the grinding tools. The height of the pallets can be adjusted, so that the portion released at each 'stroke' corresponds to a standard weight. At home, the easiest way to determine a portion is by using a measuring spoon, which allows the right volume of coffee powder to be drawn from the package, or the right volume of beans to be ground each time. The volume-to-weight equivalence is not absolute, as it depends on the density of the powder, which in turn depends on the degree of compacting in the package; variability is even larger for whole beans. In practical terms, however, the error is negligible.

7.5.2 Particle size distribution

Undoubtedly, coffee granulometry is a crucial factor for espresso, but it is also one of the least investigated, and probably least understood, variables of percolation. Particle size refers to the measurement of the geometric characteristics of a population of particles constituting a powder, and to its representation. The most studied characteristic is size distribution, while particle shape is also a matter of great interest. Common ways of measuring it have been described in 5.3.

Size distribution of a powder sample may be represented by a histogram or, when many measure points are used, by a chart showing a grading curve. The curve is plotted on a diagram (see Figure 5.4) in which the axis of the abscissa, usually on a logarithmic scale, carries the 'equivalent diameter' variable: this variable can be defined differently according to the technique adopted, but it is generally linked with particle size. The percentages of particles belonging to each size group are plotted on the axis of the ordinates in differential (frequency chart) or integral form (cumulative curve). The percentage may refer to the number of particles, to their volume or to their surface, or even to their mass or weight. By calculation, it is possible to pass from one form of chart to the other, once the relationships between particle shape and density factors have been determined. The comparison of ground coffee batches obtained from different grinding equipment shows differences that characterize each ground coffee by its own grading curve, as fingerprints identify a person.

A fundamental parameter to obtain a correct percolation, the so-called 'fineness', refers to the average particle diameter and depends on the adjustment of the grinder. The most suitable particle size distribution for espresso percolation is not, however, a single-size grading; consequently, neither the average diameter nor the mode of the curve are sufficient to describe the correct size distribution.

An empirical compromise reached by grinder manufacturers resorts to an intermediate distribution, which can be modelled either by power-law or log-normal distribution (Vicsek, 1991), occasionally with a typical bimodality or even trimodality. Such a complex characteristic of particle size is believed to produce a double effect: on the one hand, it forms a coarse fixed structure, which allows the correct flow through the cake; on the other hand, it forms a large quantity of fines of high specific surface, which permit the extraction of a large amount of soluble and emulsifiable material. Furthermore, the coffee cake appears to behave as a self-filtering structure, retaining the fines that have been displaced by the flow.

7.5.3 Cake porosity

Theoretically, the most suitable particle size distribution is the one that exposes the maximum surface area to the action of water. However, such a distribution would not permit any porosity of the cake capable of maintaining the hydraulic resistance within limits that allow a correct flow. While in aggregations of particles having the same shape the fraction void (see 7.1.2) does not vary with different powder fineness, it can be profoundly modified when size distribution is changed (for example, by transforming a monomodal into a bimodal distribution: as is the case when sand is added to gravel in order to render it less permeable). The same statement of bimodality holds true for the shape of particles, because irregular particles (such as fines made up of fragments of fractured cells) may easily sneak through regular-shaped particles and increase specific surface.

Another factor affecting cake porosity is compacting. The bed of loose ground coffee poured into the filter by dropping must be compacted, usually by hand. The compacting force may vary from a few kilos, for a vertical upward thrust (a tamping plate is usually built-in with the grinder), to approximately 20 kgf for downward compacting by a hand tool, the tamper: in the latter case, the force of the thrust may be increased by part of the operator's own weight.

Compacting influences percolation even when exerted weakly, that is, the difference in hydraulic resistance between a loose bed and a weakly compacted one is large, but there is only a minor variation between weakly or more forcefully compacted beds. A plausible explanation for this phenomenon could be the formation of small localized channels in the upper layer of a loose bed which, when struck by a jet of boiling water, does not resist to impacting water as firmly as a compacted cake does.

Compacting forces much larger than those just mentioned (40 kgf or more) are employed in the industrial production of a patented system of single portions of ground coffee, sealed between two layers of filter paper (Illy, 1982). These portions (also called pods or servings) are percolated in especially designed semi-automatic coffee machines aimed at the preparation of an espresso at home, where they free the operator from manual dosing of ground coffee and from most possible errors in preparation.

7.5.4 Cake shape

The shape of the coffee cake, determined by the filter's shape, is important when considering peculiar forms that deviate from the traditional ones, universally accepted by current coffee machines. The classical shape of the double dose is cylindrical, around 12 mm in height by 60 mm in width, the exact dimensions depending on the machine. If this height-to-diameter ratio of about 0.2 is lowered, very fine grounds are required for coffee to percolate within a standard time; this choice would degrade the reproducibility of percolation, because of the creation of localized channels by water. A larger height-to-diameter ratio would give the cake the shape of a column; in this case, optimal extraction would demand an excessively high pressure (as in liquid-phase chromatographic columns). Conversely, for normal pressure values the grind would be so coarse as not to offer a sufficient number of fractured cells, resulting in a low extraction. Consequently, only minor variations from the optimal form of the established filter can be recommended.

An important phenomenon to be taken into account when considering the shape of the cake is the expansion of the bed due to the swelling of coffee particles when wetted. This is a phenomenon of colloidal imbibition, where a chemical hydration of the organic materials (such as the polysaccharides) takes place. The water involved in wetting the coffee grounds is bound, therefore, in an aggregation state differing not only from the ordinary liquid state, but also from that of interstitial water, which is subject to the laws of capillary phenomena. This is shown by the fact that imbibition water cannot be eliminated by oven heating at 100 °C for a protracted length of time, but only by raising the oven temperature by several degrees. During expansion, wet coffee grounds exert a pressure comparable to that of the wooden wedge used in the past to cleave marble blocks, whose force of expansion, once it is driven into the block and wetted, prevails over the cohesive forces of the stone and splits the block into slabs. On account of this behaviour, an empty space is left over the ground coffee cake inside the extraction chamber. The actual expansion of the cake varies with blend, roasting degree and dose, and determines the exact headspace (around 5–6 mm) needed to prevent over-compacting.

A final shape characteristic worth mentioning is the filter perforation. In 7.2.1, the term filter was mentioned as inappropriate for indicating the perforated plate that constitutes the base of the cup containing the ground coffee bed. Actually, this plate does not act as a filter but as a supporting structure for the compacted coffee cake. The diameter of the holes (usually 0.25 mm) is designed to prevent the coarse particles that break away from the lower layer of the cake from passing into the cup. In order to prevent solidified caramelized residue from obstructing the holes, they are triangular in vertical cross-section: the coffee powder comes into contact with the smaller sized opening of the hole, beneath which there is an open truncated-conical cavity. The filter contribution to the overall hydraulic resistance is small, and becomes negligible when the holes get worn.

7.5.5 Water quality

The second most important ingredient of espresso – after the coffee itself – is water, constituting more than 95% of any coffee beverage. However self-evident this statement may seem, the role of water in coffee preparation must be taken into due consideration. It would be otherwise senseless to devote utmost care and attention to only one of the two ingredients, while neglecting the other. Yet, this regularly occurs, partly because the choice of the proper water is left to the end user and partly because that person (espresso bartender, home user) has an extremely limited influence over the choice of the water. In fact, very few people brew coffee using mineral water, whose characteristics are stated on the label and are supposed to be constant. Most consumers trust tap water quality, which they rarely purify in any way.

It must be stressed that the quality of treated water coming from public waterworks is perfectly suitable for human consumption (WHO, 1993) and further purification or disinfecting treatments are unnecessary. However, treatment of drinking water may turn out to be beneficial in two critical cases; one is linked to sensory perception, and the other concerns the use of the espresso machine.

In the first case, the beverage must be free from any unpleasant foreign flavours left over in the water by generalized disinfecting treatments, such as chlorination. Salts that release free chlorine (hypochlorites) lend the beverage their peculiar taste, which is perceptible when the initial concentration of Cl_2 exceeds 0.5 mg/l. A system most commonly used for removing chlorine taste involves passing the liquid through a bed of granules of activated carbon, which has the ability to adsorb many odour-bearing substances. A drawback in the use of activated carbon is that saturation cannot be easily detected, so that it is necessary to regenerate or replace the bed regularly.

The second critical case is related to water hardness, namely to the water's calcium and magnesium content. On heating, these cations produce insoluble salts (mainly carbonates, but also sulphates and silicates). These salts tend to precipitate in the form of compact plaques (scaling), particularly on the heated surfaces. Such scales form a coating on the surfaces, which detrimentally affect the heat exchange coefficient. This may lead to serious technical troubles with the espresso machine, in particular causing a reduction in heat transfer effectiveness with a possible consequent widening of the fluctiations in temperature (too hot/too cold) and risk of failure in electrical resistance due to overheating.

Potable water may be nearly always defined as hard, that is, having a total hardness that exceeds 15 French degrees, corresponding to a concentration of 150 mg/l of $CaCO_3$. In some waterworks in Italy, for instance, water hardness may reach a peak of 37 French degrees. This is not a reason for health concern, because the hardness for drinking water recommended by Italian law (DPR 236, 24/5/88) ranges from 15 to 50 French degrees. It is also worth remembering that mineral waters may exceed 100 French degrees.

Direct use of hard mains supply water in both professional and home espresso machines produces unacceptable deposits in 1–3 months. Manufacturers of home-use machines advise removing these deposits by periodic washing (see 7.5.10) with weak acid solutions: citric or formic acid, even vinegar, will do. In professional machines, scaling is prevented by fitting the machine upstream of a cleanup system specifically active on calcium and magnesium ions, but nearly as active on other metal ions. Two types of treatments are commonly available: softeners or demineralizers.

A softener consists of a bed of ion exchange resins that traps calcium ions and replaces them with sodium ions (which do not affect hardness, and do not form deposits) to a hardness depending on the saturation of the resin. Optimal water hardness to avoid scales and have good percolation is around 9 French degrees, as recommended by manufacturers. The resins must be periodically regenerated by washing with NaCl so as to restore their effectiveness. In the case of treatment of very hard water, the amount of sodium released (in combination with other ionic species present) may become perceptible to taste and increase water pH on heating. Two troubles may be encountered with softeners:

- A functional problem, inasmuch as localized channels form in the resin bed after some time, diminishing the effect of ion exchange. This can be avoided by stirring the resins from time to time.
- A hygiene risk, namely the danger of a possible proliferation of microorganisms in the bed. This can be averted by using beds containing a bacteriostatic additive.

Demineralizers eliminate Ca^{++} ions without introducing any foreign ions. Ion exchange of the demineralizer employs either separate beds of resins, fitted in series, or one single bed in which resins have been mixed but that can be separated by floating. One resin is of the so-called cationic type and captures Ca^{++} and Mg^{++} by replacing them with H^+; the other resin is of anionic type and captures the negative ions (Cl^-, HCO_3^-, SO_4^{--} and others) by replacing them with hydroxyls, which neutralize hydrogen ions forming H_2O. These types of resins must be regenerated separately with HCl and with NaOH, respectively. A more modern type of demineralizer is based on the principle of reverse osmosis. Pressure energy is employed to filter water through a semi-permeable membrane capable of retaining ions, which can be discarded as concentrated solution, while letting the smaller H_2O molecules through. With the use of a suitable membrane or of a graded bypass, water can be obtained at the desired degree of hardness.

The main influence of hardness on beverage quality is an increase in percolation time, compared to soft water below 2 French degrees. However, only minor differences occur when water hardness is raised above 8 French degrees to typical mains water values (Rivetti *et al.*, 2001). Hence, it is not recommended to soften or demineralize water below this value, otherwise a coarser grind should be used to compensate for the increase in percolation time. Water pH also affects percolation time even if in a less predictable way, probably in association with hardness. The influence of hardness and pH on percolation might be explained by the change, due to Ca and H ions, in foam-producing and emulsifying properties of the natural surfactants present in roasted coffee (see 8.1.2.4). It is unclear, though, whether the increase in percolation time is to be ascribed to increased foam and emulsion viscosity, with respect to pure water, or to the interstitial formation and precipitation of insoluble calcareous soaps or other pH-dependent material.

7.5.6 Temperature

Quantitatively, this relationship between the rate a reaction proceeds and its temperature is determined by the Arrhenius equation:

$$k = A \, e^{-E/RT}$$

where:

k = reaction rate
A = constant
E = activation energy
R = gas constant (0.082 litre atm/mole °K)
T = temperature

Extraction of coffee, like any dissolution reaction of a chemical species in water, depends on the constant of equilibrium and on the rate. Higher temperatures are matched by increased percentage of extracted material. It is disputable whether the additional fraction of substances extracted at higher temperature improves the taste of the brew, or enhances any desirable property. The appeal of an espresso prepared under optimal hydraulic conditions is known to increase with temperature, up to a maximum at 92 °C. The extraction of bitter and astringent substances at higher temperatures reduces the overall sensory appreciation.

Unfortunately, in practice, temperature cannot be easily singled out from the other percolation variables, because hydraulic resistance of the cake depends on extraction temperature. For instance, reference has been made in 7.3 to the fact that cold water flows are larger than hot water ones, when obtained through the same cakes and under the same conditions. This difference becomes noticeable when trimming the canonical temperature of 92 °C down to 70 °C; the effect is slighter, yet significant, at temperatures around the optimal value (Andueza et al., 2001). This makes temperature regulation one of the most important operations in designing an espresso machine (see 7.4.2).

The control of espresso water temperature in professional machines is established by varying steam pressure in the boiler, where water is heated passing through an immersed coil acting as a heat exchange surface. The temperature of steady water in equilibrium with its own steam is univocally dependent on pressure. Espresso water heated this way has a reasonably constant temperature, thanks to over-dimensioned exchange surfaces.

Home-use machines, when not equipped with a boiler, are fitted with a thermostat controlling the temperature of the metallic block, which acts as a heat exchanger. This thermostat is often built in on the unit, and fixed; in models where it is adjustable its fine-tuning can usually be done using a small screwdriver until the temperature of the water dripping from the empty extraction chamber reaches 90 ± 2 °C.

The temperature of the water is certainly the main thermal factor in percolation, but not the only one. Also important is the cake temperature, induced by both the mobile and the fixed equipment constituting the extraction chamber. In order to homogenize temperature in professional machines, a small quantity of circulating boiler water constantly heats up the fixed equipment (the block). When using home espresso machines, it is recommended to pre-heat the extraction group by letting a small quantity of hot water drip through before inserting the cake. The same operation may also be used to pre-heat the empty cup, filling it up with hot water before percolating the coffee; an excellent espresso is always served in a warm cup. To this purpose, professional machines are equipped with a cup-warming grid.

7.5.7 Pressure

The role of water pressure on percolation is widely acknowledged, but the influence of the method of exerting pressure on the dry cake has not been adequately studied and deserves more attention. Pressure does not remain constant during the process of percolation (as often assumed), but varies as a function of the characteristics of the hydraulic circuit above the cake. This can be verified by continual pressure sensors: the recordings yield a function $p = p(t)$, which exhibits an initial transient rise, increasing more or less slowly. The dry coffee bed still lacks adequate cohesion and is susceptible to resettlement; therefore the initial phase appears to be decisive in attaining the definitive microstructure of the cake. After wetting, ground coffee particles swell up, interpenetrating each other firmly, and percolation can follow its stationary course. This effect may explain why, as already mentioned in 7.3, the flow is not constant during percolation, rather showing a non-monotonic dependence on the average pressure and an asymptotic decay.

The pressure value universally applied in professional espresso preparations, where a centrifugal pump is employed, is 9 relative atmospheres or, technically, 9 kg/cm^2 (equivalent to 10 ata, i.e. absolute technical atmospheres). It is easily adjusted by regulating the bypass valve located downstream of the pump, which, as already mentioned in 7.4.1, can supply a flow exceeding percolation needs.

This value results from a series of 'trial and error' attempts in the early years of espresso machine technology, whose results were measured by consumer satisfaction. It is interesting to point out how an alert espresso operator is in a good position to test any slight variation in coffee preparation procedures, judging immediately from the consumer's face (a genuine sensor) and drawing useful indications for further trials. In the course of time, such a method has established the optimal pressure level that minimizes failure possibilities in obtaining a perfect espresso.

Hydrodynamic experiments (Baldini and Petracco, 1993) demonstrate that the average flow ceases to depend linearly on pressure in the proximity of 9 atm, and that an increase in pressure actually causes a decrease in the average flow despite the already discussed law of Darcy. This suggests that the combined machine/cake system is capable of self-adjusting to overcome pressure variations. Such a phenomenon may lead to cups of coffee that are apparently (hydraulically) similar, though originating from completely different extraction conditions, and thus potentially having different brew qualities.

In home espresso machines, which are driven by a small vibration pump without bypass valve, pressure cannot be adjusted and is settled by the hydraulic resistance of the cake. This makes the grind even more important: it should be adapted to each machine model, because the wrong particle size does not only influence percolation time, but extraction pressure as well.

7.5.8   Percolation time

Percolation time is the variable that most depends on the person who prepares the espresso. Nevertheless, in consumers' perception, extraction time is not an independent variable, but rather the result of the other preset parameters (pressure and temperature, grind and tamping/compacting). The real variable allowed to consumer's choice is actually the volume of brew in the cup, as it is the easiest thing to check: once the cup is full, percolation is stopped without minding the prescribed percolation time.

The volume in the cup is therefore left to the personal taste of the consumer, who chooses it according to the local food tradition and personal eating habits. Optimal results are attained with approximately 25–30 ml volume reached in 25–30 seconds, common in Southern Europe (Italy, Spain, Greece).

Very short extraction times (less than 10 s) would produce a brew that is under-extracted, tasting dull and diluted.

Local tastes may ask for a volume up to 70, 90 or even 120 ml or more, particularly in northern Europe (Germany, Holland, Scandinavia) and in the USA. However respectable this criterion may be, it appears quite insidious because it does not set any rule for the duration of percolation, which can become too long (over 60 s). The resulting brew is then over-extracted, tasting bitter and astringent.

It is therefore recommended that the volume in the cup be considered as a consequence of percolation, and that the extraction time be controlled as the main variable. Long expertise produces a consensus on a percolation time in the optimal range of 25–35 seconds, obtained by commanding other variables, particularly grinding. If a larger brew volume is desired, a better cup is prepared by increasing the particle size of the grounds instead of prolonging extraction time. However, its body and flavour cannot be compared to those of a full-fledged Italian espresso, prepared following the full set of rules (see 8.3).

7.5.9   Cake moistening

The moistening of the ground coffee cake is a term that indicates a specific operation, also called pre-infusion, for which every espresso bartender has their own personal method. It consists of pouring a certain quantity of hot water into the extraction chamber so as to wet the ground coffee, starting the actual percolation a few seconds later. The resulting beverage is claimed to have a richer body and better flavour, which could derive from an improved extraction caused by the swelling of the bed during the pre-infusion period.

Dynamics similar to pre-infusion can also be observed in the common straightforward preparation. Water never enters the extraction chamber instantaneously, not even when the hydraulic circuit consists simply of a tube linking the pump to the chamber; this is because the start-up transient of the pump produces a wave of pressure that reaches the ground coffee surface in a finite time. Moreover, a number of espresso machines are equipped with an empirical device – a choke nozzle (also known by its French term 'gicleur' or jet), which somewhat renders the pre-infusion operation automatic by slowing down the water entering into the chamber. The opening of the device, inserted in series on the hydraulic circuit just above the chamber, is usually about 0.1 mm wide.

7.5.10 Machine cleanliness

Let us conclude the list of variables influencing percolation by dealing with cleanliness – truly irksome and yet also a delight for *baristas* and home users alike. Following the same hygiene rules recommended for other kitchenware, percolation equipment must be kept clean to prevent foreign matter from being dragged into the cup and to eliminate any possibility of proliferation of microorganisms. Nevertheless, it must be stressed that the organoleptic qualities of the brew may also be negatively influenced by excessive cleanliness. An established tradition dictates that coffee pots (and espresso machines too) must be washed with warm running water, avoiding soap or common household detergents to avert the risk of off-flavours from residues.

This rule stems from two considerations, the first of which is easy to explain: the presence of metal filters densely arranged along the water and coffee path may hold back some cleansing agent in spite of the most careful rinsing. The residual detergent is then gradually released, spoiling the flavour of the cups brewed soon after washing. The second reason to forbid using common detergents is subtler and concerns the surface state of the metallic parts in contact with coffee. After the first few percolations, a thin layer of coffee oil and extremely fine solid particles deposits on the metallic parts, acting as a passivator of the metal surface. Excessively clean coffee machines produce a beverage having the typical metallic flavour of new material, probably produced by the catalytic activity of the metal (often copper or its alloys, but also aluminium, steel, chrome) or, possibly, by metal ions released into the water. This activity is apparently inhibited by the formation of a protective layer of coffee matter, which can be washed away if strong or abrasive detergents are used. Mild detergents, devoid of scents or foaming additives, have been developed specifically for coffee machines; their regular use is recommended to prevent the accumulation of sediments in critical sections of the equipment.

A particular aspect of cleanliness regards the removal of calcium deposits formed by hard water, as detailed in 7.5.5. This is a common occurrence in home-use espresso machines, where periodical decalcification is recommended in order to prevent serious damage to the valves and heating resistor. It is performed by re-circulating for a few minutes a water solution of a weak acid (acetic, formic, citric), or any decalcification liquid available on the market, through the looped hydraulic circuit of the machine. Thorough rinsing must always follow washing.

REFERENCES

Andueza S., Maeztu L., De Peña M.P. and Cid C. (2001) Influence of extraction temperature in the final quality of the Colombian coffee cups. *Proc. 19th ASIC Coll.*, CD-ROM.

Baldini G. (1992) Filtrazione non lineare di un fluido attraverso un mezzo poroso deformabile. Thesis, University of Florence.

Baldini G. and Petracco M. (1993) Models for water percolation during the preparation of espresso coffee. 7th Conference of the European Consortium for Mathematics in Industry – Montecatini (Italy), pp. 21–22.

Bandini S. and Cattaneo G. (1988) A Theory for Molecule Structures: the Molecular Ontology Theory. 2nd International Workshop on Qualitative Physics, Paris.

Bandini S., Casati R., Illy E., Simone C., Suggi Liverani F. and Tisato F. (1997) A reaction-diffusion computational model to simulate coffee percolation. *Proc. 17th ASIC Coll.*, pp. 227–234.

Barbanti D. and Nicoli M.C. (1996) Estrazione e stabilità della bevanda caffè: aspetti chimici e tecnologici. *Tecnologie Alimentari*, 1/96, 62–67.

Bear J. and Verrujt A. (1987) *Modelling Groundwater Flow and Pollution*. Dordrecht: Reidel, pp. 27–43.

Bullo T. and Illy, E. (1963) Considérations sur le procédé d'extraction. *Café Cacao Thé* 7 (4), 395–399.

Cammenga H.K., Eggers R., Hinz T., Steer A. and Waldmann C. (1997) Extraction in coffee-processing and brewing. *Proc. 17th ASIC Coll.*, pp. 219–226.

Cammenga H.K. and Eligehausen S. (1993) Solubilities of caffeine, theophylline and theobromine in water, and the density of caffeine solutions. *Proc. 15th ASIC Coll.*, p. 734.

Cammenga H.K. and Zielasko B. (1997) Kinetics and development of boiler scale formation in commercial coffee brewing machines. *Proc. 17th ASIC Coll.*, pp. 284–289.

Cappuccio R. and Suggi Liverani F. (1999) Computer simulation as a tool to model coffee brewing cellular automata for percolation processes; 2D and 3D techniques for fluid-dynamic simulations. *Proc. 18th ASIC Coll.*, pp. 173–178.

Dentan E. (1977) Structure fine du grain de café vert. *Proc. 8th ASIC Coll.*, pp. 59–72.

Fond O. (1995) Effect of water and coffee acidity on extraction. Dynamics of coffee bed compaction in espresso type extraction. *Proc. 16th ASIC Coll.*, pp. 413–421.

Heathcock J. (1988) Espresso microscopy and image analysis. Private communication.

Illy E. (1982) Coffee machine which brews coffee beverages from pods of ground coffee. *European Patent 0 006 175*.

Illy E. (2002) The complexity of coffee. *Sci. Am.* 286 (6), 86–91.

Illy F. (1935) Apparecchio per la rapida ed automatica preparazione dell'infusione di caffè. *Italian Patent 333293/26* December 1935.

Kay J.M. (1963) *Fluid Mechanics and Heat Transfer*. Cambridge: Cambridge University Press, p. 253.

Lee T., Kempthorne R. and Hardy J. (1992) Compositional changes in brewed coffee as a function of brewing time. *J. Food Sci.* 57, 1417–1421.

Macrae R. (1985) Nitrogenous components. In R.J. Clarke and R. Macrae (eds), *Coffee: Volume 1 – Chemistry*. London: Elsevier, p. 117.

Mandelbrot B. (1977) *The Fractal Geometry of Nature*. New York: Freeman & Co, p. 126.

Navarini L., Ferrari M., Suggi Liverani F., Liggieri L. and Ravera F. (2004) Dynamic tensiometric characterization of espresso coffee beverage. *Food Hydrocolloids* 18, 387–393.

Nicoli M.C., Dalla Rosa M. and Lerici C.R. (1987) Caratteristiche chimiche dell'estratto di caffè: Nota 1. Cinetica di estrazione della caffeina e delle sostanze solide. *Industrie Alimentari* 5/87, 467–471.

Peters A. (1991) Brewing makes the difference. *Proc. 14th ASIC Coll.*, pp. 97–106.

Petracco M. (1989) Physico-chemical and structural characterisation of espresso coffee brew. *Proc. 13th ASIC Coll.*, pp. 246–261.

Petracco M. (2001) Beverage preparation: brewing trends for the new millennium. In R.J. Clarke and O.G. Vitzthum (eds), *Coffee: Recent Advances*. Oxford: Blackwell Science, pp. 140–164.

Petracco M. and Suggi Liverani F. (1993) Espresso coffee brewing dynamics: development of mathematical and computational models. *Proc. 15th ASIC Coll.*, pp. 702–711.

Rivetti D., Navarini L., Cappuccio R., Abatangelo A., Petracco M. and Suggi Liverani F. (2001) Effects of water composition and water treatment on espresso coffee percolation. *Proc. 19th ASIC Coll.*, CD-ROM.

Spiro M. (1993) Modelling the acqueous extraction of soluble substances from ground roasted coffee. *J. Sci. Food Agric.* 61, 371–373.

Spiro M. and Hunter J.E. (1985) The kinetics and mechanism of caffeine infusion from coffee: the effect of roasting. *J. Sci. Food Agric.* 36, 871–876.

Spiro M. and Page C.M. (1984) The kinetics and mechanism of caffeine infusion from coffee: hydrodynamic aspect. *J. Sci. Food Agric.* 35, 925–930.

Spiro M. and Selwood R.M. (1984) The kinetics and mechanism of caffeine infusion from coffee: the effect of particle size. *J. Sci. Food Agric.* 35, 915–924.

Stauffer D. (1985) *Introduction to Percolation Theory.* London: Taylor & Francis.

Swartz N. (1997) The concepts of necessary and sufficient conditions. Class Notes, http://www.sfu.ca/philosophy/swartz/conditions1.htm.

Vicsek T. (1991) Modelling of the structure of coffee layers by random packing of spheres. Personal communication.

Voilley A. and Clo G. (1984) Diffusion of soluble substances during brewing of coffee. In B.M. McKenna (ed.), *Engineering and Food, Volume 1.* London: Elsevier, pp. 127–137.

WHO (1993) *Guidelines for Drinking-water Quality,* 2nd edn. *Volume 1 – Recommendations.* Geneva: World Health Organization, pp. 122–130.

Zanoni B., Pagliarini E. and Peri C. (1991) Modelling the aqueous extraction of soluble substances from ground roasted coffee. *J. Sci. Food Agric.* 58, 275–279.

CHAPTER 8

The cup

M. Petracco

At the beginning of this book, a first portrait of an espresso coffee serving was presented as 'a small heavy china cup with a capacity just over 50 ml, half full with a dark brew topped by a thick layer of a reddish-brown foam of tiny bubbles'. A chemist would not like such a description and prefer a less poetic but more explicit wording. Nevertheless, this definition characterizes the espresso beverage quite well, at least from a macroscopic point of view, pinpointing its main features:

- a small volume, indicating that the brew is more concentrated than usual;
- a composite physical nature, where the foam is as important as the liquid.

These features are, of course, the result of a very special percolation method used to prepare the cup, but they could be recognized and characterized even if nothing were known about the history of the cup presented to the analyst's examination. In the following sections espresso will be described in detail and characterized both from a physico-chemical and an organoleptic point of view.

8.1 PHYSICAL AND CHEMICAL CHARACTERIZATION OF THE ESPRESSO BEVERAGE

Espresso is unstable: all the parameters that are prone to fast modification after brewing must be determined as soon as possible. The most rapidly varying property is gas content. Gas release occurs owing to the sudden pressure drop undergone by the beverage when issuing out from the espresso machine (see 7.5.7).

Dissolved gases – mainly CO_2 – quickly effervesce in the cup, and bubble up to build a layer of froth. This makes espresso a composite beverage where two distinct constituents are present: supernatant foam and underlying liquid.

8.1.1 Foam

Espresso foam – also known by the Italian term of *crema* – is by itself a biphasic system composed of gas globules framed within liquid films (called *lamellae*) constituted by a water solution of surfactant. These films tend to set in a configuration of two layers of surface-active molecules facing the gas, with water molecules between them. The high molecular force in the film allows its peculiar geometry: a bubble if isolated, or a honeycomb-like structure of many bubbles growing close together (Figure 8.1).

An important characteristic of the *crema* is persistence, and it must survive at least a couple of minutes before breaking up and starting to show an uncovered spot on the dark surface of the beverage. The disappearance can be imputed to the phenomenon of drainage, which causes entrapped water to flow out of the films, leaving only surfactant molecules to bear the stress of the structure. The films eventually fail due to their limited elasticity, thus releasing the entrapped gas. Persistence can be easily measured by watching the surface of a standard cup at constant temperature and recording the first sign of weakening of the foam's layer, revealed by formation of a hole.

As espresso foam is fairly short-lived, its quantity must be measured soon after percolation. This is done by transferring the whole beverage

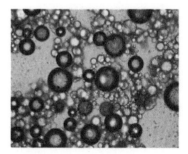 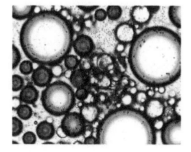

Figure 8.1 Pattern of foam in espresso: (left) at beginning of brewing; (right) after resting overnight, proteins have agglomerated

from the cup to a graduated cylinder, allowing it to stabilize and recording the volumes of both foam and liquid. Their ratio constitutes a 'foam index' that in well-prepared cups should be at least 10%.

Solidity is a noteworthy rheological property of foam, denoting its capability to bear for some time the weight of a spoonful of sugar. It can be quantified in the laboratory by loading the foam with a calibrated wire mesh ring of a standard weight and measuring the time it takes to submerge.

8.1.2 Liquid

The liquid part of the espresso beverage, where the foam stays afloat, is an even more complex system. Along with the properties of the liquid as a bulk, the presence of several dispersed phases must be taken into account to explain the peculiar characteristics of espresso. Physical and chemical parameters of the beverage are detailed in Table 8.1, which shows typical analyses of correctly prepared espresso cups (30 ml/30 s).

As already pointed out in Chapters 1 and 7, the polyphasic nature of the liquid is what makes espresso stand out in comparison with all other brews, both from a physico-chemical and from a sensory standpoint. The peculiarity of espresso beverage is the simultaneous presence of three

Table 8.1 Typical analysis of an espresso brew

Parameter	Pure arabica	Pure robusta
Density at 20 °C (g/ml)	1.0198	1.0219
Viscosity at 45 °C (mPa·s)	1.70	1.74
Refractive index at 20 °C	1.341	1.339
Refractive index of filtrate at 20 °C	1.339	1.339
Surface tension at 20 °C (mN/m)	46	48
Total solids (mg/ml)	52.5	58.2
Total solids of filtrate (mg/ml)	47.3	55.6
Total lipids (mg/ml)	2.5	1.0
Unsaponifiable lipids (mg/ml)	0.4	0.2
pH	5.2	5.2
Chlorogenic acids (mg/ml)	4.3	5.0
Soluble carbohydrates (mg/ml)	8.0	10.0
Total elemental nitrogen (mg/ml)	1.8	2.3
Caffeine (mg/ml)	2.6	3.8
Ash (mg/ml)	7.2	7.0
Elemental potassium (mg/ml)	3.2	2.9

dispersed phases coexisting within a matrix, namely a concentrated solution of salts, acids, sugars, caffeine and many other hydrophilic substances. These phases are:

1 an emulsion of oil droplets;
2 a suspension of solid particles;
3 an effervescence of gas bubbles, which evolves into a foam.

While the presence of dispersed phases is evident, for the beverage is opaque, some microscopic investigation is needed to enlighten this feature. Light microscopy, scanning electron microscopy (SEM) and transmission electron microscopy (TEM) have been employed to observe the discontinuities of the matrix. To optimize light microscopy, some form of highlighting of the particles is needed: interference techniques are quite useful to reveal bubbles, while staining media specific for lipids (e.g. osmium tetroxide or 'Oil Red O') have proved useful to examine oil droplets (Heathcock, 1989).

The study of the dispersed phases requires special care, because they differ in their stability. The effervescence, which can be observed for no longer than one or two minutes, must be tackled as soon as possible by some form of 'freezing', while the suspension remains stable for days, and the emulsion even longer. Furthermore, the last is difficult to break in order to collect the oil phase for analysis.

The relative amounts of stable particles, namely lipid droplets and solid particles, are strongly skewed in favour of droplets. An optical microscope picture of a perforated filtering sheet, retaining particles larger than $0.2\,\mu m$, shows a few cell wall fragments along with dozens of droplets of various sizes, up to $10\,\mu m$ (Figure 8.2).

Their size distribution (Figure 8.3) was evaluated by digital image processing of light micrographies (Heathcock, 1989) and shows dependence from the percolation variables used to produce the beverage (see Figure 8.3) (Heathcock, 1988).

Distribution is then one factor affecting the organoleptic appreciation of the cup. Particle size can be measured by laser scattering analysis, and average values of 2 to $5\,\mu m$ have been reported (Petracco, 1989).

8.1.2.1 Density

Density is little influenced by coffee solids. Differences with respect to pure water are limited to the second decimal digit; this could be explained by the presence of coffee lipids dispersed as an emulsion, which are less dense than water and decrease the overall density. Addition of sugar

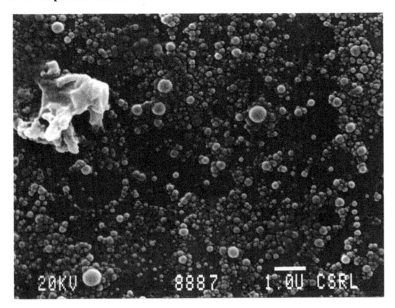

Figure 8.2 SEM of coffee lipid droplets and solid particles filtered out from their liquid matrix

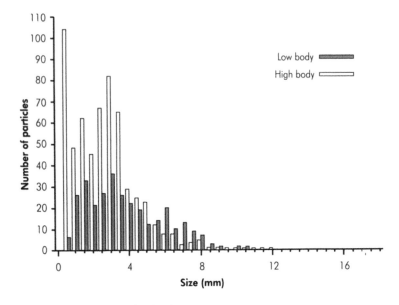

Figure 8.3 Droplet size distribution

can raise density to around 1.08 g/ml, while synthetic sweeteners like saccharin or aspartame add sweetness with little or no density increase. This changes the perception of the sensory variable 'body' between naturally and artificially sweetened espresso.

8.1.2.2   Viscosity

Viscosity is considerably higher than that of pure water, with reported values approximately double at usual consumption temperatures (Table 8.2). This property is influenced by the presence of dispersed phases, related to the amount of lipid droplets in emulsion. Increased viscosity has been associated with improved body, as shown by the comparison of a correctly prepared cup with a poorly prepared one (Figure 8.3), where viscosity differences as high as 33% have been recorded (Table 8.2) (Petracco, 1989). This may be explained by the fact that emulsions usually display higher viscosity than pure solutions.

8.1.2.3   Refractive index

The refractive index is usually determined to evaluate solute concentrations in transparent liquids. Espresso's polyphasic nature makes this method difficult to apply, and scarcely convenient to predict the body character. In comparison with the index of pure water (1.333 at 20 °C), espresso beverages display values ranging from 1.339 to 1.341. Refractive index is influenced mainly by the solute concentration, and the dispersed phases are less important as indicated by the minor differences found after filtering aliquots of espresso. Commercial refractometers specially graduated for coffee are available and correlate with total solids content.

8.1.2.4   Surface tension

Surface tension is a property of interfaces, due to the tendency of molecules to diffuse from one material to another: pure water–air

Table 8.2 Viscosity of high- and low-body espresso, compared with water (mPa·s)

°C	Pure water	Low body	High body
45	0.61	1.32	1.64
30	0.81	1.51	2.01

interfaces have a tension of $73 \cdot 10^{-3}$N/m at 20 °C. The lower values found in coffee beverages, down to $46 \cdot 10^{-3}$N/m, are due to the presence of surface-active agents, the surfactants, which accumulate at the interface between two phases building up monomolecular layers. These molecules, composed of an apolar chain with a polar head, help to form and stabilize foam and emulsion, and may also influence the fluid's behaviour during percolation through the narrow interstices of the packed coffee cake, where capillarity plays an important role. The chemical nature of natural coffee surfactants has not yet been fully clarified: several classes of complex molecules, like glycolipids or glycoproteins, might exhibit such behaviour (Nunes et al., 1997). Apparently, in espresso beverages there are two classes of compounds responsible for two different properties: foamability seems to be influenced by a proteic fraction, while foam persistency depends more on a polysaccharidic fraction (Petracco et al., 1999). An additional interesting effect of coffee surfactants, besides foam and emulsion promotion, is that they help the beverage to come into close contact with taste buds in the tongue, thus enhancing the sensory perception.

8.1.2.5 Total solids

Total solids content – the beverage's overall concentration – is the most important characteristic in the chemical composition of espresso, often perceived by lay consumers as 'strength'. This property is determined by drying the brew in an oven to constant weight without filtering. Accordingly, suspended solid material, emulsified lipids and solutes are included. Total solids concentration in the beverage is dictated by the brewing formula, namely the coffee/water ratio, but varies greatly also depending on both roasting degree and percolation temperature: a darker roasting produces more solubles, and a higher temperature extracts them more efficiently (Nicoli et al., 1991). Values as low as 20 mg/ml up to 60 mg/ml may be encountered. The true soluble fraction, typically 90% of the total solids, can be determined by oven drying after filtering the liquid.

8.1.2.6 Lipids

The emulsified lipid fraction can be determined by liquid–liquid solvent extraction. Different amounts have been reported in both espresso and non-espresso coffee beverages (Petracco, 1989; Peters, 1991; Ratnayake et al., 1993; Sehat et al., 1993). For an espresso made from a 100% arabica blend, a figure of the order of 5% of the total solids may be considered as typical (Table 8.1). This is a little less than one-tenth of the total lipids present in the roast and ground coffee portion, and represents only a minor contribution to the dietary energy intake: less than 1 kcal per cup. Within the unsaponifiable fraction of coffee lipids, two compounds belonging to the diterpene family – cafestol and kahweol – are of relevance to blood cholesterol level (see 10.3.3); present in roasted arabica beans at levels around 0.6%, they are transferred to the espresso brew in small amounts (Urgert, 1997).

8.1.2.7   Acids

Acidity of coffee beverages is an important organoleptic parameter (Woodman, 1985). The presence of acetic, formic, malic, citric and lactic acids has been detected in the brew, along with quinic and chlorogenic acids (Peters, 1991; Severini et al., 1993). The content of the last ones is of course smaller when using dark roasted coffee, in which they have largely disappeared (Clifford, 1985). A mineral acid – phosphoric – has also been found in brews (Maier, 1987), deriving from thermal degradation of phytic acid, the phosphoric ester of inositol. As already mentioned in 4.3.2.6, espresso acidity cannot be described just by pH. Measurements of this variable obtained using electrode-type instruments report values ranging from 5.2 to 5.8, showing an increase with roasting degree (Dalla Rosa et al., 1986a; 1986b) and depending on extraction time (Nicoli et al., 1987).

8.1.2.8   Carbohydrates

This broad class of compounds is present in coffee seeds, as in all plant materials, both as simple sugars and as polysaccharides (Trugo, 1985). Upon roasting, heavy reactions and transformations happen, changing the balance between soluble and insoluble carbohydrates. While only the former are relevant to the beverage, the brewing method may influence the hydrolytic degradation of insoluble carbohydrate polymers, adding to the solubles content. Typically, only small amounts of monosaccharides can be found in the brew, but no sucrose is present, because it has already been transformed on roasting into bitter tasting Maillard compounds (Trugo and Macrae, 1985). Much research has been done on coffee carbohydrates (Leloup et al., 1997), but few data on their content in coffee beverages are available: in espresso a typical figure for soluble carbohydrates is 8 g/l, corresponding to some 15% of total solids (Petracco, 1989).

8.1.2.9 Nitrogen compounds

Nitrogen-containing compounds are present in coffee beverages as transformed proteic material – grouped under the broad name of melanoidins – and caffeine, while trigonelline has largely disappeared during roasting to yield volatile aroma compounds (Macrae, 1985). A way to analyse this complex class of compounds is by determining total elemental nitrogen, subtracting the caffeine nitrogen and multiplying for the standard factor 6.25 to supply a conventional protein content value. Typical figures are in the neighbourhood of 1.4 mg/ml total N, corresponding to approximately 4 mg/ml of proteic material.

Caffeine is the most studied coffee component, due to its physiological activity (see 10.2). Its solubility in water displays a marked dependence on temperature, with a more than 30-fold increase from standard conditions to atmospheric boiling temperature, where almost quantitative extraction takes place at brewing provided enough time is allowed to cope with the kinetics (Spiro, 1993). Also water hardness seems to influence caffeine extraction to a certain extent (Cammenga and Eligehausen, 1993).

Caffeine presence in espresso beverages has not been much studied. Data suggest that caffeine extraction from coffee grounds is incomplete in espresso percolation and, as already indicated in 7.2.2, the yield is usually within the range 75–85%. This is due to the short time available with espresso brewing to extract caffeine from the cellular structure, as shown by fractionated extraction curves (Figure 8.4) (Petracco, 1989). Therefore, caffeine concentration in the brew ranges between 1.2 mg/ml up to 4 mg/ml, depending on cup size and blend composition.

Corresponding cup contents range between 60 mg for pure arabica blends, and 130 mg for pure robusta blends.

8.1.2.10 Minerals

The presence of minerals in the espresso brew is low, as can be seen from ash content scarcely exceeding 7 mg/ml. Nevertheless, they constitute about 15% of total solids and, since potassium forms a substantial part of them, they might be considered as a beneficial contribution to the diet.

8.1.3 Volatile aroma

The meaning of the substantive 'aroma' is traditionally considered to be 'a strong pleasant smell, usually from food or drink' (Procter, 1995). In the

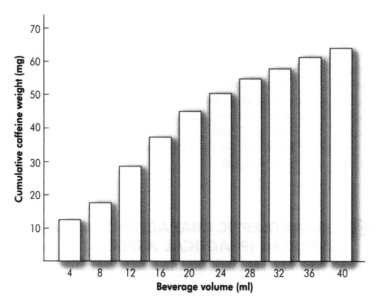

Figure 8.4 Fractionated extraction of caffeine during the brewing of a *ristretto* coffee cup

domain of food science, however, it is better addressed as 'the quality or principle of plants or other substances which constitutes their fragrance; agreeable odor; as, the aroma of coffee' (Webster, 1913). The following, more general, definition applies even better to our chemical context: 'any property detected by the olfactory system' (WordNet, 2003), with the additional specification that by 'aroma' we shall designate any set of 'odorants', i.e. chemical substances capable of reaching our olfactory organs.

There is little doubt that the aroma of any coffee beverage is dissimilar from that of the roast and ground coffee used to brew it, inasmuch as the composition of its odorants is different (Grosch, 2001). This is the result of extraction dynamics, which itemizes fast and slow extractors among the several hundreds of volatile compounds produced on roasting (Mayer *et al.*, 2000).

Further work is required to identify over-the-cup aroma compounds specific to espresso, as compared with those above a cup of filter coffee (see Figure 4.13). Along with a stronger overall concentration, espresso features a dispersed lipidic phase (see 8.1.2), which is likely to carry and release some oil-soluble volatiles to the headspace. Moreover, the presence of a thick layer of foam would certainly change the composition of the latter.

Actually, the pleasant sensation of sniffing freshly ground coffee beans is much more stimulating than breathing in the rather weak aroma exhaling from a cup of coffee. The strong sensory advantage of the beverage appears only when taking a sip in the mouth, where more complex release and hydrolysis reactions are likely to occur due to the biological environment. Thanks to new methodologies (Büttner and Schieberle, 2000), the field of the measurement of volatiles evolving *in vivo* is open to future research.

8.2 ORGANOLEPTIC CHARACTERISTICS OF ESPRESSO (PRACTICAL ASPECTS)

The word of Greek origin 'organoleptic' translates literally as 'brought by organs', and figuratively as 'learnt through perceptions', that is, by direct information coming from our organs of sense: eyes, ears, skin, nose, or mouth. According to ISO vocabulary, it denotes the attributes of a product that are perceptible by sense organs (ISO, 1992). Its implementation, called sensory analysis, is used to examine and measure objective characteristics such as taste, aroma, colour and other analogous factors of foods and beverages via an evaluation of the subjective sensations they elicit by the gustatory, olfactory, haptic (tactile), visual and auditory spheres (see also Chapter 9).

Few everyday experiences can compete with a good cup of coffee, as long as sheer sensory pleasure is considered. It is clear that most of the quality of such a beverage is determined by its overall sensory impact. In this context, espresso is the brewing method that offers the consumer the most potent experience, even if a high quality cup it is not easy to obtain: espresso's very strength – the ability to concentrate aromas – is also its weak point, because, while enhancing qualities, it shows up at the same time all the possible defects of the raw material.

A cup of well-prepared espresso is a powerful stimulator of our senses: the rich colour, the intense aroma, the intense and long-lasting taste, all provide pleasure to the nose, the mouth and the eyes. Only the sense of hearing seems to be extraneous to this multi-sensory experience (if we disregard the call of a steaming espresso machine). Sipping from this small cup creates a link between the amazing complexity of the chemical composition of coffee and the as complex sensory perception system.

8.2.1 Visual characteristics

As far as sight is concerned, the main attribute of the cup is foam. The beautiful aspect of the foam is conferred by the tiny gas bubbles framed by viscous liquid *lamellae*, where minute cell wall fragments float producing a 'tiger skin' effect (Figures 8.5 and 8.6b).

Consumers who ascribe great importance to the colour and the texture of foam are right, because perfect foam is the signature of a perfect preparation. Any error in grinding or in percolation, in temperature or extraction level, is immediately denounced by the colour, the texture and the persistence of the foam. For example, if the foam is light-coloured, inconsistent, thin and evanescent, it means that espresso has been under-extracted, probably because the grind was too coarse or the water temperature too low (Figure 8.6a).

An over-extracted espresso, on the contrary, may display either a white foam with big bubbles, if water temperature was too high, or just a white spot in the middle of the cup (Figure 8.6c).

Other extraction mistakes can derive from too low a cake porosity or from too high a coffee dose, both indicated by very dark foam with a hole in the middle.

The foam layer also acts as an aroma-sealing lid, trapping the volatile compounds responsible for the odour of espresso brew. The foam preserves them for our pleasure if the blend used to brew the cup was good, or for our torment if it was defective: it works as an amplifier, enhancing both virtues and sins.

8.2.2 Degustation characteristics

For the sake of a systematic description of organoleptic attributes, it is important to clearly show the boundaries between the multi-faceted

Figure 8.5 A perfect espresso showing (left) the depth of foam and (right) the 'tiger skin' effect on the surface of the foam

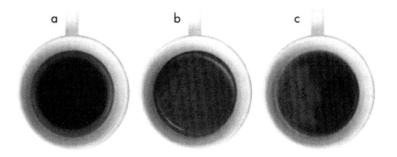

Figure 8.6   Different *espressi*: (a) under-extracted; (b) correctly extracted; (c) over-extracted

concepts of taste and after-taste, odour and flavour, body and astringency, since they all are determinants in the appreciation of a beverage such as espresso, when mouth and nose intervene. While Chapter 9 will deal in depth with these concepts from the perspective of the science of perception, in the following sections they will be addressed mostly from the hedonic point of view.

8.2.2.1   Gustatory senses: taste

Taste is perceived by the homonymous sense whose primary elements, called buds and which number several hundreds, are located on the surface of the tongue (see 9.2.1 for more on this). Specialized buds used to be associated with the perception of mainly one of the four pure taste sensations: sweet-sensitive buds were deemed to cluster on the tip of the tongue, whereas bitter-sensitive buds were to be found in its back part, with acidic and salty sensations coming mostly from the tongue's sides. However, this theory has recently been questioned (Bartoshuk, 1993) (see 9.2.4).

Sweetness, a character always positively correlated with coffee quality and price, is particularly appreciated in the espresso cup, while the contrary holds for bitterness. Nevertheless, the overall description of a fine espresso sounds like 'bitter-sweet with a initial slightly acidic note'.

Acidity is not a *desideratum* in espresso because it tends to give an unbalanced feeling amplified by the high concentration of the brew. Consumers in different countries show marked differences in the appreciation of the ratio of bitterness to acidity. In southern European countries, consumers prefer an espresso with a bitter dominance (and consider body very important), while in the northern ones a more balanced taste is preferred and excess bitterness is considered a defect. This stems also from the concentration in the espresso cup, very small and concentrated in the south, where it is drunk without milk; more diluted and frequently with milk or dairy cream in the north. Unfortunately, the preparation of an espresso of more than 50 ml without over-extracting is very difficult: a large water volume also extracts more of the less soluble, and less pleasant, components. For this reason, in the north, consumers asking for their favourite rather diluted espresso shot often receive a bitter-tasting, woody and sometimes astringent one. Espresso percolation resembles a fractional extraction, with an increase in less soluble substances as extraction proceeds. Alas, not all that is soluble is agreeable: one must know when to close off.

Both blend formula and roasting level can be used to change the bitterness–acidity ratio. Fine washed coffees, frequently having a strong acidic tone, are rewarding in the preparation of filter coffee since they will withstand hydrolysis, and keep their quality during the time elapsing between preparation and consumption (Feria-Morales, 1989). In an espresso blend they will help with the aroma, but a blend of pure washed coffees will taste excessively acidic and lacking body. Sun-dried coffees will, on the contrary, enhance both body and balance. Also, the roasting intensity can be used to change the bitterness–acidity ratio, as seen in 4.3.2.6. Caution must be used when roasting washed coffees: if roasted too dark they will develop a pungent, burnt character usually found unpleasant. Another objectionable acidic, astringent and metallic taste is the consequence of very fast roasting, as already discussed in 4.2.2.

8.2.2.2 Gustatory senses: after-taste

The rheology of espresso is rather special compared to other coffee brews. Its strong concentration is responsible for high density and viscosity, whereas the presence of natural surfactants lowers surface tension. These apparently contrasting characters could explain the intense taste and the long-lasting after-taste (and after-flavour) sensations, namely those felt for a longer period after emptying the mouth. When drunk, espresso first soaks the surface of the tongue, colouring it; the brew is then trapped by the taste buds, and oil droplets fix themselves to the mucous membrane. There they slowly release the volatile substances dissolved, so that they are perceived for a while (up to 15 minutes) after the beverage has been swallowed.

Coating of the tongue may possibly explain a reduction in the perception of bitterness. This statement can be indirectly supported by experiments in which an expert cup-testers' panel is presented with a quinine solution, whose bitterness is rated 100. Then a colloidal polysaccharide is added to the bitter solution at a dilution of 1% and, surprisingly, bitterness is rated at 40. The conclusion is that colloids are able to reduce the perception of bitterness by receptor blockade. Droplets of the emulsified oil in espresso probably behave like colloids, and they act on perceived bitterness owing to their dimensions, which are smaller than 10 μm.

This theory may also explain the phenomenon of increased perceived bitterness when diluting an espresso. Dilution normally reduces the intensity of any taste, but in this case the molecules responsible for the bitter taste are present in the liquid in several billions, and the oil droplets in several orders of magnitude less. The probability of a bitter molecule catching an unblocked receptor is increased by the dilution of the beverage, so that diluted espresso tastes more bitter.

8.2.2.3 Olfactory senses: odour

Odour here is meant as the olfactory sensations caused by inhaling. Whereas the aroma of freshly ground coffee is a powerful stimulus, coffee beverages do not release a large amount of volatile compounds into the environment that surrounds the cup, and this is even more so for espresso, as already explained in speaking about the foam's role in trapping aroma.

On the contrary, many volatile molecules of aroma are released within the mouth after drinking and reach the olfactory receptors in the nose by retro-diffusion, namely the movement of gaseous molecules from the mouth through the pharynx up to the nose. A simple experiment that can prove the existence of this phenomenon is to munch and swallow a bite of apple while pinching one's nostrils. The missing sensation is exactly what will be hereinafter called flavour (but which – just to add to the confusion of terms – is referred to as 'arome' in French, at least according to ISO 5492 vocabulary:

> 3.33 aroma (noun): NOTE – The sense of the terms 'aroma' in English and 'arome' in French is not exactly equivalent. (1) French sense: Organoleptic attribute perceptible by the olfactory organ via the back of the nose when tasting. (2) English sense and French informal language: An odour with a pleasant connotation. (ISO 1992)

8.2.2.4 Olfactory senses: flavour

To make the distinction between aroma and flavour even clearer, it is useful to explain that the human sense of olfaction perceives the presence of odorous volatile molecules – the ensemble of the ones given off by coffee is, in this context, called aroma – by means of thousands of receptors located in the inner *mucosae* of the nose. Olfaction is in charge of both odour, by direct inhaling of the molecules arising from the roast and ground coffee or from the cup, and flavour, the nasal perception of the volatile substances evolving in the mouth and reaching the nose cavity by the pharyngeal pathway (Petracco, 2001). While evaluation of pure taste is quite easy by closing up the nostrils and excluding in this way the olfactory side, the opposite is unfortunately not feasible (at least without nerve surgery...): no retro-pharyngeal flow can occur without actually passing the sample through the mouth. Probably for this reason, most authors, even in normative papers (ISO, 1992; Civille and Lyon, 1996), define flavour as a complex combination, in which olfactory sensations merge with gustatory and somatic (haptic and trigeminal) ones. This concept is shared by other contributors to this book (see 9.4.3). Contrarily, the present author's opinion calls for reserving the term 'flavour' to the pure olfactory component of this overall sensation, while the latter is felt sadly to lack a specific term: 'palatability' could be a good candidate, if freed from its current hedonic meaning (taken to be positive by default).

Among the organoleptic characters of coffee, flavour is one of the most debated (Wrigley, 1988). Both intensity and quality of the olfactory component of taste are important, especially when espresso is concerned. More than 800 volatile species have been identified so far in coffee headspace, and many of them are liposoluble. Since espresso brewing extracts a substantial quantity of coffee oil as a dispersion of small droplets, it also transports into the cup the liposoluble volatile substances dissolved in the lipid phase of roast coffee during roasting and storage. After brewing, these gaseous molecules tend to escape from the liquid and finally reach the nasal receptors of the consumer. This complex mechanism explains the difference between the flavour of espresso and filter coffee: not only a matter of concentration, but a distinct aromatic spectrum as well.

Terms to describe flavour are usually borrowed from everyday life, and recall the world of flowers, fruits and fresh or baked foods. Unfortunately, there is sometimes a need to describe negative flavours or taints, where terms like 'stinker', describing a rotten vegetable tang, or 'mouldy', reminding of fruits infected by mildew, or 'peasy', reminiscent of pea pods, are used (see 3.7.2.10). A consumer-oriented list has been developed by the International Coffee Organization (ICO, 1991), but each expert panel usually has its own terminology.

8.2.2.5 Haptic senses: mouthfeel

Mouthfeel is a tactile sensation perceived by buccal mucous membranes, along with the thermal response due to the temperature of the beverage. Most of its nature is related to small movements of the tongue against palate and gums, which apply a shear stress to the liquid performing a sort of rheological measurement of viscosity and texture. Body, an important attribute of espresso coffee, is felt in this way and can be considered either as a synonym of mouthfeel or, with astringency, as part of its definition.

It is sometimes claimed that body is a peculiarity of robusta espresso brews, and, as a matter of fact, those beverages 'fill the mouth'. It is difficult to believe that oil droplets of colloidal size are present in higher amounts in robusta than in arabica espresso, given that the lipids content of roast and ground robusta is much lower than that of arabica (some 9% as opposed to 13%). Cup-testing experiments have been reported, indicating that freshly percolated robusta espresso has actually a stronger body than arabica, but only as a first sensation: if the sip is held in the mouth, after some seconds arabica's body is still present while robusta's fades away. Moreover, an inversion of assessment takes place if the beverage is allowed to rest for 2 minutes or so, with arabica again displaying stronger body. It has been postulated that this behaviour could derive from a transient viscosity rise induced by gas bubbles of colloidal size, possibly more evident in robusta (Petracco, 1989).

8.2.2.6 Haptic senses: astringency

Another tactile sensation, astringency, reminds one of medicine and is always considered as negative. It is related to a chemical phenomenon, namely the precipitation of the proteins of saliva due to specific phenolic compounds present in some beverages and unripe fruits (Rider, 1992). In coffee, it has been ascribed to the presence of immature beans containing dicaffeoylquinic (dichlorogenic) acids (Ohiokpehai et al., 1982), which are deemed to be astringent to mucous membranes. Deplorably, immature beans are nowadays found more and more frequently, because green coffee producers seem to pay less attention to the quality of their crops due to the low price of recent periods.

8.2.3   Forecasting sensory quality

As seen so far, the traditional cup-testing approach still remains the ultimate assessment tool for coffee quality. After all, the reason why coffee has become so popular, achieving the position of the second most largely consumed beverage after water, is its flavour or, even better, its overall impact on our senses. Sensory evaluation, which used to be considered the magic because 'taste is a matter of taste', is nowadays earning the status of a highly respected analytical tool, able to produce key information with good reliability (Meilgaard *et al.*, 1999).

8.2.3.1   Espresso cupping practice

In industrial coffee routine, some form of objective evaluation is needed to ascertain product overall quality, along with the constancy of that quality in time and in varying process conditions. For a product that is consumed for pleasure, there is little doubt that the final word must be left to the sensorial sphere. While the general aspects of organoleptic laboratory activity will be dealt in 9.4.3, we present here some practical guidelines to approach the assessment of espresso.

The 'tool' commonly used is a panel of assessors, who may be either coffee experts (professional cup-testers) or naive consumers after a basic training. The reason for employing more than one person is obvious: by averaging responses, the risk of incorrect judgement due to a possible indisposition of one person is minimized. Another panel potential is the synergy that can be gained by debating coffee characteristics among the assessors during open sessions: this procedure may extract more information, since individual sensitivities and perception thresholds may be different.

Sensory tests used in espresso evaluation may be grouped in three basic types, listed by increasing difficulty for the panel members:

■ Trio tests, used simply to determine whether any perceivable difference exists between two samples. In this configuration the two beverages are split in two cups each, but only three of the four samples are presented to the panel: the assessors are requested to tell which is the 'foreign' single cup, as opposed to the pair of 'sister' cups.
■ Duo tests, where two or more single cups are presented to the panel, who are asked to rank them in relationship to one sensory variable. When more than one variable is to be determined, a pre-filled card proves to be useful to summarize the evaluations.

■ Absolute tests, in which some complex variable, like aromaticity or overall merit, is to be determined by comparison with a mental paradigm present in each assessor's memory based on previous experience. Coffee aroma profiling (ICO, 1991) can also be included in this type, since it is based on assessors' recall of variegated flavour knowledge.

In regular day-to-day espresso cup-testing sessions in support of the purchasing activity, the panel should be presented with a maximum of a dozen samples, each served in three different cups according to three fundamental preparation techniques:

■ Infusion: it is a brewing method that is widely used in northern Europe and in the USA, in which boiling water is poured on coarsely ground coffee powder and allowed to rest for a given period before filtering away the spent grounds from the remaining clear beverage. The concentration of the beverage is low (below 20/l) and only the soluble substances may pass into the cup, giving an aromatic pattern typical of filtered coffee beverage.
■ Espresso: the classic espresso cup is prepared under standardized and thoroughly controlled conditions, allowing the panel to evaluate its foam and body along with the taste and flavour characteristics.
■ Diluted with hot water up to reach the concentration of the infusion method: in this way the high concentration of regular espresso does not hinder the evaluation of a weaker aroma's nuances, and the difference between the aromatic pattern of the solution and the emulsion can be determined.

Espresso cup-testing sessions cannot be too long nor too frequent during the day because some fatigue develops after the first dozen or so espresso cups, due to the lingering after-taste deriving from the sticking of coffee oil droplets on the tongue and mouth membranes (Petracco, 1989). Rinsing the mouth with water, albeit necessary between each sampling, is not effective to remove the taste completely. On the contrary, cool whole milk seems to act better to this purpose, perhaps because, being itself an oil-in-water emulsion, it can displace coffee oil droplets from the tongue by dilution.

8.2.3.2   Instrumental testing

Unfortunately, it is fairly evident that implementing the cup-testing practice as explained in the previous section is neither simple nor inexpensive. As a consequence, industry strives to take advantage of sensory data collections, using them as raw experimental data to calibrate instrumental screening methodologies.

A good instance of such a rapid, non-destructive fingerprinting technique is near-infrared spectroscopy. It is based on absorption measurements of scanned monochromatic near-infrared light (wavelength between 1100 nm and 2500 nm), whose energy is dissipated in rotational and vibrational movements of the molecular bonds of the material under examination, and ultimately transformed into heat (Murray and Williams, 1987). Energy absorption patterns contain a lot of implicit information about molecular response to specific wavelengths and to their combinations.

Spectrophotometers may operate by filters or by scanning monochromators: the latter present the advantage that a higher amount of information can be collected at the same time, and are therefore most useful for research and new applications. Spectra are obtained by recording either the transmitted part of the incident light (NIT, used mainly for transparent materials like solutions) or the reflected one (NIR, applied to opaque materials like grains or powders), and by plotting its reciprocal logarithm against the wavelength. The procedure, generally employed to analyse solid materials opaque to visible light, is called diffuse reflectance. Since infrared radiation is able to penetrate some millimetres under the surface of solid materials, the term near infrared transflectance (NIRT) may better describe the technique as applied for the analysis of coffee beans.

All spectra of vegetal materials do not differ that much at a first glance: this is the reason for the need for sophisticated statistical elaboration of the raw data in order to correlate spectra with the reference analysis data of interest. The branch of chemistry studying the best fit of data from such secondary analytical methods with the primary ones is called chemometrics. It uses computerized regression methods like principal components analysis or partial least squares to calibrate, that is, to build an equation able to convert spectra into chemical previsions. Calibration sets of some dozens of analysed samples are needed, and the robustness of the result is linked to their careful choice.

NIRT has been shown suitable for supplying simultaneous forecasts of many chemical characteristics of the sample examined, provided that a good calibration has been previously obtained by statistical correlation with traditional, time-consuming analytical methods. This secondary method has widely been used with agricultural products (Shenk, 1992) and sometimes in the specific domain of coffee too.

Applications to coffee have been described for both green beans and roast and ground products: they include prediction of chemical characters like moisture and caffeine content (Guyot *et al.*, 1993), trade features like pureness (Downey and Boussion, 1996) or arabica/robusta content (Davrieux *et al.*, 2001), and also for modelling sensory data (Feria-Morales, 1991).

Several instrument brands, among are which NIRSystems, PerTen and Buhler, sell equipment with extensive autodiagnostics and powerful software: some claim to have developed specific calibrations for coffee too. It would be wise nevertheless for users to check the equations on the specific products of interest, and be prepared to develop their own calibrations.

8.3 ESPRESSO DEFINITION AGAIN

At the end of Chapter 1, a tentative definition of Italian espresso was proposed to the reader. Let us discuss it once again:

> *Italian espresso is a small cup of concentrated brew prepared on request by extraction of ground roasted coffee beans, with hot water under pressure for a defined short time.*

At this point, after having analysed and characterized espresso as carefully as possible, we are able to complete our definition. The above features, related to 'espresso style' and to 'espresso method', can be merged with those characteristics of the beverage which assure the fulfilment of consumer's expectations:

> *Italian espresso is a polyphasic beverage, prepared from roast and ground coffee and water alone. It is constituted by a foam layer of small bubbles with a peculiar tiger-tail pattern, on top of an emulsion of microscopic oil droplets in an aqueous solution of sugars, acids, protein-like material and caffeine, with dispersed gas bubbles and colloidal solids.*

The analysis of the beverage is:

Viscosity at 45 °C	>1.5 mPa·s
Total solids	20–60 g/l
Total lipids	2 mg/ml
Droplet size count (90%)	<10 μm
Caffeine	<100 mg/cup

The distinguishing sensory ensemble of characteristics of Italian espresso includes a rich body, a full and fine aroma, a balanced bitter-sweet taste with an acidic note and a pleasant lingering after-taste. It must be exempt from unpleasant flavour defects, such as stinking, mouldy, grass-like or other.

Owing to its instability, Italian espresso must be prepared on request from roasted and ground coffee beans, by means of a specific brewing method defined as:

Italian espresso is a brew obtained by percolation of hot water under pressure through a compacted cake of roasted ground coffee, where the energy of the water pressure is spent within the cake.

The variables ranges are:

Ground coffee portion	6.5 ± 1.5 g
Water temperature	90 ± 5 °C
Inlet water pressure	9 ± 2 bar
Percolation time	30 ± 5 s

the volume in the cup is to be left to the personal taste of the consumer inside the range 15 to 50 ml, with an optimal outcome at 25 to 30 ml.

8.4 ESPRESSO–MILK MIXES

A chapter about espresso coffee beverages cannot ignore the enormous number of coffee-with-milk cups consumed every day. Caloric effects of espresso, from a nutritionist's point of view, are insubstantial (see 8.1.2.5 and 8.1.2.6). Nevertheless, coffee does play a significant role in the diet when drunk together with other foods such as milk or sugar, inasmuch as its cup behaves as a carrier of nutrients not inherent to coffee.

The nutritional importance of milk in everybody's diet is not to be understated: a complete feed for newborns on its own, it helps adults to keep in good health mainly via its calcium content. Therefore, blending coffee drinks with milk may be seen as a great way to increase milk consumption and foster calcium intake, to the benefit of bone strength. Conversely, it must be remembered that the caloric content of milk is rather high (60–80 kcal/100 ml for whole milk, around 40 for skim milk), and may contribute seriously to the total energy intake, exceeding by far the effect of coffee when they are mixed together.

A plethora of recipes are present in various cultures, ranging from a drop of coffee in a glass of milk, just to add some flavour, to a drop of milk to discolour a coffee cup. Milk adds to appearance, to texture and to after-taste persistence thanks to its fat content, which contributes mouthfeel and flavour too, and helps to distribute fat-soluble flavour components of coffee, principally when used in espresso drinks. It may be added cold or hot, and some are now even questioning the priority: is milk to be added to coffee, or coffee to milk?

Perhaps the most attractive marriage between the two products happens when milk is added in the form of foam, originating the family of *cappuccino*-like drinks. The chemistry of milk foam formation is at least as complex as that of coffee, ensuing from casein–lipid interactions mediated by phospholipids (Goff and Hill, 1993). Both the protein and fat contents in milk are critical to foam development: skimmed milk contains the greatest percentage of protein and foams better than low-fat or whole milk, while fat content helps to maintain foam stability. The freshest milk – coming directly from the cow – is seldom used for hygienic reasons. Pasteurized milk is favoured because it can be kept up to 14 days in refrigerated storage with no noticeable change in foaming properties. It can be produced either by conventional process (heated to 64 °C for 30 min) or by HTST process (72 °C for 15 seconds) (Hinrichs and Kessler, 1995). Aseptically processed milk (the so-called UHT, heated up to 144 °C for 4 seconds) can be kept sealed at room temperature for several months. From a practical standpoint, it has many advantages: less refrigeration is needed in a business that is usually carried out in a limited space, and less time and temperature is needed for heating a product already at room temperature. However, it is not widely used by *baristas*.

The rule-of-thumb for making a good *cappuccino* is 'thirds':

- first, make a standard espresso shot in a larger cup, where espresso should take about one third of its volume;
- then add a third of hot liquid milk; and
- a third of steamed, frothed milk.

Any variation on the above prescription is permitted, producing beverages that have been baptized with fancy, often exotic-sounding names like *café au lait, café crème, macchiato, latte, frappuccino* and *mochaccino*.

Several pitfalls may spoil *cappuccino*: off-flavour water can taint the steamed milk; damaged or unclean steaming tools can add a burnt or scorched flavour; lengthy steaming can produce milk temperatures above 70 °C, imparting a cooked flavour. Typically, more than 75% fresh (unsteamed) milk should be present when producing the foam. Failure to drain steam pipes before frothing can dilute milk thus reducing mouthfeel and creaminess, as well as decreasing foaming capacity.

A final crucial warning: unlike roast and ground coffee, milk is a 'perfect' food not just for people but for microorganisms too, so when handled incorrectly it is prone to spoilage. Keeping it refrigerated as long as possible and maintaining all relevant tubing and vessels clean is mandatory.

REFERENCES

Bartoshuk L.M. (1993) The biological basis of food perception and acceptance. *Food Qual. Pref.* 4 (1/2), 21–32.

Büttner A. and Schieberle P. (2000) Exhaled odorant measurement (EXOM)? A new approach to quantify the degree of in-mouth release of food aroma compounds. *Lebensm. Wiss. Technol.* 33, 553–559.

Cammenga H.K. and Eligehausen S. (1993) Solubilities of caffeine, theophylline and theobromine in water, and the density of caffeine solutions. *Proc. 15th ASIC Coll.*, p. 734.

Civille G.V. and Lyon B.G. (1996) *Lexicon for Aroma and Flavor Sensory Evaluation: Terms, Definitions, References and Examples*. ASTM Data Series Publication 66. West Conshohocken, PA: American Society for Testing and Materials.

Clifford M.N. (1985) Chlorogenic acids. In R.J. Clarke and R. Macrae (eds), *Coffee: Volume 1 – Chemistry*. London: Elsevier, pp. 153–202.

Dalla Rosa M., Barbanti D. and Nicoli M.C. (1986a) Produzione di caffè tostato ad alta resa. Nota 2. Qualità della bevanda di estrazione. *Industrie Alimentari* 7–8/86, 537–540.

Dalla Rosa M., Nicoli M.C. and Lerici C.R. (1986b) Caratteristiche qualitative del caffè espresso in relazione alle modalita' di preparazione. *Industrie Alimentari* 9/86, 629–633.

Davrieux F., Laberthe S., Guyot B. and Manez J.C. (2001) Prediction of arabica content from ground roasted coffee blends by near infrared spectroscopy. *Proc. 19th ASIC Coll.*, CD-ROM.

Downey G. and Boussion J. (1996) Authentication of coffee bean variety by near infrared reflectance spectroscopy of dried extract. *J. Sci. Food Agric.* 71, 41–49.

Feria-Morales A. (1989) Effect of holding-time on sensory quality of brewed coffee. *Food Qual. Pref.* 1 (2), 87–91.

Feria-Morales A. (1991) Correlation between sensory evaluation data (taste and mouthfeel) and near infrared spectroscopy analyses. *Proc. 14th ASIC Coll.*, pp. 622–630.

Goff H.D. and Hill A.R. (1993) Chemistry and physics. In Y.H. Hui (ed.), *Dairy Science and Technology Handbook – Volume 1: Principles and Properties*. New York: VCH Publishers, pp. 1–82.

Grosch W. (2001) Chemistry III: volatile compounds. In R.J. Clarke and O.G. Vitzthum (eds), *Coffee Recent Advances*. Oxford: Blackwell Science, pp. 80–82.

Guyot B., Davrieux F., Manez J.C. and Vincent J.C. (1993) Détermination de la caféine et de la matière sèche par spectrométrie proche infrarouge. Applications aux cafés verts robusta et aux cafés torrefiés. *Proc. 15th ASIC Coll.*, pp. 626–636.

Heathcock J. (1988) Espresso microscopy and image analysis. Personal communication.

Heathcock J. (1989) Staining techniques for lipid and protein in espresso coffee. Personal communication.

Hinrichs J. and Kessler A.G. (1995) Thermal processing of milk – processes and equipment. In P.F. Fox (ed.), *Heat-induced Changes in Milk*, 2nd edn. Brussels: International Dairy Federation, pp. 9–21.

ICO (1991) *Sensory Evaluation of Coffee*. London: International Coffee Organization, p. 74.

Leloup V., DeMichieli J.H. and Liardon R. (1997) Characterisation of oligosaccharides in coffee extracts. *Proc. 17th ASIC Coll.*, pp. 120–127.

London ISO (1992) Sensory analysis – Vocabulary. ISO Standard 5492:1992. Geneva: International Organization for Standardization.

Macrae R. (1985) Nitrogenous components. In R.J. Clarke and R. Macrae (eds), *Coffee. Volume 1 – Chemistry*. London: Elsevier, pp. 115–152.

Maier H.G. (1987) The acids of coffee. *Proc. 12th ASIC Coll.*, pp. 229–237.

Mayer F., Czerny M. and Grosch W. (2000) Sensory study on the character impact aroma compounds of coffee beverage. *Eur. Food Res.Technol.* 211, 212–276.

Meilgaard M., Civille G.V. and Carr B.T. (1999) *Sensory Evaluation Techniques*, 3rd edn. Boca Raton, FL: CRC Press.

Murray I. and Williams P.C. (1987) Chemical principles of near-infrared technology. In P. Williams and K. Norris (eds), *Near-infrared Technology in the Agricultural and Food Industries*. St. Paul, MI: American Association of Cereal Chemists, pp. 17–34.

Nicoli M.C., Dalla Rosa M. and Lerici C.R. (1987) Caratteristiche chimiche dell'estratto di caffè: Nota 1. Cinetica di estrazione della caffeina e delle sostanze solide. *Industrie Alimentari*, 5/87, 467–471.

Nicoli M.C., Severini C., Dalla Rosa M. and Lerici C.R. (1991) Effect of some extraction conditions on brewing and stability of coffee beverage. *Proc. 14th ASIC Coll.*, pp. 649–653.

Nunes F.M., Coimbra M.A., Duarte A.C. and Delgadillo I. (1997) Foamability, foam stability, and chemical composition of espresso coffee as affected by the degree of roast. *J. Agric. Food Chem.* 45, 3238–3243.

Ohiokpehai O., Brumen G. and Clifford M.N. (1982) The chlorogenic acids content of some peculiar green coffee beans and the implications for beverage quality. *Proc. 10th ASIC Coll.*, pp. 177–186.

Peters A. (1991) Brewing makes the difference. *Proc. 14th ASIC Coll.*, pp. 97–106.

Petracco M. (1989) Physico-chemical and structural characterisation of espresso coffee brew. *Proc. 13th ASIC Coll.*, pp. 246–261.

Petracco M. (2001) Beverage preparation: brewing trends for the new millennium. In R.J. Clarke and O.G. Vitzthum (eds), *Coffee: Recent Advances*. Oxford: Blackwell Science, pp. 140–164.

Petracco M., Navarini L., Abatangelo A., Gombac V., D'agnolo E. and Zanetti, F. (1999) Isolation and characterization of a foaming fraction from hot water extracts of roasted coffee. *Proc. 18th ASIC Coll.*, pp. 95–105.

Procter P. (ed.) (1995) *Cambridge International Dictionary of English.* Cambridge: Cambridge University Press.

Ratnayake W.M.N., Hollywood R., O'Grady E. and Stavric B. (1993) Lipid content and composition of coffee-brews prepared by different methods. *Food Chem. Toxicol.* 31, 263–269.

Rider P.J., DerMarderosian A. and Porter J.R. (1992) Evaluation of total tannins and relative astringency in teas. In Ho Chi-Tang, Chang Y. Lee and Huang Mou-Tuan (eds), *Phenolic Compounds in Food and Their Effects on Health I. Analysis, Occurrence and Chemistry.* Washington, DC: American Chemical Society, pp. 93–99.

Sehat N., Montag A. and Speer K. (1993) Lipids in the coffee brew. *Proc. 15th ASIC Coll.*, pp. 869–872.

Severini C., Pinnavaia G.G., Pizzirani S., Nicoli M.C. and Lerici R.C. (1993) Etude des changements chimiques dans le café sous forme de boisson pendant l'extraction et la conservation. *Proc. 15th ASIC Coll.*, pp. 601–606.

Shenk J.S. (1992) NIRS analysis of natural agricultural products. In K.I. Hildrum, T. Isaksson, T. Naes and A. Tandberg (eds), *Near-Infrared Spectroscopy, Bridging the Gap between Data Analysis and NIR Applications.* New York: Ellis Horwood, pp. 235–240.

Spiro M. (1993) Modelling the aqueous extraction of soluble substances from ground roasted coffee. *J. Sci. Food Agric.* 61, 371–373.

Trugo L.C. (1985) Carbohydrates. In R.J. Clarke and R. Macrae (eds), *Coffee: Volume 1 – Chemistry.* London: Elsevier, pp. 83–114.

Trugo L.C. and Macrae R. (1985) The use of the mass detector for sugar analysis of coffee products. *Proc. 11th ASIC Coll.*, pp. 245–251.

Urgert R. (1997) Health effects of unfiltered coffee. Thesis, Agricultural University of Wageningen (Holland), p. 50.

Webster N. (ed.) (1913) *Webster's Revised Unabridged Dictionary.* Springfield, MA: G. & C. Merriam.

Woodman J.S. (1985) Carboxylic acids. In R.J. Clarke and R. Macrae (eds), *Coffee: Volume 1 – Chemistry.* London: Elsevier, pp. 266–289.

WordNet 2.0 (2003) by Princeton University, available online at: www.wkonline.com/cgi-bin/Dict.

Wrigley G. (1988) *Coffee.* London: Longman Scientific & Technical, p. 491.

CHAPTER 9

Physiology of perception

R. Cappuccio

9.1 INTRODUCTION

When we take a sip of *espresso*, we can have a hedonic reaction of pleasure, caused by the rich and smooth aromatic notes, the balanced acidity and bitterness and the intense body of the beverage, or a rejection reaction, caused by a low quality blend or an incorrect extraction process. These hedonic reactions are consequences of a chemical reception process, which occurs in the mouth and retronasal cavity, and a transduction process, which transforms the chemical signals into electrical ones. The information is then transmitted to the central nervous system (CNS), where an integration process occurs, giving that overall complex mixture of sensory inputs – composed of taste (gustation), smell (olfaction) and the tactile/trigeminal sensations (temperature and mouthfeel) – which we call *flavour* (Heath, 1988). This integrated sensation is evaluated, and a response in terms of acceptance and liking is given. Comprehension of how sensory information is processed and integrated is important in understanding the evaluation process, even if we do not fully understand all the steps (Stone and Pangborn, 1968).

Three semi-independent molecular sensors to check the external environment have been identified in mammals:

■ The vomeronasal organ (VNO) in the nasal septum is specialized in the detection of particular chemical signals, called pheromones, which are correlated to mating and aggressive behaviours. Recently, evidence of this organ has been found also in humans – to the great joy of perfumers (Meredith, 2001; Thorne *et al.*, 2002).
■ The main olfactory epithelium (MOE) in the back of the nasal cavity detects volatiles. Its main features are an extraordinary sensitivity; the

human threshold of detection can be as low as 10^{-11}M for trichloroanisole – rio corky flavour (see 3.7.2.6). Humans are thought to be able to discriminate between several thousands of different odours, even if in theory the number is billions, but bloodhound dogs have been shown to detect some odours at a concentration of 10^{-17}M (Dulac, 2000a), and to possess an extremely large discriminatory power.

■ The taste sensory epithelium of the mouth responds principally to water-soluble chemicals present in the oral cavity. Chemicals from food termed tastants dissolve in saliva and contact the taste cells through the taste pores, to give a fast response, in terms of acceptance or rejection, to food to be ingested.

These chemical senses are semi-independent because, even if they are located in different zones and their reception/transduction mechanisms are different, they often work together to give the organism a unique answer. The perception of coffee quality involves both olfactory and gustative sensory inputs: after drinking and swallowing, volatiles released in the throat reach the nasal cavity and stimulate the olfactory epithelium.

9.2 GUSTATION

Gustation has been studied for many years, but only recently has progress in genetics and molecular biology enabled greater understanding of its fundamental characteristics, namely reception and transduction mechanisms. Lately many reviews have appeared, among which should be mentioned (in alphabetical order) Dulac (2000a), Kinnamon (2000), Lindemann (2001) and Smith and Margolskee (2001).

Even though people often use the word 'taste' to mean 'flavour', in the strict (and correct) sense, the word is applicable only to the sensations coming from specialized cells in the oral cavity. Gustation or taste is the sensory system whose scope is essentially an evaluation of the food to be ingested, and in fact all organisms, be they bacteria or mammals, perform this check by chemoreceptive evaluation. Taste evolved in order to help organisms to detect hazardous substances, such as acids or alkaloids, and to recognize nutritionally important compounds like sugars, salts and aminoacids (Lindemann, 2001). Actually, even a few days' old baby is able to discern between sweet and bitter and react with a smile when tasting a sweet substance, or with a cry when tasting a bitter one (Ganchrow et al., 1983).

While the olfactory system can recognize and discriminate a huge number of different odours in humans, the taste system discriminates only a few basic taste qualities: sweet, salty, sour, bitter and umami – the sensation elicited by monosodium glutamate.

Taste receptor cells (TRCs) supply information for the recognition and response to the nutrients we need. The sweet taste of sugars, for example, is a stimulus for eating carbohydrate-rich foods. Salty and acid tastes are markers for ions in food: evidence is reported that humans and other animals with sodium and other dietary deficiencies will seek foods rich in sodium and minerals, as well as vitamins (Smith and Margolskee, 2001). Umami can be considered a signal coming from protein-rich food, whereas the perception of bitterness is fundamental for its protective role, since many harmful substances, like strychnine and other alkaloids, often present a strong bitter taste (Dulac, 2000a). The sour taste of spoiled foods also contributes to their avoidance. All animals, including humans, generally reject acids and bitter-tasting substances at all but the weakest concentrations. And in fact thresholds of perception are very different, ranging from about 10^{-10} M for denantonium benzoate (benzyldiethyl [(2,6-xylylcarbamoyl)methyl] ammonium benzoate), the most bitter substance known, to 8×10^{-4} M for citric acid and 9×10^{-3} M for sucrose.

Taste signals also evoke physiological responses, such as insulin release, that aid in preparing the body to use the nutrients effectively. Individual taste stimuli, even within one quality, often differ greatly in molecular size, lipophilicity and pH (Kinnamon, 2000). For example, as far as sweet taste is regarded, the molecular weights can be as low as 92 for glycerol, up to 180 for glucose, 205 for saccharin, 342 for sucrose, but also as high as 11 233 for monellin or 22 204 for thaumatin.

It is easy to understand that there may be multiple mechanisms for transducing even a single taste quality such as sweetness. Moreover, there is a wide range of variation in people's sensitivities to compounds belonging to the same 'taste-category' (Delwiche et al., 2001), and in fact some researchers provocatively assert that there may not be any 'fundamental' tastes (Ishii and O'Mahony, 1987; Delwiche, 1996). The division of responses to taste stimuli into five types might be nothing but a useful categorization that humans do, according to the particular function that the molecules associated to those chemical stimuli have.

9.2.1 Reception – taste buds

A field of taste research is devoted to the comprehension of how molecules corresponding to the taste qualities become electrical signals that carry information to the brain about what substance is in the mouth. This process is called transduction and starts in the oral cavity.

Human TRCs can be found as single cells or packed in onion-shaped structures called taste buds (Figure 9.1), which consist of up to one hundred cells. Taste buds are located predominantly on the tongue, but can be found also in the soft palate, in the epiglottis. The taste marker molecule gustducin can be found even in nasal mucosa and in the stomach (Lindemann, 2001).

Gustducin is a taste-specific G-protein: the name 'G-protein' derives from the fact that the activity of such proteins is regulated by a chemical called guanosine triphosphate (GTP). G-proteins are made up of three subunits: an α subunit that can be considered the active portion, and β and γ subunits that regulate. When a tastant binds to G-protein coupled receptors (GPCRs) on a taste cell's surface, the α subunit dissociates from the $\beta\gamma$ portion, and activates a nearby enzyme. The enzyme starts a transduction cascade, ending with a firing of an electrical impulse, which is sent to the brain. The same transduction mechanism can be found in olfaction.

In humans the tongue contains around 5000 taste buds, but the number of buds can vary considerably from one person to another. Most of taste buds on the tongue are located within specialized structures called papillae, which give the tongue its velvety appearance (Smith and Margolskee, 2001).

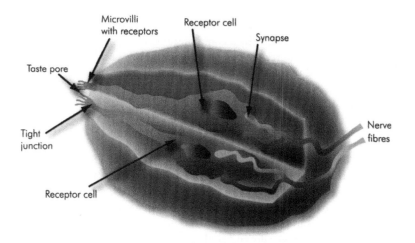

Figure 9.1 Tastebud with two receptor cells (Lindemann, 2001). On the top surface of the bud, *microvilli* are noticeable; on the bottom part the synapses are shown. Tastants enter the bud through the pore, are processed in the cells, and a signal is sent to the brain by the nerve fibres. Reproduced with permission.

Taste buds in the anterior two-thirds of the tongue reside in fungiform papillae, each containing one or at most a few taste buds. The posterior region of the tongue contains the *papillae circumvallatae* and the *papillae foliatae*, each of which contains hundreds of taste buds (Dodd and Castellucci, 1991). The most numerous papillae on the tongue, called 'filiform', lack taste buds and are involved in tactile sensation.

9.2.2 Taste coding in the brain

Many research papers on animals and humans demonstrate that there is not always a strict correlation between taste quality and chemical class, particularly for bitter and sweet tastants. Bitter compounds can be salts like magnesium sulphate, aminoacids like L-phenylalanine, alkaloids such as caffeine, or glycosides of phenolic compounds (naringin in grapefruit). Many carbohydrates are sweet but some are not. Furthermore, very disparate types of chemicals can evoke the same sensation: the chemical structures of artificial sweeteners aspartame and saccharin have nothing in common with sugar. On the contrary, the compounds that elicit salty or sour tastes are less diverse and are typically ions.

Coding in the periphery of the taste system could occur at two levels: taste receptor cells and afferent fibres. In principle, a taste receptor cell could be tuned to a single modality (e.g., sweet, sour, bitter, or salty), or to more than one modality. Likewise, subsets of cells having similar response profiles could be innervated by a common fibre, or single fibres may carry information from different types of cells. Whether and how individual taste cells respond to multiple chemical stimuli is still a matter for debate.

9.2.3 Transduction mechanisms

Given a so widely varying class of compounds eliciting a particular taste, much effort has been devoted to understanding the molecular components of the taste transduction mechanism, leading to the conclusion that multiple reception–transduction systems occur for different tastes and even for different tastants of the 'same' taste.

Ionic stimuli, such as salts and acids, can depolarize taste cells by direct interaction with particular ion channels. More complex stimuli, such as aminoacids, sugars and most bitter-tasting compounds, bind to surface receptors that trigger a series of signals within the cell, the result of which is the opening and closing of ion channels.

9.2.3.1 Salty and acid tastes

Two tastes can detect ions in the oral cavity: salty and sour.

Salty taste can regulate the assumption of NaCl and other minerals, and it is therefore very important for homeostasis processes; it is elicited by many ionic species, but the component that is probably most important, and surely most studied, is the one due to the presence of sodium ions. Sodium ions penetrate into TRCs through Na-channels, which are located on the top or lateral surface. The accumulation of sodium ions leads to an electrochemical change, called depolarization, which eventually is transformed into an electrical impulse that is sent to the brain through nerve cells. TRCs then repolarize, that is, reset.

Na^+-channels play a role in NaCl taste transduction, as well as in the transduction process of other salts, like KCl or NH_4Cl. Thus salty taste detection and transduction can both depend on the use of certain combinations of common and specific transcellular pathways (Lundy *et al.*, 1997).

Acidity can be considered a quality in many foods and beverages when of low intensity, and it is in fact a marker for good quality coffee, but it becomes unpleasant as intensity grows (Ganchrow *et al.*, 1983). Its role for the organism could be to give an alarm for an unripe fruit or spoiled food. Acids taste sour because they dissociate into hydrogen ions (H^+) in solution. Those ions act on a taste cell in three ways:

- By directly entering the cell as ions or undissociated weak acid.
- By blocking potassium ion (K^+) channels on the microvilli: K^+-channels are responsible for maintaining the cell membrane potential at a hyperpolarized level. Block of these channels causes a depolarization, Ca^{2+} entry, transmitter release and increased firing of electrical signals.
- By binding to and consequently opening channels and thus allowing other positive ions to enter the cell. These positive charges eventually depolarize the cell and an electrical signal is then sent to the brain.

A fundamental issue surrounding sour taste reception is the identification of the sour stimulus. It must be pointed out that no strong correlation has been found between the pH of a solution (e.g., coffee) and the perceived sourness, even if to some extent the intracellular pH of taste cells is correlated with extracellular changes in pH (Figure 9.2). Recently a hypothesis has been successfully tested, that acids induce sour taste perception by penetrating plasma membranes as H^+ ions or as

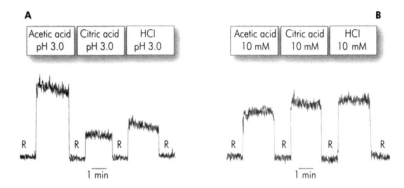

Figure 9.2   Recorded nerve signals from responses to different acid stimuli with (a) same pH and (b) same concentration (Reproduced with permission from Lyall *et al.*, 2001)

undissociated molecules and decreasing the intracellular pH of TRCs. The results evidence that taste nerve responses to weak acids (like acetic acid) are independent of stimulus pH but correlate strongly with the intracellular acidification of polarized TRCs (Lyall *et al.*, 2001).

With regard to coffee, there is so far no clear consensus (Woodman, 1985; Heath, 1988; Balzer, 2001) as to which acids are the most important in determining acidity by pH measurement and perceived acidity, and whether pH or titrable acidity (Maier, 1987) is a good chemical estimator for perceived acidity. Extending the above-mentioned results to humans, titrable acidity at pH 6.5, which is the average pH of human saliva, can be considered the optimal chemical measure of perceived acidity.

9.2.3.2 Bitter and sweet tastes

The sensations of bitter and sweet tastes derive from the interaction of tastants with GPCRs in the apical membranes of TRCs. The top surface of TRCs, which makes contact with the oral cavity, is rich in microvilli containing GPCRs, ion channels and other transduction elements. The lateral region of TRCs contains ion channels and synapses with afferent taste nerves (Margolskee, 2002).

Like acidity, **bitterness** is also acceptable when of low intensity, but it becomes absolutely disagreeable when intense. It is usually associated with the idea of something harmful, and in fact many toxic substances (as nicotine and strychnine) taste bitter. Bitterness is elicited by a number of chemically heterogeneous substances, but they present the same transduction mechanism, through GPCRs and second messengers.

By scanning mouse genomic databases, a group of new GPCRs, the T2R family, was discovered in 2000 by different research groups (Adler *et al.*, 2000; Chandrashekar *et al.*, 2000; Matsunami *et al.*, 2000). This family of receptors is very large, consisting of 40–80 genes, thus explaining the large number of different bitter-tasting compounds. In humans, the T2R family comprises more than 20 genes coding for GPCRs.

Sweet taste is mainly caused by soluble carbohydrates, and it may represent a stimulus for the intake of calorie-rich food. However, a large number of non-carbohydrate molecules are also sweet. Sweet taste elicits a strong hedonic effect (Ganchrow *et al.*, 1983). Like bitter stimuli, sweet tastants also bind to GPCRs on a taste cell's surface and trigger the downstream process of secondary messengers, which eventually send the electrical impulse to the brain. The candidate receptor for sweet taste is T1R3, a GPCR, which was found simultaneously by four laboratories in 2001, by means of the latest tools and techniques of genetics and molecular biology (Nelson *et al.*, 2001; Montmayeur *et al.*, 2001).

Ultimately, the study of sweet taste perception should help us explore the hedonic aspects of taste and perhaps understand why 'a spoonful of sugar helps the medicine [and the cup of coffee] go down' (Sherman and Sherman, 1964).

9.2.3.3 Umami taste

Umami, a term derived from the Japanese *umai* (delicious), corresponds to a taste sensation that is qualitatively different from sweet, salty, sour and bitter. Umami is a dominant taste of food containing L-glutamate, like chicken-broth, meat extracts and ageing cheese. The biological signifi-cance of this basic taste, discovered about 100 years ago, is comparable to that of sweet taste, being elicited by protein-rich food.

A taste receptor for L-glutamate could be linked to one of the receptors for glutamate already known from neuronal synapses. Starting from this hypothesis, Roper and colleagues (Chaudhari *et al.*, 2000) found a subset of gustative cells which presents a structure adapting the receptor to the concentrations of glutamate present in food: thus a receptor has been identified which responds to aminoacids.

Other laboratories (Matsunami *et al.*, 2000; Nelson *et al.*, 2002), have identified and characterized a mammalian aminoacid taste receptor, which respond to L-aminoacids and monosodium glutamate, belonging to another family of genes, sharing many analogies with T2Rs, recognized as receptors for bitter compounds.

Umami stimuli are known to bind to G-protein-coupled receptors and to activate second messengers. But the intermediate steps between the second messengers and the release of neurotransmitters are unknown.

9.2.4   Challenging the tongue map myth

One of the greatest myths about taste – which is unfortunately still found in textbooks – is the so-called 'tongue map', showing a regional specialization of the tongue for a particular taste quality, indicating that sweetness is detected by taste buds on the tip of the tongue, bitterness at the back and saltiness and acidity on the sides.

The misconception arose early in the twentieth century as a result of misinterpretation of a German work (Hänig, 1901), and still resists in the literature. In the actual research it had been, on the contrary, determined that each chemoreceptive area of the human tongue responds to each of the qualities of sweet, sour, salty and bitter taste. Only minor differences in subjective sensitivities, computed as the inverse of the detection threshold, were noted across areas (Lindemann, 2001). The misinterpretation and the lack of check of the original reference led to an interesting graphical 'evolution' of those images, providing the impression of specialized areas for taste reception. In reality, taste buds responsive to each category of taste stimuli are found in all regions of the oral cavity, even if with minor differences; moreover, each individual taste bud contains taste cells with sensitivity to the different types of tastants. Taking bitter compounds as an example, some of them are perceived more intensely on the tip of the tongue, others on the back of the tongue and others in the throat after swallowing. The lack of specificity of individual taste buds across the tongue is directly confirmed by the large distribution of the T2R genes in taste buds of all types of papillae (Dulac, 2000a).

9.2.5   PROP sensitivity as a marker for food choice

There are both cultural and genetic reasons why we perceive foods differently, and have different habits as far as eating is regarded. For example, thresholds of perception and sensitivity to a bitter compound called 6-N-propylthiouracil (PROP) vary enormously across individuals. The differences are so sharp that scientists tend to categorize people into 'non-tasters', 'tasters' and 'super-tasters', according to their sensitivity to PROP (Bartoshuk, 2000). The general dislike of bitterness led early investigators to suspect that 'tasters' and 'super-tasters' would tend to dislike bitter foods, coffee included.

However, other scientists are more sceptical (Mattes, 2002). While significant differences in bitter taste perception have been observed under laboratory conditions, their impact on actual food choices and eating habits may be limited. The research has not reached a definitive answer.

As far as coffee is concerned, out of respondents who consumed drip coffee, tasters were more likely to drink coffee with milk/cream and sweetener than non-tasters were. Conversely, a greater percentage of non-tasters preferred their coffee black as opposed to tasters (Ly and Drewnowski, 2001).

9.3 OLFACTION

Despite the aesthetic and emotional role to which humans have confined olfaction, for most animals it represents a fundamental sense, on which they rely in order to recognize predators or preys, food, kin and mates. Today's big challenge of olfaction research is to understand how scents are recognized and how recognition of odours in the nose is transferred to the brain to provide interpretation. Coffee, for example, emits a specific combination of some 1000 different odour molecules (Flament, 2002): therefore the central olfactory system has to integrate signals from a huge number of odorant receptors (OR) in order to provide recognition and interpretation, which in the end is translated into a smile of appreciation for a lovely sensory experience or a grimace for unsatisfied expectations.

Olfaction is the chemical sense whose scope is to inform the organism about the chemical composition of the environment. The breakthrough for understanding the mechanisms of this sense came in 1991 with Buck and Axel's discovery of the genes encoding ORs in mammals (Buck and Axel, 1991), and since then new findings appear regularly in scientific journals, adding to our understanding of this sense. Among the latest reviews on the subject are (in alphabetical order): Axel (1995), Doty (2001), Firestein (2001), Leffingwell (2002), Mombaerts (2001a).

The human nose is an extremely sophisticated device of molecular recognition, still far superior to the electronic noses found on the market (Mombaerts, 2001a). A common misconception is that humans can smell some 10 000 odours; this view arises from confusion between detection, discrimination and identification of odours. Theoretically there is no upper limit to the number of possible detected smells. It is true, however, that humans are very good at detecting odours, but lack the ability to identify them. In other words, our lives are not ruled by smells, whereas odours and chemical signals in general play a fundamental role in the life of most animals, from basic tasks like finding foods to complex functions, like social behaviour.

When we smell an odour out of its normal context (e.g. in a laboratory vial), it is very common to experience the 'tip-of-the-nose' phenomenon: we are familiar with it, but just cannot remember its name (Pines, 2001). That is, it is easier to recognize the scent of coffee in a coffee shop than in a theatre, because the context helps to narrow down the list of potential odours. Moreover, the emotional role of smell is high.

Memories associated with odours are linked to emotions; conversely they appear to play a special role for memories: it is the 'Proust effect'. In A la recherche du temps perdu, French novelist Marcel Proust described his feelings after consuming a spoonful of tea in which he had soaked a madeleine:

> The taste [meaning, of course flavour] was that of a little piece of madeleine which on Sunday mornings in Combray ... my Aunt Leonie used to give me, dipping it first in her own cup of tisane ... Immediately the old grey house on the street, where her room was, rose up like a stage set ... and the entire town, with its people and houses, gardens, church, and surroundings, taking shape and solidity, sprang into being from my cup of tea.

What if he had been drinking an espresso?!?

9.3.1 What are odours?

As far as odorants are concerned, olfaction scientists have to face a complex problem: there is no 'smell scale', and odorous molecules vary widely in chemical composition and three-dimensional shape (Pines, 2001): there is no comparison with sight, where wavelengths and corresponding colours can be placed on a linear scale. Represented in the olfactory repertoire are aliphatic and aromatic molecules with varied carbon backbones and diverse functional groups, including (in alphabetical order) alcohols, aldehydes, alkenes, amines, carboxylic acids, esters, ethers, halides, imines, ketones, nitriles, sulphides and thiols (Firestein, 2001). The requirements a molecule must fulfil in order to be an odorant are that it should be volatile and hydrophobic. These requirements can be due to the fact that molecules have to reach the nose and have to pass through membranes. Moreover, their molecular weight should be less than approximately 300 daltons (one dalton is one-twelfth the weight of an atom of ^{12}C). Actually, the largest known odorous molecule is labdane, which has a molecular weight of 296. There is no definitive explanation of this requirement, although there may be some clue in that vapour pressure (and therefore volatility) decreases rapidly with molecular size. Conversely, some of the strongest odorants present high molecular weights. Moreover, 'the cut-off point is very sharp: for example, substitution of the slightly larger silicon atom for a carbon in a benzenoid musk causes it to become odourless' (Turin and Yoshii, 2003). Another hint of the correlation between dimension and chemoreception is that occurrences of specific anosmias are correlated with the molecular weight. Anosmia (from the Greek; *a* [no] *osmia* [smell]), or, more correctly hyposmia, is the decreased sensitivity to some or all odorants. Everybody is anosmic to some compounds, and there are molecules to which up to one-third of the human population is hypo-osmic, as, for example, benzyl salicylate.

9.3.2 Physiology: olfactory sensory systems, olfactory sensory neurons, olfactory epithelium, *glomeruli*

Receptors of three different neural systems can be found in most land mammals. They are located in the nasal cavity, and their role is to check the external environment. The main olfactory system is the sensor of the environment, responsible in large part for the flavour of foods and beverages. Additionally, this system serves as an early warning system for spoiled food and noxious or dangerous environmental chemicals.

A second olfactory system (the vomeronasal system) has developed for the specific task of finding a receptive mate (Dulac, 2000b). The nature of vomeronasal sensations, if any, is unknown to humans (Doty, 2001), even if chemical communication appears to occur (Meredith, 2001).

The trigeminal system provides a response to both chemical and non-chemical stimuli, inducing somatosensory sensations (e.g. cold, hot, spicy and tickling), and provoking reactions, secretion of mucus and halting of inhalation, in order to minimize possible damage to the organism.

Odorant stimuli are detected by olfactory sensory neurons (OSNs) located in the olfactory epithelium (OE), which is situated in the dorsal part of the nasal cavity. The OE is only a few square centimetres wide and it is composed of some millions of OSNs, which are generated *in situ* from supporting stem cells. Like in other epithelia, cell renewal persists throughout adult life to replace OSNs, which have a lifespan of weeks to months (Mombaerts, 2001a). The OSNs are bipolar neurons with a single dendrite that reaches up to the surface of the tissue and ends in a knob-like swelling with some 20–30 very fine projecting cilia. These cilia, which actually lie in the thin layer of mucus covering the tissue, are now known to be the site of the sensory transduction apparatus. A thin axon projects from the cell directly to higher brain regions (Dodd and Castellucci, 1991) (Figure 9.3a).

The OE is a complex melange of different OSN populations randomly distributed, with the aim of maximizing the probability of interaction of an odour molecule with its receptor. All axons of OSNs which express a particular receptor, no matter what their position is in the OE, converge to a single 'target' in the olfactory bulb (Mombaerts, 1999), called glomerulus. Glomeruli are spherical specialized synaptic arrangements, in which thousands of axons from OSNs converge (Firestein, 2001). Olfactory glomeruli have been viewed as functional units since they were first described in the nineteenth century, but their role was unknown until the discovery of odorant receptor genes. A century after its anatomical description, molecular biology has proved that a glomerulus receives the axonal input from OSNs that express a particular OR gene, thus providing for integration of information and for an increase in the signal-to-noise ratio (Mombaerts, 2001a). The concept of glomerular convergence is consistent with a widely accepted view of olfactory coding, in which the particular combination of activated glomeruli informs the brain on what the nose is smelling (Buck, 1996).

Sensory information is processed within the bulb and relayed to the olfactory cortex and several other centres in the brain, where it gives rise to the olfactory perception of, for instance, coffee. The olfactory pathway is unique among the senses because in olfaction the cell detecting the stimulus, the OSN, is a neuron that projects its axon directly to the brain (Mombaerts, 2001b).

9.3.3 Olfactory receptor genes

Scientists have long been convinced that receptor proteins capable of recognizing and binding odorants are on the cilia in the OE, but the discovery happened only in 1991, when Buck and Axel found the family of transmembrane proteins that were believed to be the ORs and some of the genes that encode them, thus giving a partial answer to how the OS responds to the thousands of molecules of different shapes and sizes, i.e. the odorants, and how the brain makes use of these responses to discriminate between odours (Buck and Axel, 1991). They cloned and characterized 18 different members of an extremely large multigene family

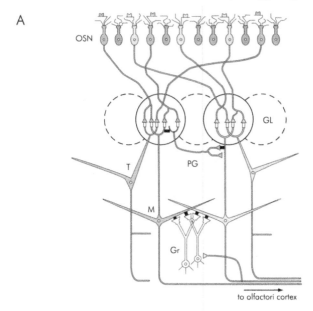

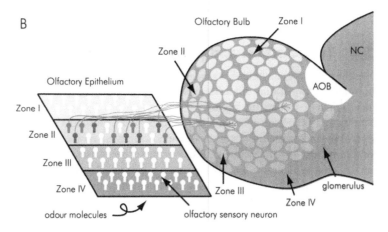

Figure 9.3 Schematic synaptic organization circuit, from the olfactory sensory neurons (OSNs) to higher regions in the brain (Mori *et al.*, 1999). The olfactory epithelium is mapped onto the olfactory bulb in the brain with a convergent topography. Axons from OSNs expressing the same odorant receptor converge to defined *glomeruli*. (Reprinted with permission from Mori *et al.*, 1999. Copyright © 2005 American Association of the Advancement of Science)

that encodes the seven transmembrane proteins whose expression is restricted to the OE. All the proteins they found contained sequence similarity to other members of the 'G-protein' coupled receptor family (Leffingwell, 2002).

A second giant step in understanding the olfactory reception process happened in 1998, when Firestein and colleagues (Zhao *et al.*, 1998) proved by experiment, measuring the electrical activity in the neurons, what Axel and Buck had found were actually ORs. In 2001, Lancet and co-workers reported a first draft of the human OR repertoire through a data-mining process (Glusman *et al.*, 2001). This dataset includes 906 human OR genes, of which more than two-thirds seem to be pseudogenes (i.e., unexpressed genes) (Mombaerts, 2001b). This extremely high number of genes for ORs in the mammalian genome (it is always approximately 1000) makes it by far the largest family of GPCRs, and probably the largest gene family in the entire genome (Firestein, 2001). This family represents 1% of the genome, which means that, at least in animals like dogs or rats, where almost all genes are expressed, 1 out of every 100 genes is used for the detection of odours. This astounding number of genes reflects the crucial importance of smell to animals (Pines, 2001).

9.3.4 Transduction and processing in the brain

Once the receptor has bound an odour molecule, a cascade of events is initiated that transforms the chemical energy of binding into a neural signal, which is sent to the brain.

The ligand-bound receptor activates a G-protein, causing a cascade of events that ends with the firing of an electrical signal to the brain. The central olfactory system receives the odour molecule information through axons of sensory neurons. The information is processed and integrated as the olfactory quality of objects. The human perception of an odour is characteristic in that it is usually associated with pleasant or unpleasant emotions (Mori *et al.*, 1999). In order to understand the rules for interpreting the olfactory stimuli, it is useful to use a 'molecular receptive range' for OSNs (Figure 9.3b). Individual OSNs express only one type of OR gene out of a repertoire of up to 1000 genes. It appears that to match the ability of organisms to detect far more than 1000 odours, the odours must participate in some kind of combinatorial processing: that is, one receptor must be able to interact with several discrete odorants. Olfactory sensory neurons should respond to many odorants with varying affinities. Conversely, an odour molecule should interact with more receptors. In other words, an individual odour will activate multiple glomeruli in the olfactory bulb, and different odours activate overlapping but non-identical patterns of glomeruli. One particularly striking case involves enantiomers (molecules that are optical isomers, or mirror images, of one another).

It seems that receptors that recognize similar odours tend to stimulate the same area of the olfactory bulb (Figure 9.3b) (Firestein, 2001). Recent studies have proved that the olfactory axon projection follows a specific convergence pattern towards glomeruli (Mori et al., 1999). There are specific zones of activity in the olfactory bulb, caused by odour-specific patterns. These zones seem to correspond to individual glomeruli, since there is a correspondence between the size distribution of the discrete responses and the size distribution of glomeruli (Bozza and Mombaerts, 2001).

Despite these findings, the basic understanding of the functional organization of the axonal projection to the olfactory cortex and of the neuronal circuits in the olfactory cortex is still unknown. The next fundamental step will be therefore the comprehension of the neuronal mechanisms by which the olfactory cortex combines or compares signals from all different glomeruli (Mori et al., 1999).

9.3.5 How many odours can we detect?

Different figures are given in the literature on the number of odours humans can detect, ranging from approximately 2000 to more than 100 000. Theoretically there could be billions, based on the possible combinations of 1000 receptors. In fact, the question is probably irrelevant, just as it makes little sense to ask how many colours or hues we can see (Firestein, 2001). Perfumers, chefs, sommeliers, coffee cup-tasters or highly trained animals can discriminate more odours than the average, but this is due to experience, not to a difference in the 'measurement instrument'. Physical chemistry may be the primary limiting factor, as odour chemicals must possess a certain volatility, solubility and stability to act on the nasal sensory tissue.

The variable sensitivity of individual glomeruli produces distinct maps for different odorant concentrations. It seems that odorants and their concentrations can be encoded by distinct spatial patterns of glomerular activation (Rubin and Katz, 1999). For example, methanethiol, which smells like rotten cabbage at high concentrations, has a sweet coffee-like aroma at low concentrations (Flament, 2002). However, most odours remain constant in their quality over orders of magnitude of concentration. Amylacetate, a pleasant fruity smelling substance also found in coffee aroma, can be easily identified at concentrations from 0.1 μM to 10 mM (Firestein, 2001).

9.3.6 Cross-cultural differences

Social and cultural cues provide another interpretation of non-sensory influence on perception, and are potent modulators of responses to odours and foods. Cross-cultural studies show how cultural experience can affect expectations, thereby altering sensory perception (Cristovam et al., 2000; Hübener et al., 2001; Prescott et al., 2002).

Much work has been carried out in the past decade, with increasingly accurate and standardized psychophysical tests leading to the following broad results, reported by Doty (2001):

■ Women, on the average, have a better sense of smell than men – superiority that goes beyond cultures and is noticeable already at the age of four; they also retain the ability to smell longer than men.
■ There is a substantial genetic influence on the ability of humans to identify odours.
■ Starting about in the fourth decade of life, many individuals begin to notice a gradual decline in their ability to smell, affecting both sensitivity and ability to identify odours. Because olfactory stimulation makes up a major component of food and beverage flavours, people with olfactory decrements often report a diminished ability to taste. A major loss of olfactory function occurs after age 65, with over half of those between 65 and 80, and over three-quarters of those 80 years of age and older, having such loss.
■ The decrement in olfactory function associated with smoking is present in former smokers and recovery to pre-smoking levels, while possible, can take years, depending on the duration and amount of past smoking.
■ The olfactory function is compromised in urban residents and workers in various industries.

9.4 HUMAN CHEMOSENSORY PSYCHOPHYSICS

So far, reception and transduction mechanisms for chemical senses have been described. When we drink a cup of espresso, the chemical stimuli coming from taste and smell are transduced, together with tactile sensations, into electrical signals, which are sent to the brain. Taste, smell and trigeminal systems are separate systems with separate functions, as discussed above. As in vision, where movement, position and identification of objects are perceived separately, without, however, perceiving them as independent properties, the experience of flavour integrates information from smell, taste and the trigeminal system. The orbitofrontal cortex may be the brain structure that performs this integration (Abdi, 2002), with an analytical and hedonic reaction and a final response in terms of recognition and liking of the characteristics of the coffee (or whatever food or beverage).

How can this response be correlated to the physico-chemical characteristics of the coffee and the chemoreception process? Psychophysics is the branch of research that interprets perception, that is, the psychological process that provides the link between the physical stimulus and our everyday experience of odour, taste, tactile sensations and sight. Psychophysics uses specific behavioural methods to determine the relationship between the physical stimuli and the perceptions they elicit, that is, people's subjective experience. It is a method for measuring some characteristics of an object, specifically those features eliciting a reaction from our senses, which goes beyond the '*I like it, I do not like it!*' judgement.

In an analytical measurement method, like, for example, headspace gas chromatography, a sample is prepared and injected into the column, and the detector gives, as an output, the aroma profile of the sample, in terms of quality and quantity of volatile compounds. When dealing with sensory analysis, the detectors are our senses of gustation and olfaction. The output is a profile in terms of quality and quantity of sensory attributes.

Among the literature on best sensory analysis practice and methods is the pioneer work of Amerine and Pangborn (Amerine *et al.*, 1965), and more recent works by Stone and Sidel (1985), Moskowitz (1988), Piggot (1988), Lawless and Klein (1991) and Meilgaard *et al.* (1999).

9.4.1 Psychophysical scales

An important issue, which has to be faced in sensory analysis, is the use of scales. Scales are psychophysical metrics used to evaluate the intensity of a stimulus. Usually five types of scales are used:

- Category scale (e.g., barely detectable, weak, moderate, strong, very strong), which is easy to use but has a fixed number of categories and is sensitive to bias.
- Discrete numerical scale, the numerical counterpart of the category scale. The most commonly used is the 9-point scale (Peryam and Girardot, 1952).
- Visual analogue scale, that is a straight line (with or without semantic or numeric anchors) on which the intensity of the sensation has to be rated. This scale presents an infinite number of categories and it is easy to use, but it does not allow between-subject comparison, because different assessors have their own 'mental' scale.
- Ratio scale, in which the assessors evaluate the ratio between two stimuli. It provides information about the relative intensity of stimuli, but it requires considerable training and does not allow between-subject comparisons.
- The category-ratio scale, like the Labelled Magnitude Scale (Green *et al.*, 1993; 1996), combines features of ratio and category scales, allowing comparisons between subjects and providing absolute intensity estimates. However, when evaluation of a complex stimulus is involved, like in coffee, it becomes quite cumbersome.

No scale is correct or incorrect, in an absolute sense: the use of one or the other depends on the task being performed, on the differences looked for, and on the panel being used. When comparing very different items, such as a robusta with an arabica, the category scale, or numeric discrete scale, would be sufficient; but in the comparison between, for example, two lines of the same cultivar, then an analogous scale or a ratio scale should be used, to look for fine differences. Non-linear anchored scales can be used in some situations, like log scales or power scales. The problem that the use of anchored scales brings along is the tendency of assessors to stick to the anchors.

In a preference test with consumers, a 5-point or 9-point scale for the degree of liking (1 = not at all ... 5 or 9 = very much) would be sufficient: a more detailed scale could lead to confusion for the consumer.

9.4.2 Stimuli-reaction categorization: type of sensory tests

The reactions to the presentation of a sensory stimulus can be divided into three types:

■ qualitative perception of intrinsic characteristics – analytical;
■ quantitative perception of the intensity of the stimulus – analytical;
■ hedonic reaction, i.e. the degree of pleasure caused by the stimulus – hedonic.

A sensory stimulus (e.g., an espresso coffee) is described first in terms of its characteristics (e.g., acid or bitter taste, smooth mouthfeel, chocolate and toasted bread aroma notes), then these characteristics can be quantified (on a given scale), and finally the degree of pleasure caused by the stimulus is noted. This last, being a hedonic judgement, is not an attribute of the product itself. It can vary with culture, gender, age, and it represents the basis of consumer segmentation science. The two types of assessments (analytical or hedonic) should not be performed together by the same panel, because trained assessors are not representative of consumers' preferences, and, moreover, they are particularly critical of the product. On the other hand, consumers are unable to describe correctly the sensations they experience, for lack of training (Majou, 2001).

However, statistical techniques, like preference mapping, allow the drawing of correlations between sensory characteristics and consumers' preferences.

9.4.2.1 Analytical tests

Analytical tests can be essentially of two types: discriminative or descriptive.

Discriminative tests are used to check whether two products can be differentiated. For example, the roasting process is changed, but the sensory characteristics of the coffee are not to be affected, or, conversely, one ingredient of the blend is changed, in a way to be perceived by consumers. For these tasks discriminative tests are be used, with a trained panel, not necessarily composed of experts, who are asked to identify which are the identical products. Commonly used tests are the paired comparison test, the triangle test and the 'two-out-of-five' test. What changes is the probability of a correct guess, and therefore the number of judges and correct answers required in order to reach a result with a certain degree of confidence.

Descriptive tests are used to characterize the sensory profile of a product, by describing its perceived sensory features. The characterization of coffees from different origins, as well as the profiles of different brewing methods, is achieved through such tests.

Descriptive analysis techniques involve a panel to specify the intensities of specific attributes. In order to obtain reliable results, a trained panel of expert judges is required, who should be trained with reference standards.

The test methods are classified according to the result they provide. There can be qualitative tests, like the flavour profile, and quantitative tests, like the Quantitative Descriptive Analysis (QDA). The panel must rate the perceived intensity of each attribute based on a psychophysical model for subjective intensity, which accepts the proposition that individual ratings vary in subjective intensity as a function of stimulus concentration. The result of these tests is usually visualized in 'spider' plots.

It must be noticed that it is usually assumed that each descriptor is unrelated to the others (i.e., they are orthogonal in mathematical terms) and it is therefore necessary for an exhaustive profile of the product. Moreover, it is assumed that they are perceptually separable features that we can perceive distinctly, which is not the case for complex stimuli (Lawless, 1999) like coffee, which consists of taste qualities and a mixture of hundreds of different, distinct odours.

The intensity-based descriptive approach can bring with it several hidden problems, and therefore it can be dangerous to assume a perfect correspondence between the descriptive data that are obtained and the sensations actually experienced by the panellists.

These considerations do not imply that the profile data cannot be used to differentiate and characterize products, but it is scientifically more honest to state that 'ratings for coffee A were higher or lower than for coffee B', rather than 'Coffee A was perceived as more intense in its bitter and smoky notes and less flowery'. Such an inference about perception would be unjustified or even incorrect (Lawless, 1999).

9.4.2.2 Hedonic tests

Hedonic tests evaluate the liking or acceptance of a product. They should be carried out with an untrained panel of consumers, which must be representative of the population; quite large numbers are therefore required. In the sensory testing sequence, acceptance testing usually follows analytical tests, which have reduced the number of possible alternatives. If one ingredient of the blend is changed, to the extent that consumers will likely perceive it, and the sensory profile of the new blend is established, a check on the acceptance of the product must be performed before launching it into the market.

Hedonic tests can be preference tests, where two or more products (e.g., the old blend and the new blend) are compared, or acceptance tests, where the degree of liking (for example on a 9-point scale) is evaluated.

9.4.3 Espresso cupping technique (fundamental aspects)

Espresso cupping has been dealt with in 8.2.2 from a practical point of view. Here more theoretical aspects will be considered.

Cupping methods followed in the industry are quite heterogeneous. In the following some general rules will be given for a correct cupping procedure. These are valid for all brewing methods, espresso included. Cupping should be performed alone, without external stimuli, in order to minimize noise and bias introduced by interaction with other assessors.

A sip of some 10–15 ml of coffee should be taken. Salivary-induced changes in pH of sour taste stimuli affect taste perception, and therefore smaller volumes of espresso would be perceived as less sour than larger volumes. Saliva induces an increase of solution pH, and a small quantity (4 ml) can be totally buffered by saliva (Christensen et al., 1987), thus distorting the perception of acidity.

In order to avoid the consequences of caffeine intake, when evaluating many samples a sip-and-spit procedure is recommended. Flavour perception during eating or drinking depends on the movement of odorants released in the mouth via the retronasal cavity to the olfactory epithelium. An accurate analysis of the swallowing process, by means of real-time magnetic resonance imaging (Büttner et al., 2002) has proved (what already was well-known in medicine) that during mastication and swallowing there is no connection to the dorsal oropharynx, the nasopharynx and the airways, preventing leaking and aspiration of food. When the bolus has left the pharynx, the upper oesophageal sphincter is closed and the larynx, epiglottis and velum palatinum return to their original position, followed by a short pulse of respiration, the so-called 'swallow breath', by which volatiles reach the retronasal cavity (Figure 9.4). A simple proof of this phenomenon can be performed by drinking an espresso with a nose-clip: only 5% of the population would recognize it as coffee.

The 'slurping' technique (Heath, 1988), that is the deliberate opening of the *velum palatinum* border when coffee is in the oral cavity, is used to avoid swallowing and yet perceive the aroma. In this way the amount of volatiles reaching the olfactory epithelium ranges from 30% to 130% of the values observed with swallowing, depending on how effective each panellist is in the control of exhaling (Büttner and Schieberle, 2000).

Frequent rinsing has been prescribed to minimize adaptation of salivary influence to taste perception, as well as to eliminate or minimize saturation or carryover effects, and to enhance discrimination among samples. Palate cleansers are commonly employed in sensory evaluation to rinse the mouth

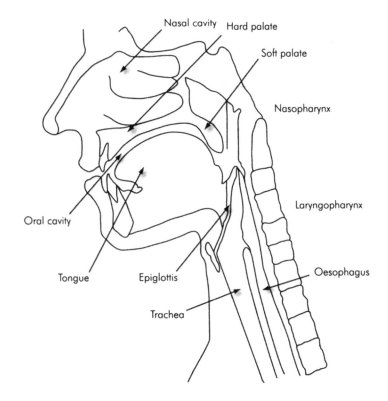

Figure 9.4   View of the mouth and retronasal cavities (adapted from Büttner et al., 2000). When the bolus has left the pharynx, volatiles can reach the retronasal cavity, thanks to the so-called 'swallow breath'

before and in between samples. Despite the widespread use, little literature regarding the effectiveness of these materials or methods is available, and some studies report that no significant differences between palate cleansers were found on any measure (Johnson and Vickers, 2002), unless in very particular test conditions.

When tasting coffee, several main aspects have to be evaluated and their definitions ought to be stated unequivocally. Unfortunately, between areas of investigation there is a wide range of definitions for the same terms used for describing a sensory experience. In the following, a brief survey of most commonly used definitions will be given with a stress on our own interpretation, which tends to align with the consensus in the field of sensory science.

9.4.3.1 Appearance

Appearance, mainly that of the *crema*, attracts consumer attention. It has been said that espresso is consumed first by sight. This is true also for other beverages, like wine, where wine tasters create perceptual illusions, describing the odours of a wine according to its colour (Morrot *et al.*, 2001). As discussed in 8.2.1, the *crema* should be persistent (some minutes), and made up of very small, hazelnut-coloured bubbles.

9.4.3.2 Gustatory sensations

Gustatory sensations, produced by the beverage's taste, encompass acidity, bitterness and sweetness (no salt or umami sensations are reported in coffee literature), whose perception-transduction mechanisms have already been discussed (see 9.2).

The stimulus eliciting **acidity/sourness** sensations acts on the same receptors – and sensory scientists use them as synonyms. Nevertheless, the reaction to it can have both positive and negative connotations, due to the fact that this stimulus is often coupled with other sensations. Therefore, two distinct terms are commonly used in the coffee field, according to the hedonic reaction that the stimulus causes:

■ Acidity, a basic taste produced by the solution of an organic acid. The International Coffee Organization defines it as 'A desirable sharp (producing a strong physical sensation as if of cutting) and pleasing taste is particularly strong with certain origins' (ICO, 1991), which applies, for instance, to Central American wet-processed coffees.
■ Sourness, a different basic taste descriptor that 'refers to an excessively sharp, biting and unpleasant flavour' (ICO, 1991). It is used when associated to an astringent or metallic sensation, typical of fast roasted coffees where the short time does not allow the extensive elimination of chlorogenic acids (Ohiokpehai *et al.*, 1982).

As in the case of acidity, the definition of **bitterness** has hedonic connotations. Bitterness is 'considered desirable up to a certain level' (ICO, 1991). When it brings about an unpleasant reaction, the attribute 'harsh' (rough or sharp in a way that is unpleasant, especially to the senses) is associated with the product.

In coffee cupping, **sweetness** is a marker of good quality coffee, and it is often associated with aroma descriptors such as fruity, flowery, 'chocolatey' and caramel-like (ICO, 1991).

9.4.3.3   Olfactory sensations

Olfactory sensations, produced by the beverage's aroma *sensu lato*, are commonly described by the terms odour and/or aroma.

Aroma can be defined as 'the property of certain substances, in very small concentrations, to stimulate chemical sense receptors that sample the air or water surrounding an animal' (*Encyclopaedia Britannica*, 2003). According to the ASTM, it is the 'perception resulting from stimulating the olfactory receptors; in a broader sense, the term is sometimes used to refer to the combination of sensations resulting from stimulation of the nasal cavity' (ASTM E253-03). ISO does not give a unique definition, describing it as 'an odour with a pleasant connotation', according to English sense and French informal language, or, according to French sense, as an 'organoleptic attribute perceptible by the olfactory organ via the back of the nose when tasting' (ISO, 1992a). Other sources characterize aroma as 'a sensory perception based on one's olfactory senses; i.e., sense of smell' (IFT, 2003), or prefer to define it from a chemical point of view, as illustrated in section 8.1.3, leaving the term odour to describe olfactory perception. One author in the coffee field states that it is due to the volatiles, perceived retronasally after swallowing, or by the 'slurping' technique (*see above*) (Heath, 1988). In other fields of research the definition is applied to pleasant sensations perceived by direct inhalation in non-specialized language (i.e. orthonasally), whereas, technically speaking, aroma is referred to as 'pleasant sensations perceived indirectly (i.e. retronasally) by the olfactory organ when tasting food or beverages' (Consejo Oleicola International, 1987). We accept the last definition, which brings about a difference between odour and aroma, as volatiles perceived orthonasally or retronasally. The latter has been called 'flavour' elsewhere in this book (see 8.2.2.4).

Odour is the sensation arising from the 'sniffing' of the cup (i.e., the volatiles smelled orthonasally). Sensory scientists tend to use the term odour to indicate an 'organoleptic attribute perceptible by the olfactory organ on sniffing certain volatile substances' (ISO, 1992a).

9.4.3.4   Flavour

Flavour is by far the most debated term and it varies according to the field of research. An early sensoric definition was 'the sensation realized when a food or a beverage is placed in the oral cavity. It is primarily dependent upon reactions to taste and olfactory receptors to the chemical stimulus. However, some flavours also involve tactile, temperature and pain receptors' (Beidler, 1958). Several standards have accepted it: for example, 'complex combination of the olfactory, gustatory and trigeminal sensations perceived during tasting. The flavour may be influenced by tactile, thermal, painful and/or kinaesthesic effects' (ISO, 1992a). The definition by ASTM is similar: 'Flavour is the complex effect of the basic tastes, olfactory sensations – perceived retronasally – and chemical feeling factors all stimulated by foods and/or other materials in the mouth' (Civille and Lyon, 1996). Or, 'flavour is the combined perception of mouthfeel, texture, taste and aroma' (British Standards Institute, 1975). A recent definition (*Encyclopaedia Britannica*, 2003) is: 'in sensory perception, attribute of a substance (apart from its texture and temperature) that is produced by the senses of smell, taste and touch and is perceived within the mouth. These sensations help to identify substances and are sources of enjoyment when eating and drinking'. One independent author proposes to reserve the term flavour for a pure olfactory sensation (Petracco, 2001): as elaborated in 8.2.2.4, this would ask for a new term to cover the overall feeling produced by aliments in the mouth. Summarizing, we could take as our definition the conclusion from a recent survey (Delwiche, 2003), defining flavour as a complex reaction to chemical stimuli, generated in the orbitofrontal cortex. Taste and smell are considered to be essential in the perception of flavour.

9.4.3.5   After-flavour

After-flavour is the sensation produced by the lingering of taste and aroma for a while (up to 15 minutes in the case of espresso) after having swallowed a beverage. It might be ascribed to the mouth-coating effect of the espresso, which has been related to the wetting properties of the beverage on the oral cavity, deriving from its surface and interfacial behaviour (Navarini *et al.*, 2002).

9.4.3.6   Astringency

Astringency, a tactile sensation belonging to 'mouthfeel' with a negative connotation for espresso coffee, is described as 'the shrinking of the epithelium probably related to precipitation of proteins of the saliva, and somehow related to pH viscosity and anion species' (Smith and Noble, 1997; Sowalsky and Noble, 1998). It is characteristic of an after-taste sensation consistent with a dry feeling in the mouth (ICO, 1991). A typical case is tannic acid, which precipitates gelatine to give an insoluble compound used to perform the task of tanning leather.

9.4.3.7 Texture

Texture is primarily the response of the tactile senses to physical stimuli that result from contact between the oral cavity and the food. It is a multi-parameter attribute, which derives from the structure of the food and is detected by several senses (Szczesniak, 2002). This is another difficult term to define since it means different things to different people (Bourne, 2002). Texture refers mostly to solid foods. When liquids are studied, other terms are used, like body and mouthfeel.

Body (see also 3.8.8) is defined by ASTM as 'the quality of a food or beverage relating either to its consistency, compactness of texture, fullness, flavour, or to a combination thereof' (ASTM E253-03). In the coffee field this attribute descriptor is used to characterize the physical properties of coffee beverages (ICO, 1991). We would tend to eliminate olfactory sensation from its definition, and describe the body of an espresso coffee as the integration of tactile sensations due to the interaction between the beverage and the oral cavity, the tongue and the palate. This definition cannot easily be applied to other brewing methods, where differences in tactile properties are minimal and not supported by scientific literature. However, with the term 'body' having so wide a range of definitions, and espresso being a typically Italian brewing method, we could propose the Italian term *corpo* to describe it. Adapting Szczesniak's definition of texture (Szczesniak, 2002), *corpo* is a sensory property and, thus, only a human being can describe it. It is a multi-parameter attribute not just oiliness or creaminess but a set of characteristics; it derives from the structure of the beverage (molecular, microscopic or macroscopic) and is detected by the senses of touch, pressure and sight (Navarini, 2003).

Mouthfeel is often explained as 'the mingled experience deriving from the sensations of the skin in the mouth during and/or after ingestion of a food or beverage. It relates to density, viscosity, surface tension, and other physical properties of the material being sampled' (Bourne, 2002).

The literature on the subject is of little help, apart from the pioneer work of Szczesniak (1979), who performed numerous studies in order to define and measure mouthfeel for a vast range of food and beverages (coffee included). In her survey, commonly used definitions for mouthfeel are categorized. Starting from Szczesniak's categorization, a set of descriptors, which apply to espresso coffee, can be found. They are divided into viscosity-related terms (thick, viscous), feel on soft tissue surfaces (smooth, creamy), body-related terms (heavy, light, watery), chemical effect (astringent), coating of oral cavity (mouth-coating), resistance to tongue movement (syrupy), after-feel in the mouth (lingering). Body thus represents a category of mouthfeel, relating to sensations apparently caused by the denseness of the rheological structure, therefore purely tactile. However, most terms in this group can be interpreted as having both flavour and mouthfeel characteristics (Szczesniak, 1979).

When dealing with coffee in general many definitions are found, most of them relating mouthfeel with olfactive properties connected to the intensity of the aroma perceived (Pictet, 1989), and describing it with rather vague terms such as 'rich' or 'round'. But, according to the International Coffee Organization, the descriptor mouthfeel comprises only body 'to describe the physical properties of the beverage', and astringency, 'characteristic of an after-taste sensation consistent with a dry feeling in the mouth' (ICO, 1991).

With regard to espresso, little is found in the scientific literature, where body is linked with physical parameters like viscosity or dry matter (Petracco, 1989). The body of an espresso coffee is, in our opinion, better described as a purely tactile sensation due to the interaction between the beverage and the oral cavity.

To conclude, it must be pointed out that words that have different meanings to different panellists (i.e., words that a panel might have difficulty reaching agreement on) should be avoided at all costs. Therefore, an accurate definition of the vocabulary must be developed, on which every panellist should agree.

9.4.4 Psychophysical measurement: from stimulus to perception

Psychophysical studies require essentially two parameters: the variation of specific aspects of a physico-chemical property of the stimulus, and the measurement of a dimension of the subject's response. Therefore a fundamental issue of sensory studies is the correct delivery of the stimulus. Only in this way can we hope to get a reliable answer from an instrument (in our case, a panel) that is characterized by a low signal-to-noise ratio and presents problems with offset, gain and drift with time.

It is highly probable that different judges will have different thresholds of perception for chemical compounds (both gustative and olfactive) (Bi and Ennis, 1998; Bartoshuk, 2000; Linschoten et al., 2001); also, thresholds and sensitivity change with age (Doty, 1989). The rating of two assessors on an intensity scale will differ, even if the same stimulus is applied. Furthermore, perception scales are non-linear, meaning that when the concentration of the stimulus is doubled, a double intensity on a perception scale will seldom be rated. Finally, the repetition of the delivery of the stimulus over time (e.g., coffee brewed from the same blend in successive days) will not lead to the same answer from the panel (Vanne *et al.*, 1998).

These problems lead to the following considerations. The number of panellists involved in an evaluation must be high enough to assure an averaging of the responses, representative of the population. Relying on the judgement of just one or a few assessors for determining the sensory properties of one product can be misleading because of specific anosmias (see 9.3.1) or hypersensitivity of the judge. The number can be determined by simple algorithms, starting from a binomial distribution, in the case of discriminative tests. Once the significance of the result is set; as far as profiling is concerned, the assumption is that the scores given by the judges would distribute around an average value with normal distribution. Thus, analysis of variance (ANOVA) procedures can be followed to evaluate the characteristics of the product. In this case the central limit theorem states that the minimum number of measures for a distribution to be approximated to a normal distribution should be 10. This requirement can be met in a university laboratory performing basic tasks, but it is seldom fulfilled in industry's everyday experience. However, the risk of reducing the number of assessors in a panel too far must be kept in mind. A cost–risk analysis should be performed before planning the campaign.

In order to perform an analytical task, such as discrimination between products or profiling, the panel must be recruited, checked for sensitivities, then calibrated and trained, in order to share a common vocabulary and provide homogeneous results. Continuous monitoring of the panel should be done to check for possible outliers and to verify when recalibration might be needed (McEwan, 2001; Stucky *et al.*, 2001).

A set of ISO Standards has been developed in order to assure maximum reliability of the results. They deal with the design of the test rooms (ISO, 1988a), the recruitment and training of the panel (ISO, 1992b, 1993, 1994a) and propose the descriptive and discriminative tests (ISO, 1983a, 1983b, 1985a, 1985b, 1987a, 1987b, 1988b, 1991a, 1991c, 1992a, 1994b, 1994c, 1999a, 1999b, 2002a) to be performed. As far as coffee is concerned, there is one standard on sample preparation and one on vocabulary (ISO, 1991b, 2002b).

9.4.5　Statistical analysis of sensory data

Once data have been collected, they have to be 'transformed' into results. Because of the complexity of the stimulus and the low signal-to-noise

Table 9.1 Techniques used in the statistical analysis of sensory data

Technique	Scope
Analysis of variance	(ANOVA) Comparison of multiple attribute-samples-judges in case of normal distributed data
Friedman test	ANOVA for non-normal data; ranking
Procrusted analysis	Interpretation of consensus configuration; detection of outliers
Principal component analysis	Reduction of dimensionality for multivariate data, showing the relationship between groups of variables and between objects
Multidimensional scaling	Identification of similarities among stimuli by picture mapping
Discriminant analysis	Classification of an item into one of several mutually exclusive groups

ratio, sophisticated statistical techniques must be employed, in order to obtain reliable results. Table 9.1 gives a short list of the most commonly used tests with their application.

REFERENCES

Abdi H. (2002) What can cognitive psychology and sensory evaluation learn from each other? *Food Qual. Pref.* 13, 445–451.

Adler E., Hoon M.A., Mueller K.L., Chandrashekar J., Ryba N.J.P. and Zuker C.S. (2000) A novel family of mammalian taste receptors. *Cell* 100, 693–702.

Amerine M.A., Pangborn R.M. and Roessler E.B. (1965) *Principles of Sensory Evaluation of Food*. New York: Academic Press.

ASTM E253-03 *Standard Terminology Relating to Sensory Evaluation of Materials and Products*. West Conshohocken, PA: American Society for Testing and Materials.

Axel R. (1995) The molecular logic of smell. *Sci. Am.* 273, 154–159.

Balzer H.H. (2001) Acids in coffee. In R.J. Clarke and O.G. Vitzthum (eds), *Coffee: Recent Developments*. Oxford: Blackwell Science, pp. 18–32.

Bartoshuk L.M. (2000) Comparing sensory experiences across individuals: recent psychophysical advances illuminate genetic variation in taste perception. *Chemical Senses* 25, 447–460.

Beidler L.M. (1958) Physiological basis of flavor. In A.D. Little (ed.), *Flavor Research and Food Acceptance*. New York: Rheinhold, pp. 1–28.

Bi J. and Ennis D.M. (1998) Sensory thresholds: concepts and methods. *J. Sensory Studies* 13, 133–148.

Bourne M.C. (2002) *Food Texture and Viscosity: Concept and Measurement.* Food Science and Technology International Series. New York: Academic Press.

Bozza T.C. and Mombaerts P. (2001) Olfactory coding: Revealing intrinsic representations of odors. *Curr. Biol.* 11 (17), R687–R690.

British Standards Institute (1975) Glossary of Terms Relating to Sensory Analysis of Food. BS 5098:1975.

Buck L.B. (1996) Information coding in the vertebrate olfactory system. *Ann. Rev. Neurosci.* 19, 517–544.

Buck L. and Axel R. (1991) A novel multigene family may encode odor recognition: a molecular basis for odor recognition. *Cell* 65, 175–187.

Büttner A. and Schieberle P. (2000) Exhaled odorant measurement (EXOM) – a new approach to quantify the degree of in-mouth release of food aroma compounds. *Lebensm. Wiss. Technol.* 33, 553–559.

Büttner A., Beer C., Hannig C., Settles M. and Schieberle P. (2002) Quantitation of the in-mouth release of heteroatomic odorants. In G.A. Reineccius and T.A. Reineccius (eds), *Heteroatomic Aroma Compounds.* ACS Symposium Series 826. Washington, DC: American Chemical Society, pp. 296–311.

Chandrashekar J., Mueller K.L., Hoon M.A., Adler E., Feng L., Guo W., Zuker C.S. and Ryba N.J.P. (2000) T2Rs function as bitter taste receptors. *Cell* 100 (6), 703–711.

Chaudhari N., Landin A.M. and Roper S.D. (2000) A metabotropic glutamate receptor variant functions as a taste receptor. *Nat. Neurosci.* 3, 113–119.

Christensen C.M., Brand J.G. and Malamud D. (1987) Salivary changes in solution pH: a source of individual differences in sour taste perception. *Physiol. Behav.* 40, 221–227.

Civille G.V. and Lyon B.G. (1996) *Lexicon for Aroma and Flavor Sensory Evaluation.* ASTM Data Series Publication 66. West Conshohocken, PA: American Society for Testing and Materials.

Consejo Oleicola International (1987) COI/ T.20/ Doc. no. 4, 18 June 1987.

Cristovam E., Russell C., Paterson A. and Reid E. (2000) Gender preference in hedonic ratings for espresso and espresso-milk coffees. *Food Qual. Pref.* 11 (6), 437–444.

Delwiche J.F. (1996) Are there 'basic' tastes? *Trends Foodsci. Technol.* 7, 411–415.

Delwiche J. F. (2003) Attributes believed to impact flavor: an opinion survey. *J. Sensory Studies* 18 (4), 347–352.

Delwiche J.F., Buletic Z. and Breslin P.A. (2001) Covariation in individuals' sensitivities to bitter compounds: evidence supporting multiple receptor/ transduction mechanism. *Percept. Psychophys.* 63, 761–776.

Dodd J. and Castellucci V.F. (1991) Smell and taste: the chemical senses. In E.R. Kandel, J.H. Schwarz and T.M. Jessel (eds), *Principles of Neural Science.* New York: Elsevier Science.

Doty R.L. (1989) Age-related alterations in taste and smell function. In J.C. Goldstein, H.K. Kashima, C.F. Coopmann Jr (eds), *Geriatric Otorhinolaryngology*. Burlington, BC: Decker, pp. 97–104.

Doty R.L. (2001) Olfaction. *Ann. Rev. Psychol.* 52, 423–452.

Dulac C. (2000a) The physiology of taste, vintage 2000. Minireview. *Cell* 100, 607–610.

Dulac C. (2000b) Sensory coding of pheromone signals in mammals. *Curr. Opin. Neurobiol.* 10, 511–518.

Encyclopaedia Britannica (2003) http://www.britannica.com.

Firestein S. (2001) How the olfactory system makes sense of scents. *Nature* 413, 211–218.

Flament I. (2002) *Coffee Flavor Chemistry*. New York: Wiley & Sons, pp. 336–337.

Ganchrow J.R., Steiner J.E. and Daher M. (1983) Neonatal facial expressions in response to different qualities and intensities of gustatory stimuli. *Infant. Behav. Dev.* 6, 189–200.

Glusman G., Yanai I., Rubin I. and Lancet D. (2001) The complete human olfactory subgenome. *Genome Res.* 11, 685–702.

Green B.G., Dalton P., Cowart B., Shaffer G., Rankin K. and Higgins J. (1996) Evaluating the 'Labeled Magnitude Scale' for measuring sensations of taste and smell. *Chem. Senses* 21, 323–334.

Green B.G., Shaffer G.S. and Gilmore M.M. (1993) Derivation and evaluation of a semantic scale of oral sensation magnitude with apparent ratio properties. *Chem. Senses* 18, 683–702.

Hänig D.P. (1901) Zur Psychophysik des Geschmackssinnes. *Philosophische Studien* 17, 576–623.

Heath H.B. (1988) The physiology of flavour: taste and aroma perception. In R.J. Clarke and R. Macrae (eds), *Coffee: Volume 3 – Physiology*. London: Elsevier Applied Science, pp. 141–170.

Hübener F., Laska M., Kobayakawa T. and Saito S. (2001) Cross-cultural comparison of olfactory perception in Japanese and Germans. Abstract: Fourth Pangborn Sensory Science Symposium 2001: A Sense Odyssey. Dijon.

ICO (1991) Consumer-oriented vocabulary for coffee. In *Sensory Evaluation of Coffee*. London: International Coffee Organization, pp. 72–95.

IFT (2003) Glossary of Terms. http://www.ift.org.

Ishii R. and O'Mahony M. (1987) Taste sorting and naming: can taste concepts be misinterpreted by traditional psychophysical labelling systems? *Chemical Senses* 12, 37–51.

ISO (1983a) Sensory Analysis – Methodology – Triangular Test. ISO 4120:1983. Geneva: International Organization for Standardization.

ISO (1983b) Sensory Analysis – Methodology – Paired Comparison Test. ISO 5495:1983. Geneva: International Organization for Standardization.

ISO (1985a) Sensory Analysis – Methodology – General Guidance. ISO 6658:1985. Geneva: International Organization for Standardization.

ISO (1985b) Sensory Analysis – Methodology – Flavour Profile Methods. ISO 6564:1985. Geneva: International Organization for Standardization.

ISO (1987a) Sensory Analysis – Methodology – Evaluation of Food Products by Methods Using Scales. ISO 4121:1987. Geneva: International Organization for Standardization.

ISO (1987b) Sensory Analysis – Methodology – 'A'–'not A' test. ISO 8588:1987. Geneva: International Organization for Standardization.

ISO (1988a) Sensory Analysis – General Guidance for the Design of Test Rooms. ISO 8589:1988. Geneva: International Organization for Standardization.

ISO (1988b) Sensory Analysis – Methodology – Ranking. ISO 8587:1988. Geneva: International Organization for Standardization.

ISO (1991a) Sensory Analysis – Methodology – Method of Investigating Sensitivity of Taste. ISO 3972:1991. Geneva: International Organization for Standardization.

ISO (1991b) Green Coffee – Preparation of Samples for Use in Sensory Analysis. ISO 6668:1991. Geneva: International Organization for Standardization.

ISO (1991c) Sensory Analysis – Methodology – Duo–Trio Test. ISO 10399:1991. Geneva: International Organization for Standardization.

ISO (1992a) Sensory Analysis – Vocabulary. ISO 5492:1992. Geneva: International Organization for Standardization.

ISO (1992b) Sensory Analysis – Methodology – Initiation and Training of Assessors in the Detection and Recognition of Odours. ISO 5496:1992. Geneva: International Organization for Standardization.

ISO (1993) Sensory Analysis – General Guidance for the Selection, Training and Monitoring of Assessors – Part 1: Selected Assessors. ISO 8586-1:1993. Geneva: International Organization for Standardization.

ISO (1994a) Sensory Analysis – General Guidance for the Selection, Training and Monitoring of Assessors – Part 2: Experts. ISO 8586-2:1994. Geneva: International Organization for Standardization.

ISO (1994b) Sensory Analysis – Identification and Selection of Descriptors for Establishing a Sensory Profile by a Multidimensional Approach. ISO 11035:1994. Geneva: International Organization for Standardization.

ISO (1994c) Sensory Analysis – Methodology – Texture Profile. ISO 11036:1994. Geneva: International Organization for Standardization.

ISO (1999a) Sensory Analysis – General Guidance and Test Method for Assessment of the Colour of Foods. ISO 11037:1999. Geneva: International Organization for Standardization.

ISO (1999b) Sensory Analysis – Methodology – Magnitude Estimation Method. ISO 11056:1999. Geneva: International Organization for Standardization.

ISO (2002a) Sensory Analysis – Methodology – General Guidance for Measuring Odour, Flavour and Taste Detection Thresholds by a Three-Alternative Forced-Choice (3-AFC) Procedure. ISO 13301:2002. Geneva: International Organization for Standardization.

ISO (2002b) Coffee and Its Products – Vocabulary. ISO 3509:2002. Geneva: International Organization for Standardization.

Johnson E. and Vickers Z. (2002) The effectiveness of palate cleansers. Abstract. Proceedings IFT 2002 Annual Meeting and Food Expo, Anaheim, California.

Kinnamon S.C. (2000) A plethora of taste receptors. Neuron 25, 507–510.

Lawless H.T. (1999) Descriptive analysis of complex odors: reality, model or illusion? Food Qual Pref. 10, 325–332.

Lawless H.T. and Klein B.K. (eds). (1991) Sensory Science Theory and Applications in Foods. New York: M. Dekker.

Leffingwell J.C. (2002) Olfaction – a review. http:// www.leffingwell.com/ olfaction.htm.

Lindemann B. (2001) Receptors and transduction in taste. Nature 413, 219–225.

Linschoten M.R., Harvey L.O. Jr, Eller P.M. and Jafek B.W. (2001) Fast and accurate measurement of taste and smell thresholds using a maximum-likelihood adaptive staircase procedure. Perception Psychophysics 63 (8), 1330–1347.

Lundy Jr R.F., Pittman D.W. and Contreras R.J. (1997) Role for epithelial Na1 channels and putative Na+/H+ exchangers in salt taste transduction in rats. Am. J. Physiol. 273, R1923–R1931.

Ly A. and Drenowski A. (2001) PROP (6-n-propylthiouracil) tasting and sensory response to caffeine, sucrose, neohesperidin dihydrochalone and chocolate. Chemical Senses 26, 41–47.

Lyall V., Alam R.I., Phan D.Q., Ereso G.L., Phan T.T., Malik S.A., Montrose M.H., Chu S., Heck G.L., Feldman G.M. and Desimone J.A. (2001) Decrease in rat taste receptor cell intracellular pH is the proximate stimulus in sour taste transduction. Am. J. Physiol. Cell Physiol. 281, C1005–C1013.

Maier H.G. (1987) The acids of coffee. Proc. 12th ASIC Coll., pp. 229–237.

Majou D. (ed.) (2001) Sensory Evaluation – Guide of Good Practice. Paris: Actia.

Margolskee R.F. (2002) Molecular mechanisms of bitter and sweet taste transduction. J. Biol. Chem. 277, 1–4.

Matsunami H., Montmayeur J.-P. and Buck L. (2000) A family of candidate taste receptors in human and mouse. Nature 404, 601–604.

Mattes R.D. (2002) PROP taster status: Dietary modifier, marker or misleader. Abstract. Proceedings ECRO 2002, Erlangen (Germany).

McEwan J. (2001) Proficiency testing for sensory profile panels: issues for measuring panel performance. Abstract. Fourth Pangborn Sensory Science Symposium 2001: A Sense Odyssey. Dijon.

Meilgaard M., Civille G.V. and Carr B.T. (1999) Sensory Evaluation Techniques, 3rd edn. Boca Raton, FL: CRC Press.

Meredith M. (2001) Human vomeronasal organ function: A critical review of best and worst cases. Chemical Senses 26 (4), 433–445.

Mombaerts P. (1999) Seven-transmembrane proteins as odorant and chemosensory receptors. *Review*. *Sci.* 286 (5440), 707–711.

Mombaerts P. (2001a) How smell develops. *Nature Neurosci.* 4, 1192–1198.

Mombaerts P. (2001b) The human repertoire of odorant receptor genes and pseudogenes. *Ann. Rev. Genomics Human Genet.* 2, 493–510.

Montmayeur J.P., Liberles S.D., Matsunami H. and Buck L. (2001) A candidate taste receptor gene near a sweet taste locus. *Nature Neurosci.* 4 (5), 492–498.

Mori K., Nagao H. and Yoshihara Y. (1999) The olfactory bulb: coding and processing of odor molecule information. *Science* 286, 711–715.

Morrot G., Brochet F. and Dubourdieu D. (2001) The colors of odors. *Brain Lang.* 79 (2): 309–320.

Moskowitz H.R. (1988) *Applied Sensory Analysis of Foods – Volumes I and II*. Boca Raton, FL: CRC Press.

Navarini L. (2003) Personal communication.

Navarini L., Ferrari M., Liggieri L., Ravera F. and Suggi Liverani F. (2002) Dynamic tensiometric characterization of espresso coffee beverage, Abstract. Proceedings 16th Conference of the European Colloid and Interface Society, Paris, p. 337.

Nelson G., Chandrashekar J., Hoon M.A., Feng L., Zhao G., Ryba N.J.P. and Zuker C.S. (2002) An amino-acid taste receptor. *Nature* 416, 199–202.

Nelson G., Hoon M.A., Chandrashekar J., Zhang Y., Ryba N.J.P. and Zucker C.S. (2001) Mammalian sweet taste receptors. *Cell* 106, 381–390.

Ohiokpehai O., Brumen G. and Clifford M.N. (1982) The chlorogenic acids content of some peculiar green coffee beans and the implications for beverage quality. *Proc. 10th ASIC Coll.*, pp. 177–186.

Peryam D.R. and Girardot N.F. (1952) Advanced taste-test method. *Food Eng.* 24 (7), 58–61 and 194.

Petracco M. (1989) Physico-chemical and structural characterisation of 'espresso' coffee brew. *Proc. 13th ASIC Coll.*, pp. 246–261.

Petracco M. (2001) Beverage preparation: brewing trends for the new millennium. In R.J. Clarke and O.G. Vitzthum (eds), *Coffee: Recent Advances*. Oxford: Blackwell Science, pp. 140–164.

Pictet G. (1989) Home and catering brewing of coffee. In R.J. Clarke and R. Macrae (eds), *Coffee: Volume 2 – Technology*. London: Elsevier Applied Science, p. 238.

Piggot J.R. (1988) *Sensory Analysis of Food*, 2nd edn. New York: Elsevier Applied Sciences.

Pines M. (2001) Seeing, hearing, smelling and the world: new findings help scientists make sense of our senses. A report from the Howard Hughes Medical Institute, pp. 49–55.

Prescott J., Young O., O'Neill L., Yau N.J.N. and Stevens R. (2002) Motives for food choice: a comparison of consumers from Japan, Taiwan, Malaysia and New Zealand. *Food Qual. Pref.* 13 (7–8), 489–495.

Rubin B. D. and Katz L. C. (1999) Optical imaging of odorant representations in the mammalian olfactory bulb. *Neuron* 23, 499–511.

Sherman R.M. and Sherman R.B. (1964) 'A spoonful of sugar', in *Mary Poppins*. The Walt Disney Company, Burbank, CA.

Smith A.K. and Noble A.C. (1997) Effects of increased viscosity on the sourness and astringency of aluminum sulfate and citric acid. *Food Qual. Pref.* 9 (3), 139–144.

Smith D.V. and Margolskee R.F. (2001) Making sense of taste. *Sci. Am.* 284, 32–39.

Sowalsky R.A. and Noble A.C. (1998) Comparison of the effects of concentration, pH and anion species on astringency of organic acids. *Chemical Senses* 23, 343–349.

Stone H. and Pangborn R.M. (1968) Intercorrelations of the senses. In *Basic Principles of Sensory Evaluation*. ASTM STP 433 30–46. West Conshohocken, PA: American Society for Testing and Materials.

Stone H. and Sidel J.L. (1985) *Sensory Evaluation Practices*. New York: Academic Press, New York.

Stucky G., Lundahl D. and Hutchinson D. (2001) Panelist monitoring and tracking principal component analysis and segmentation with self-organizing maps. Abstract: Fourth Pangborn Sensory Science Symposium 2001: A Sense Odyssey. Dijon.

Szczesniak A.S. (1979) Classification of mouthfeel characteristics of beverages. In P. Sherman (ed.), *Food Rheology and Texture*. New York: Academic Press, pp. 1–20.

Szczesniak A.S. (2002) Texture is a sensory property. *Food Qual Pref.* 13, 215–225.

Thorne F., Neave N., Scholey A., Moss M. and Fink B. (2002) Effects of putative male pheromones on female ratings of male attractiveness: influence of oral contraceptives and the menstrual cycle. *Neuroendocrinol. Lett.* 23, 291–297.

Turin L. and Yoshii F. (2003) Structure–odor relations: a modern perspective. In R. Doty (ed.), *Handbook of Olfaction and Gustation*, 2nd edn. New York: M. Dekker.

Vanne M., Tuorila H. and Laurinen P. (1998) Recalling sweet taste intensities in the presence and absence of other tastes. *Chemical Senses* 23, 295–302.

Woodman J.S. (1985) Carboxylic acids. In R.J. Clarke and R. Macrae (eds), *Coffee: Volume 1 – Chemistry*. London: Elsevier Applied Science, pp. 266–289.

Zhao H., Ivic L., Otaki J.M., Hashimoto M., Mikoshiba K. and Firestein S. (1998) Functional expression of a mammalian odorant receptor. *Science* 279, 237–242.

CHAPTER 10

Coffee consumption and health

M. Petracco and R. Viani

The history of coffee starts with legends about its discovery as a food. The story is sometimes told of herdsmen in Abyssinia who, following the example of their goats consuming the seeds of some plant, chewed either raw cherries or cooked, smoked beans (Burton, 1961) (Figure 10.1). It is difficult to imagine, for those who have tried to munch on a raw bean, that the success of coffee, which is today one of the world's most traded food commodities, could ever have come about if such primitive habits had persisted, without proper product segmentation.

It seems logical, on the one hand, that the attractiveness of coffee to those early pioneers of food science derived basically from their experiencing a stimulant condition that proved to be beneficial to their activities (Ellis, 1824). Put differently, the primal cause for coffee consumption must have been the physiological effects of caffeine on the human organism, which are nowadays well substantiated (Weinberg and Bealer, 2001).

On the other hand, it is fairly obvious why coffee has become so popular, attaining the status of the second most largely consumed beverage after water. As seen in Chapter 9, it is a question of flavour, or better still, of overall impact on our senses. Sensory evaluation, which used to be considered as in the realm of 'magic' because 'taste is a matter of taste', is nowadays earning the status of a highly respected analytical tool (cup-testing being its technical name), able to produce key information with good reliability (Meilgaard *et al.*, 1999).

Thus, it is no surprise that coffee is seen by many consumers as a mild stimulant that additionally gives pleasure. In this perspective, coffee drinking is often denigrated and deemed guilty by association with social habits that give raise to harmful effects, as tobacco smoking and alcohol drinking undoubtedly do. In fact, the medical community does not see

Figure 10.1   Smoked-toasted coffee, as is still consumed in East Africa (Uganda); each banana leaf pouch contains around 20 cherries

coffee as a real worry: only a couple of areas show some reasons for concern that could advise moderation in coffee consumption. There are, by contrast, several rational motives to consider coffee as beneficial to health, acting as a protective agent against specific pathologies.

10.1   CONSUMPTION PATTERNS

An often-underestimated question should be addressed first when designing a clinical or epidemiological experiment: 'How much coffee (and how much caffeine) do my subjects take in?' Typically, just the number of cups per day is reported, with no mention as to cup size and coffee bean type. Of course, it is often difficult to make subjects (especially patients) produce accurate reports of their drinking habits in distant periods of their life (Bracken *et al.*, 2002). It is yet more awkward to arrive at an estimation of the quality of coffee they used to drink, with its related chemical composition. Thus, most of the time, subjects are simply categorized in two or three classes: abstainers, moderate drinkers and heavy drinkers (Bunker and McWilliams, 1979).

The answer is only apparently simple when dealing with intervention studies, where the actual food and beverages intake can be controlled, or even prescribed (Lelo *et al.*, 1986): the content of caffeine and other coffee materials in a cup depends both on cup size and extraction efficiency, and on coffee composition. As discussed in Chapter 3, large chemical variations are to be foreseen with different proportions of pure varieties in the blend. Also the size of a single serving is highly variable, ranging from 15 ml of concentrated Italian espresso (low in caffeine) to over 250 ml in the English-speaking countries. Moreover, a coffee serving can be derived from the more or less efficient brewing of a roasted ground coffee portion as small as 5 g, up to 15 g or more, and the variation in active constituents, such as caffeine and diterpenes, can be very high depending on blending and brewing technique (Table 10.1).

In population studies, reference is usually made to average national consumption. This is calculated from raw coffee production, import and disappearance data, which may not be universally agreed upon and may not represent an accurate portrait of the local coffee market situation (just think about re-export streams and non-official parallel markets). The most reliable statistics are those published on a quarterly basis by the International Coffee Organization (ICO, 2003), along with those occasionally produced by market research companies (for instance LMC, 2002).

As an international trade commodity, global coffee production ranks second in value only to petroleum, and totals more than 6 billion kg/year. Cultivation occurs in some 50 tropical countries (Figure 10.2). As many of these are less developed countries, coffee often represents the chief hard currency income.

Coffee consumption is widespread throughout the world, especially in Europe, the USA and Japan. The largest coffee-consuming country is the USA, at 16% of the world total, followed by Brazil (which is also the largest producer country), with 11%. National associations keep track of their domestic consumption trends, divided per outlet segment (home, restaurant, bar) and product type – traditional, gourmet, espresso, etc. (for instance NCA, 2003).

The type of related beverages and the pattern of consumption are strictly associated with social habits and culture of the single countries. Variations in raw bean composition, in roasting conditions and in the extraction procedures used to prepare coffee brews result in a great diversity of the final product, the cup of coffee. A remarkable difference between coffee and all the other beverages is indeed the extraordinary variety of brewing techniques that have been developed and used traditionally in different countries: decoction methods (boiled coffee, Turkish coffee, percolator coffee and vacuum coffee), infusion methods (drip filter coffee and *Napoletana*), and the original Italian pressure methods (*Moka* and espresso) (Petracco, 2001).

In Europe, the highest *per capita* consumption is found in the Nordic countries (Figure 10.3), where mainly light-roasted wet-processed arabica

Table 10.1 Caffeine[1] and diterpenes[2] content of various coffees

Brew	Brewing technique[3]	Cup size (ml)	Caffeine (mg/cup)*	Diterpenes (mg/cup)
Boiled	Coarsely ground roasted arabica coffee (50–70 g/l) is boiled in water	150–190	75–135	Very high (13)
Filter	Boiling water is poured over finely ground light to dark roasted coffee (30–80 g/l) in a paper filter or automatic drip machine	150–190	110–200	Negligible
Percolated	Boiling water is re-circulated through coarsely ground light to medium roasted coffee (30–60 g/l)	150–190	100–190	Not available
Plunger	Hot water (875 ml) is poured onto coffee (50 g); brew and spent grounds are separated by pushing down a perforated plunger	150–190	90–180	High
Instant	Instant coffee is dissolved in hot water	150–190	45–180	Negligible
Moka	Just overheated water is forced through very finely ground medium to dark roasted coffee (5–8 g/cup)	40–50	50–190	Low (2)
Espresso	Very finely ground medium to dark roasted mainly arabica coffee is extracted with hot water under pressure (see Chapter 8)	25–35	60–80 (130)	Low (2)
Middle-Eastern	Extremely fine ground medium to dark roasted coffee (4–6 g) in water (50–60 ml) and sugar (5–10 g) is brought to a gentle boil	30–50	35–60	Very high (13)
Decaffeinated	All brewing techniques	Depending on brewing technique	1–5	Depending on brewing technique

*Caffeine content per cup can be estimated from:
Blend (arabica 1.3%, robusta 2.4%)
Ratio coffee grounds/brewing water (lowest in espresso and Middle-Eastern coffees)
Extraction time (shortest in espresso and *Moka*)
Physical separation of grounds (lacking in Middle-Eastern and boiled coffees)
It varies between 100% for soluble coffee, 90–98% for filter and percolated coffees, down to 70–80% for espresso.
Sources: [1]Petracco, 1989; Peters, 1991. [2]Gross et al., 1997; Urgert and Katan, 1997. [3]D'Amicis and Viani, 1993; Petracco, 2001

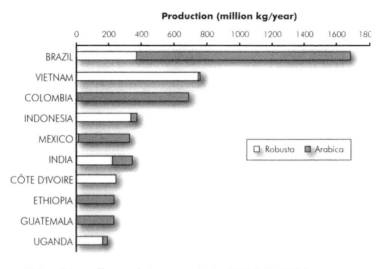

Figure 10.2   Major coffee producing countries in 2001 (ICO, 2003)

coffee is consumed. Both robusta content in the blend and roasting degree increase going southwest: medium in Germany, the Netherlands and Austria, very dark in France, Spain and Portugal, darker in Italy.

East European countries drink mainly robusta, while countries with a tea tradition, the UK, Ireland, the Russian Federation, Asian and South

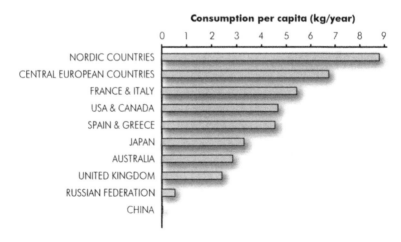

Figure 10.3   Major coffee consuming countries in 2000 (ICO, 2003)

Pacific countries, consume mostly soluble coffee. In the USA, where brewing of thinner and thinner cups of poor quality blends had drastically reduced consumption, the introduction of 'gourmet' blends, and fancy preparations such as espresso and *cappuccino*, have lately begun to reverse consumption trends.

10.2 COFFEE IS MORE THAN CAFFEINE

Not all is yet known with certainty about the physiological aspects of coffee, but a bewildering array of health and mood effects are attributed to it. For the reasons mentioned above, any figure quoted in quantitative assessments of coffee consumption must be taken with due caution, notwithstanding the plethora of epidemiological studies on the effect of coffee that have been published in the past 20 years (Debry, 1994), whose conclusions did not always lead to clear-cut answers.

Over 90% of the research carried out on the physiological properties of coffee has been devoted to caffeine, an alkaloid that stimulates the central nervous system (Stavric, 1992). Caffeine, the most used pharmacologically active substance, is also present in other common beverages like tea, colas and chocolate, besides both prescription and non-prescription drugs (Spiller, 1998; Brice and Smith, 2002a). Its daily consumption by the vast majority of the human population, while a generic habit, has never been substantiated as imputable to any addictive properties (Nehlig, 1999), even if, when caffeine intake is stopped abruptly, some individuals can experience temporary headache, fatigue or drowsiness. In any case, coffee is much more than caffeine. Its complex composition and the presence of other substances as yet unidentified (see 10.3.1), but with evident physiological effects, indicate that further research is needed to demonstrate both the wholesomeness of coffee consumption as well as the favourable effects this beverage can have in humans.

10.3 COFFEE IS BENEFICIAL TO HEALTH

Fruits and vegetables are often highlighted for their role in the prevention, or delay, of chronic human diseases by protective mechanisms that are not yet fully elucidated, albeit many plausible ones have been proposed and are currently under investigation. In this context, coffee appears to be a very interesting food plant, with a long history of use. In contrast to most other traditional foods, it has been the subject of extensive scientific research addressing its potential impact on human health (Gray, 1998).

This section will review information extracted from the large body of literature on coffee and health, giving evidence to support a direct link between coffee intake and positive health effects. Human epidemiology strongly suggests the possibility of coffee-mediated protective actions on physiology, whose putative mechanisms are proposed in the light of recent research conducted in experimental *in vivo* and *in vitro* models (Schilter *et al.*, 2001). Although further scientific confirmation is clearly needed, it can be argued that, just as for other food plants, available data on the health effects of coffee are compatible with a contribution of this common beverage to a wholesome and balanced diet.

10.3.1 Effects generically linked with coffee consumption

The quantity of coffee consumed is one of the questions commonly considered in epidemiological and clinical studies and as a consequence a mass of data has been generated (but see also 10.1). The presence in the brew of various physiologically active compounds – mutagenic compounds formed in trace amounts during roasting (hydrogen peroxide, glyoxal, etc.) on the one hand and antioxidants like chlorogenic acids and melanoidins on the other – is sometimes used to try and correlate the findings of the epidemiological approach. Interesting effects of coffee consumption have been reported on behaviour and brain activity, on metabolic activity, on chemoprotection and related prevention of cancer and of degenerative diseases, on protection of the digestive system and on the alleviation of various miscellaneous symptoms.

10.3.1.1 Behaviour and brain activity

While a very large body of evidence shows the effects of coffee consumption on behaviour, most of the experimental studies deal with placebo-controlled administration of pure caffeine, and will be discussed in 10.3.2.1.

A few researchers have investigated the activity of coffee beverages *per se* on improving alertness and performance during the day and night (Smith *et al.*, 1993) and on the speeding of information processing in adults (Bättig and Buzzi, 1986; Van Boxtel and Jolles, 1999) and older people (Riedel *et al.*, 1995a; Johnson-Kozlow *et al.*, 2001). There is a strong inverse correlation with tendency to suicide (Kawachi *et al.*, 1996), even though an increased risk has been observed among heavy coffee drinkers (Tanskanen *et al.*, 2000). An exciting possibility has been proposed by some scientists, who have observed a preventive activity of coffee on depression and against addiction to heavy drugs (Boublick et al., 1983; Wynne et al., 1987; Knapp et al., 2001).

10.3.1.2   Metabolic activity

Coffee acts on metabolism in different ways: it has long been used to enhance endurance during exercise and has been shown to display a mild anti-obesity activity. In sport, the ethical aspects may range from reducing the athlete's freedom of diet choice by limiting intake of cups of coffee, to whole question of doping (IOC, 2000), recently solved by the elimination of caffeine from the list of banned substances. The studies dealing with the effects of caffeine on athletic performance will be discussed in 10.3.2.2.

A mild anti-obesity action may be ascribed to caffeine and explained by its thermogenic effect, i.e. an increased expenditure of energy at doses associated with moderate coffee consumption (Costill et al., 1978; Acheson et al., 1980). Furthermore, coffee has been recently shown to abate uricaemia, therefore protecting from gout (Kiyohara et al., 1999).

10.3.1.3   Chemoprotective activity

By chemoprotection we imply all phenomena linked with the capacity of coffee to protect against the onset or aggravation of neoplastic (cancer) or degenerative pathologies (Parkinson's and Alzheimer's).

Recent research suggests that coffee contains anti-toxic components, displays anti-oxidant activity and may act as an anti-inflammatory. While some of these studies are actually focused on caffeine (Devasagayam et al., 1996; Varani et al., 1999, 2001; Foukas et al., 2002; Lu et al., 2002), much interest is raised by the observation that several coffee components, other than caffeine, display protective activity in vitro and in vivo.

From a biochemical point of view, several studies have investigated the antioxidant activity of coffee, scavenging free radicals and protecting against their neoplastic or degenerative action (Singhara et al., 1998; Daglia et al., 2002; Natella et al., 2002). Special attention has been paid to coffee polyphenols (Kato et al., 1991) and to other classes of compounds present in coffee, like melanoidins (Del Castillo et al., 2002; Morales and Babbel, 2002), or in tea (Hertog et al., 1993), or in both (Lekse et al., 2001; Richelle et al., 2001). Recently, a new coffee compound with activity on the detoxifying system has been identified as methyl pyridinium (Somoza et al., 2003).

Other authors have studied the anti-genotoxic effects of coffee (Abraham, 1996; Abraham and Singh, 1999) and of coffee diterpenes

(Kono, 1994; Huggett and Schilter, 1995; Poikolainen and Vartiainen, 1997; Honjo *et al.*, 1999; Huber *et al.*, 2002), where the finding that coffee consumption enhances the level of the protective system of glutathione and its transferase is particularly interesting (Scharf *et al.*, 2001; Esposito *et al.*, 2003).

Much attention has been devoted to cancer onset, an area where the protective action of coffee consumption may be inferred from statistical data. Particularly evident is the case of colorectal cancer, where a large body of evidence on its protective effect has now been collected. The identified, or at least putative, effects of coffee on prevention of human carcinogenesis are listed in Table 10.2.

Coffee consumption does not emerge as evidently linked with cancer; however, a few cancer locations might still be adversely linked with coffee intake. The weak association found with urinary tract and pancreatic cancers might be due to residual confounding factors (Tavani and La Vecchia, 2000).

Degenerative diseases like Parkinson's and Alzheimer's mainly affect the elderly population, and regular coffee consumption may play a preventive role, thus reducing their social burden. Parkinson's disease has recently been studied in some detail with respect to its correlation with coffee, and preliminary conclusions indicate a neuroprotective effect of caffeine (and nicotine) intake (Benedetti *et al.*, 2000; Honig, 2000; Ross *et al.*, 2000; Ascherio *et al.*, 2001; Ross and Petrovitch, 2001; Checkoway *et al.*, 2002; Hernán *et al.*, 2002; Ragonese *et al.*, 2003), with proposed models of action (Chen *et al.*, 2001; Schwarzchild *et al.*, 2002).

Alzheimer's disease has also been the object of recent attention (Lindsay *et al.*, 2002), with studies on its relationship with caffeine (Maia and de Mendonça, 2002). The potential role of caffeine and adenosine has been stressed in reducing inflammation (Montesinos, 2000; Dall'Igna *et al.*, 2003), which may be beneficial in atherosclerosis (Libby, 2002).

10.3.1.4   Digestive tract

In spite of popular belief in some countries that in some way coffee upsets digestion, there is evidence that in reality its regular consumption may act in a beneficial manner on various organs linked with food assimilation.

Updated research shows its helpful role on the liver, where the risk of hepatic cirrhosis is reduced with increased quantity and length of coffee consumption, while other caffeinated beverages have no effect (Klatsky and Armstrong, 1992; Corrao *et al.*, 1994, 2001; Gallus *et al.*, 2002a; Tverdal and Skurtveit, 2003). Moreover, it has been reported that coffee

Table 10.2 Cancer protection by coffee

Cancer	Findings/explanation	Comments	Reference
Bladder	Small increased risk in heavy drinkers, but unrelated with dose or length of exposure	Residual confounding by smoking?	Sala et al., 2000; Zeeger et al., 2001
Breast	No link in females; protective effect in males attributed to polyphenols		Johnson et al., 2002; Michels et al., 2002
Colon and rectum	Reduced risk/inhibition of bile acids by diterpenes Antimutagenic compounds in coffee Increased colon motility Avoidance of coffee by high-risk individuals		Urgert and Katan, 1997; Favero et al., 1998; Giovannucci, 1998
Liver	Reduced risk		Gallus et al., 2002b; Yagasaki et al., 2002
Lung	Reduced risk in women		Kubik et al., 2001
Oesophagus	No or reduced risk		Inoue et al., 1998
Ovary	Inadequate evidence or no association		Tavani et al., 2001
Pancreas	Reduced risk at low doses; possible increased risk at high doses for males	Confounding factors?	Nishi et al., 1996; Lin et al., 2002
Prostate	No association or protection due to boron content of coffee		Cui et al., 2004
Testicle	Correlation with coffee and pork consumption attributed to ochratoxin A (OTA) contamination of coffee	Lack of correlation observed with cereal intake (the major source of OTA) would exclude the link between OTA and testicular cancer	Schwartz, 2002
Thyroid	No association		Mack et al., 2002

decreases the level of gamma-glutamyltransferase (GGT), an enzyme that is a marker of liver damage induced by alcohol abuse (Sharp and Benowitz, 1990; Casiglia et al., 1993). Unfortunately, an explanation for both effects and the identity of the active compound are still missing. Coffee consumption has been linked to a reduction of the risk of stone formation in the gallbladder (Leitzmann et al., 1999, 2002) and in the kidneys (Curhan et al., 1996, 1998): all these associations might, however, be explained by a reduction of coffee drinking in subjects with symptoms related to the pathology.

Furthermore, coffee seems to act on sugar metabolism, reducing the risk of developing diabetes (Van Dam and Feskens, 2002; Rosengren et al., 2004; Salazar-Martinez et al., 2004), which could be explained by the fact that caffeine enhances patients' perception of hypoglycaemia, helping them to react accordingly (Kerr et al., 1993; Debrah et al., 1996; Watson et al., 2000; Keijzers et al., 2002). Recently, the quinic fraction of coffee has also been studied for its effects on glucose metabolism (Johnson et al., 2003; Shearer et al., 2003). Lastly, there is even evidence of a preventive action on dental caries (Daglia et al., 2002).

10.3.2   Effects associated with caffeine consumption

The amount of caffeine present in a cup depends on the type of coffee used, with wide variation between arabica and robusta (see Table 2.5), and on its mode of processing and preparation, varying between 1–5 mg for decaffeinated coffee, up to 200 mg for filter coffee prepared from a robusta blend. The amount of caffeine in a cup of espresso, which is usually made from pure arabica or from a blend composed mainly of arabica, is low, no more than 80 mg, usually less (see Table 10.1).

Within minutes after ingestion, caffeine is absorbed and distributed throughout the body tissues. It is excreted, mostly metabolized, with a half-life of 2–6 hours in healthy adults. Half-life increases in pregnant women and in subjects with an impaired liver function, and is shortened in smokers. Consumption of caffeine delays falling asleep and reduces sleep duration.

10.3.2.1 Behaviour and brain activity

Caffeine has been known long since to modify human reactions, helping to cope with awkward situations (Liebermann, 1987). Agreed beneficial effects on behaviour associated with moderate caffeine intake are an increase in alertness and reduction of fatigue, an improvement in performance on vigilance tasks and tasks requiring sustained response

(Smith and Brice, 2000; Smith, 2002), and on mood (Brice and Smith, 2002b). The area of prevention of driving accidents, where caffeine might bring considerable alleviation, has been particularly investigated (Horne and Reiner, 1996, 1999; Reyner and Horne, 1997, 1998, 2000; Brice and Smith, 2001; De Valck and Cluydts, 2001).

Complex novel molecular mechanisms have been recently proposed to explain the effect of caffeine on brain functions (Acquas et al., 2002; Lindskog et al., 2002). The hypothesis that caffeine enhances short-term, possibly also long-term memory, has been advanced by several authors (Di Chiara et al., 2001; Nguyen Van Tam and Smith, 2001), and a mechanism of action has been proposed (Riedel et al., 1995b).

While there is nowadays a large literature on the beneficial effects of caffeinated coffee on performance and mood, a debate has started as to whether the effects of caffeine merely reflect the removal of the negative effects of caffeine withdrawal (James, 1994), or if they correspond to a real benefit (Smith, 1994).

10.3.2.2   Metabolic activity

At doses associated with moderate coffee consumption, caffeine ingestion has been found to increase energy expenditure (Hollands et al., 1981; Astrup et al., 1990), which is confirmed in both younger and older women (Arciero et al., 2000). The use of caffeine as a treatment for obesity therefore appears very inviting, and has been much investigated (Acheson et al., 1980; Dulloo et al., 1989; Daly et al., 1993; Yoshida et al., 1994; Caraco et al., 1995). Unfortunately, thermogenesis and lipid oxidation are less stimulated in obese than in lean subjects (Bracco et al., 1995).

Caffeine enhances athletic performance in several sport disciplines (MacIntosh and Wright, 1995; Pasman et al., 1995; Jackman et al., 1996; Ferrauti et al., 1997); it has been found to increase the maximal anaerobic power and to bring the effects of fatigue down, both by affecting by-products metabolism (Anselme et al., 1992) and by lowering its perception threshold (Cole et al., 1996). Ergogenic effects of caffeine, which occur at doses far lower than the limit accepted by the International Olympic Committee Medical Code (12 mg/l of urine) (IOC, 2000), cannot be satisfactorily explained just by thermogenesis and lipid oxidation (Graham, 2001).

Another property of caffeine is its mild diuretic effect, which, however, at moderate doses does not produce dehydration and fluid-electrolyte imbalance during physical exercise (Stookey, 1999; Armstrong, 2002). Further, caffeine appears to influence the resynchronization of hormonal rhythms after rapid transmeridian travel (jet-lag), probably through a modulation of melatonin activity (Hartter *et al.*, 2003; Beaumont *et al.*, 2004).

10.3.2.3   Alleviation of symptoms

Various diseases have been investigated, both from a statistical and a clinical standpoint, with regard to their link with coffee consumption. For some there are hints that coffee, or caffeine, can alleviate the symptomatic framework.

Caffeine has long since been known as a bronchodilator (mentioned by Marcel Proust in one of his literary works early in the last century). In the recent past, this simple treatment for soothing bronchial asthma symptoms has again been given consideration (Pagano *et al.*, 1988; Kivity *et al.*, 1990; Schwartz and Weiss, 1992; Bara and Barley, 2003; Henderson *et al.*, 1993).

The problem of hyperactive children (Schnackenberg, 1973) has seen recent developments proposing caffeine, or coffee, treatment to help in relaxing hyperkinetic behaviour in school age children (Stein *et al.*, 1996; Castellanos and Rapoport, 2002). Some reported side-effects of caffeine administration could prove useful in clinical practice, such as an inhibitory action on anaphylactic shock (Shin *et al.*, 2000), enhanced recovery after surgical intervention (Weber *et al.*, 1997), reduction of hypnotic effects of alcohol (El Yacoubi *et al.*, 2003) and even a certain amount of radioprotection (George *et al.*, 1999).

Finally, a novel hope has been raised by the suggestion of a role for caffeine as an adenosine receptor antagonist, to minimize manifestations of Huntington's chorea (Varani *et al.*, 2001).

10.3.3   Effects attributed to specific constituents other than caffeine

The coffee diterpene cafestol has been shown to raise serum total cholesterol (Weusten-van der Wouw *et al.*, 1994; Urgert and Katan, 1997; De Roos *et al.*, 2001); this reversible effect has been observed in subjects consuming large amounts of the unfiltered beverage popular in Nordic countries, the so-called 'boiled' coffee. Consumption of moderate quantities of either espresso or *Moka* type brews has no effect on total cholesterol (D'Amicis *et al.*, 1996).

Together with kahweol, cafestol also possesses anticarcinogenic activity (Cavin *et al.*, 2002), and might play a role in the protective effects of coffee. However, animal data indicate that the level at which the coffee diterpenes would be expected to exert a chemoprotective effect in humans may be unrealistically high (Huggett and Schilter, 1995).

The importance as antioxidants of the chlorogenic acids, along with their quinide derivatives formed on roasting, has lately been proposed (Natella et al., 2002). Inasmuch as they are absorbed by humans (Olthof et al., 2001; Nardini et al., 2002), their physiological value lies in possible inhibition of the human adenosine transporter (Martin et al., 2001; de Paulis et al., 2002), with consequences on brain functions (de Paulis and Martin, 2003). Further work in this field would be welcomed.

10.4 COFFEE IS NOT HARMFUL TO HEALTH

Coffee is often attributed with the nature of a pleasant, yet malicious beverage; this conviction originates perhaps from the darkness of the beans, which are roasted on a fire – the mephistophelian element. Little scientific proof is usually forthcoming to substantiate this vague negative feeling about the beverage, with the finger of guilt most often being pointed at the most physiologically active constituent in the cup – caffeine. Nevertheless, the much investigated caffeine emerges from the scrutiny of medical research as guiltless, particularly with respect to the charge of being an addictive and procreation-jeopardizing substance.

10.4.1 Caffeine absolved from blames

10.4.1.1 Caffeine is not addictive

The question of a possible dependence on caffeine has been hotly debated, particularly as it fulfils some of the established criteria for dependence (APA, 1992; WHO, 1994). The social consequences of caffeine are unknown and, if any, are certainly negligible when compared to those of nicotine or alcohol. Caffeine, at doses typically consumed in the diet, may lead to withdrawal effects and some physical dependence in adults; the prevalence of such effects being variable, their intensity is usually very low in most individuals. Caffeine withdrawal symptoms, such as irritability, sleepiness, increased fatigue and headache, have been observed in some habitual moderate coffee drinkers. They usually start within 12–24 h, with a peak after one or two days, lasting for up to one week (Van Dusseldorp and Katan, 1990; Lane, 1997). These symptoms are not dose-dependent, i.e., they are unrelated to the quantity of caffeine ingested. Non-pharmacological factors related to expectation might also explain them (Dews *et al.*, 2002). Further research is required to examine whether similar withdrawal effects and physical dependence occur in children.

In some subjects, the intake of more than 200 mg of caffeine significantly prolongs sleep latency and shortens sleep duration; this could be explained by differences in the rate of caffeine metabolism among individuals. Blood pressure increases, which occur after acute caffeine administration, show a rapid tolerance development.

Both low (75 mg or lower) and high (200 mg and higher) doses of caffeine can be discriminated by some subjects against placebo with an effect on mood. The mild reinforcing effect of caffeine varies with the dose, with higher doses tending to reduce the choice or frequency of caffeine self-administration, and very high doses leading to avoidance. Animal studies confirm this low addictive potential of caffeine:

1   At 1 mg/kg body weight or lower – equivalent to one cup of espresso – caffeine activates the nuclei involved in locomotion, mood and sleep, without activating reward circuits in the brain (Nehlig and Boyet, 2000).

2   Only at aversive doses of 10 mg/kg body weight (equivalent to 10 cups of coffee taken at one time) does caffeine induce a release of dopamine and produce an increase in glucose utilization in the *nucleus accumbens*, the area involved in addiction and reward with drugs of abuse like amphetamines, cocaine and nicotine.

It may be concluded that 'although caffeine fulfils some of the criteria for drug dependence, the relative risk of addiction to caffeine is quite low' (Nehlig, 1999). The controversy on the status of caffeine is created, rather than by actual dependence, by the lack of quantitative criteria of abuse potential and negative health consequences in the current classification schemes (Fredholm *et al.*, 1999).

10.4.1.2   *Caffeine does not affect reproduction*

Caffeine has been widely studied in relationship to reproductive functions, without convincing evidence emerging for an increased risk of reproductive adversity (Leviton and Cowan, 2002); contradictory results may be explained by erroneous reporting of caffeine exposure (Bracken *et al.*, 2002).

Findings of a reduction in conception (infertility) in one study were not confirmed by larger studies, and could be explained by the confounding effect of alcohol consumption and smoking (Wilcox et al., 1988; Olsen, 1991; Caan et al., 1998). In the issue of miscarriage, again, the small increased risk of spontaneous abortion observed for women consuming more than 100–150 mg/day during pregnancy could be due to confounders (Fenster et al., 1998; Fernandes et al., 1998; Cnattingius et al., 2000; Wen et al., 2001). A lowered birth weight was found in one study, but no association in two others (Eskenazi et al., 1999; Grosso et al., 2001; Clausson et al., 2002). Also, indication of a link between caffeine consumption during pregnancy and sudden infant death has not been confirmed by the most recent work (Ford et al., 1998; Alm et al., 1999).

10.4.1.3 Caffeine has minimal impact on other organs

Caffeine in coffee had in the past been seen as responsible for the gastric discomfort known as 'heart-burn'; recent work (Boekema et al., 1999, 2001) shows that caffeine has no effect on gastro-oesophageal sphincter pressure, with the possible exception of patients suffering from reflux disease, where a reduction in gastric reflux was observed after they switched to decaffeinated coffee (Pehl et al., 1997).

Caffeine's implication in coronary heart disease appears to be unimportant, particularly when confounding factors like diet and cigarette smoking, which often come in parallel with coffee consumption, are taken into consideration. This applies to acute myocardial infarction, where caffeine is unlikely to be a relevant factor even if there is an increased risk at high consumption (D'Avanzo et al., 1993; La Vecchia et al., 1993). Also, its effect on hypertension, if any, is minor, and a possible increase in blood pressure at high consumption, particularly in hypertensive persons, has been reported (Nurminen et al., 1999; Beilin et al., 2001; Costa, 2002; Klag et al., 2002). Recently, a role of undefined components other than caffeine has been attributed to coffee (Corti et al., 2002).

Finally, no influence at moderate doses on cardiac arrhythmias is to be ascribed to caffeine (Rosmarin, 1989; Myers, 1991).

The contribution of caffeine to osteoporosis is still unclear. Very high levels of caffeine – more than 450 mg/day – cause significant bone loss in healthy women when limited to calcium intakes below the recommended daily dietary allowance (800 mg) (Harris and Dawson-Hughes, 1994). Such findings are, however, generally not supported by more recent studies (Franceschi et al., 1996; Loyd et al., 1997, 1998; Huopio et al., 2000), although acceleration in bone loss at the spine in elderly postmenopausal women consuming more than 300 mg of caffeine per day has been recently reported (Rapuri et al., 2001). Losses can be overcome by daily milk intakes (Barrett-Connor et al., 1994).

10.4.2   Coffee itself does no harm ... but spoiled coffee might

The many favourable effects exhibited so far by a moderate and sustained coffee consumption represent an important body of evidence. However, this must not desensitize the scientific community to progress in analysing possible contaminants that might be present in coffee, as in many other commodities.

10.4.1.2   Contaminants from moulds

Mention has been made in other chapters (see 3.11.11.1 and 4.6.1) of ochratoxin A (OTA), a fungal metabolite signalling past exposure to mould attack. Its presence in coffee at levels of a few parts per billion (ng/g) has long since been reported in the literature as occasional findings linked with poor processing practices, particularly careless drying, in the producer countries.

The noxiousness of OTA has been actively monitored: it has been considered as a possible cause of Balkan nephropathy, an endemic disease in some rural areas. Since it induces renal tumours in experimental animals, the International Agency for Research on Cancer (IARC) classifies it as possibly carcinogenic to humans (Plestina, 1996). However, the genotoxicity of OTA is still under debate: recent studies have observed that OTA is unlikely to form reactive intermediates capable of binding to human DNA, because, unlike in animals, it is not metabolized by cytochrome P450. A genotoxic carcinogenic effect is therefore implausible (Zepnik et al., 2001, 2003).

It is therefore understandable that the scientific coffee community is tackling the issue of OTA prevention in coffee very actively: in this, both analytical aspects and correct sampling techniques are of great importance. Destruction of up to 85% of the amount present in green beans has been shown to occur during processing, most of it at roasting (Blanc et al., 1998). OTA exposure from coffee consumption is estimated at around 9%, behind cereals (44%) and wine (10%), and followed by beer (7%) (Miraglia and Brera, 2002).

10.4.2.2   Other contaminants

Another unpleasant travel companion of coffee is acrylamide (AA), a compound recently discovered in various baked vegetal foods (Tareke et al., 2002). Whilst initial reports on AA in food focused on cooked starch products such as potato and cereal products, AA has also been detected at low levels in a number of other foods, including coffee. On the basis of animal data, AA is considered to be a probable human carcinogen (IARC, 1994). Nevertheless, there is no scientific evidence for this effect, especially at the very low levels at which this compound is generally present in cooked food and even more so in coffee. In the meantime, the World Health Organization advises consumers that there is no reason to alter their diets, and stresses that 'People should eat a balanced and varied diet, which includes plenty of fruit and vegetables, and should moderate their consumption of fried and fatty foods' (FAO/WHO, 2002). Coffee drinkers can continue to enjoy their favourite beverage in moderate amounts without health concerns.

10.5 CONCLUSIONS

In the scientific debate on the health aspects of coffee, which has been conducted since at least the sixteenth century (Jussieu, 1715), the regular scares about the supposed dangerous effects of coffee have each time been counteracted by opposite indications. This suggests that the health effects of coffee are, to say the least, subtle.

At the doses present in a cup of espresso, the main active constituent of coffee – caffeine – has well-documented mild positive effects, which themselves explain the reasons for coffee consumption. On the other hand, the bonus constituted by other proven positive effects of clinical relevance (such as, for instance, protection against colorectal cancer) has yet to be perceived by the public. A sustained effort, founded on solid scientific evidence, is needed to raise consumer awareness that coffee, as any other nutrient, has a contribution to make to many facets of our well-being. That said, it is doubtful that people will ever base their preference for a cup of coffee on anything other than the sheer sensory pleasure they find in it. And why not?

REFERENCES

Abraham S.K. (1996) Anti-genotoxic effects in mice after the interaction between coffee and dietary constituents. *Food Chem. Toxic.* 34 (1), 15–20.

Abraham S.K. and Singh S.P. (1999) Anti-genotoxicity and glutathione S-transferase activity in mice pretreated with caffeinated and decaffeinated coffee. *Food Chem. Toxic.* 37 (7), 733–739.

Acheson K.J., Zahorska-Markiewicz B., Pittet P.H., Anantharaman K. and Jequier E. (1980) Caffeine and coffee: their influence on metabolic rate and substrate utilization in normal weight and obese individuals. *Am. J. Clin. Nutr.* 33, 989–997.

Acquas E., Tanda G. and Di Chiara G. (2002) Differential effects of caffeine on dopamine and acetylcholine transmission in brain areas of drug-naive and caffeine-pretreated rats. *Neuropsychopharmacol.* 27 (2), 182–193.

Alm B., Wennergren G., Norvenius G. *et al.* (1999) Caffeine and alcohol as risk factors for sudden infant death syndrome. Nordic Epidemiological SIDS Study. *Arch. Dis. Child.* 81, 107–111.

Anselme F., Collomp B., Mercier B., Ahmadi S. and Perfaut C. (1992) Caffeine increases maximal anaerobic power and blood lactate concentration. *Eur. J. Appl. Physiol.* 65, 188–191.

APA (1992) *Diagnostic and Statistical Manual of Mental Disorders*, 4th edn. Washington, DC: American Psychiatric Association.

Arciero P.J., Bougopoulos C.L., Nindl B.C. and Benowitz N.L. (2000) Influence of age on the thermic response in women. *Metabolism* 49, 101–107.

Armstrong L.E. (2002) Caffeine, body fluid-electrolyte balance, and exercise performance, *Int. J. Sport Nutr. Exerc. Metab.* 12, 189–206.

Ascherio A., Zhang S.M., Hernán M.A., Kawachi I., Colditz G.A., Speizer F.E. *et al.* (2001) Prospective study of caffeine consumption and risk of Parkinson's disease in men and women. *Ann. Neurol.* 50 (1), 56–63.

Astrup A., Toubro S., Cannon S., Hein P., Breum L. and Madsen J. (1990) Caffeine: a double-blind, placebo-controlled study of its thermogenic, metabolic, and cardiovascular effects in healthy volunteers. *Am. J. Clin. Nutr.* 51, 759–767.

Bara A.I. and Barley E.A. (2003) *Caffeine for asthma*. The Cochrane Library, Issue 2, 2003. Oxford: Cochrane Database System.

Barrett-Connor E., Chang J.H. and Edelstein S.L. (1994) Coffee-associated osteoporosis offset by daily milk consumption. The Rancho Bernardo study. *JAMA* 271, 280–283.

Bättig K. and Buzzi R. (1986) Effect of coffee on the speed of subject-paced information processing. *Neuropsychobiol.* 16, 126–130.

Beaumont M., Batejat D., Pierard C., Van Beers P., Denis J.B., Coste O., Doireau P., Chauffard F., French J. and Lagarde D. (2004) Caffeine or melatonin effects on sleep and sleepiness after rapid eastward transmeridian travel. *J. Appl. Physiol.* 96, 50–58.

Beilin L.J., Burke V., Puddey I.B. *et al.* (2001) Recent developments concerning diet and hypertension. *Clin. Exp. Pharmacol. Physiol.* 28, 1078–1082.

Benedetti M.D., Bower J.H., Maraganore D.M., McDonnell S.K., Peterson B.J., Ahlskog J.E. *et al.* (2000) Smoking, alcohol, and coffee consumption preceding Parkinson's disease: a case-control study. *Neurology* 55 (9), 1350–1358.

Blanc M., Pittet A., Muñoz-Box R and Viani R. (1998) Behavior of ochratoxin A during green coffee roasting and soluble coffee manufacture. *J. Agric. Food Chem.* 46, 673–675.

Boekema P.J., Samsom M. and Smout A.J. (1999) Effect of coffee on gastro-oesophageal reflux in patinets with reflux disease and healthy controls. *Eur. J. Gastroenterol. Hepatol.* 11, 1271–1276.

Boekema P.J., Samsom M., Roelofs J.M. and Smout A.J. (2001) Effect of coffee on motor and sensory function of proximal stomach. *Dig. Dis. Sci.* 46, 945–951.

Boublick J.H., Quinn M.J., Clements J.A., Herington A.C., Wynne K.N. and Funder J.W. (1983) Coffee contains potent opiate receptor binding activity. *Nature* 301, 246–248.

Bracco D., Ferrara J.M., Arnaud M.J., Jequier E. and Schutz Y. (1995) Effect of caffeine metabolism, heart rate, and methylxanthine metabolism in lean and obese women. *Am. J. Physiol.* 269, 671–678.

Bracken M.B., Triche E., Grosso L., Hellenbrand K., Belanger K. and Leaderer B.P. (2002) Heterogenity in assessing self-reports of caffeine exposure: implications for studies on health effects. *Epidemiology* 13, 165–171.

Brice C. and Smith A. (2001) The effects of caffeine on simulated driving, subjective alertness and sustained attention. *Hum. Psychopharmacol.* 16, 523–531.

Brice C.F. and Smith A.P. (2002a) Factors associated with caffeine consumption. *Int. J. Food Sci. Nutr.* 53, 55–64.

Brice C.F. and Smith A.P. (2002b) Effects of caffeine on mood and performance: a study of realistic consumption. *Psychopharmacol.* 164, 188–192.

Bunker M. and McWilliams M. (1979) Caffeine content of common beverages. *J. Am. Dietetic Assoc.* 74, 28–32.

Burton R.F. (1860, reprinted 1961) *The Lake Regions of Central Africa.* New York: Horizon Press.

Caan B., Quesenberry C.P. Jr and Coates A.O. (1998) Differences in fertility associated with caffeinated beverage consumption. *Am. J. Publ. Health* 88, 270–274.

Caraco Y., Zylber-Katz E., Berry E.M. and Levy M. (1995) Caffeine pharmacokinetics in obesity and following significant weight reduction. *Intern. J. Obes. Relat. Metab. Disord.* 19 (4), 234–239.

Casiglia E., Spolaore P., Ginocchio G. and Ambrosio G.B. (1993) Unexpected effects of coffee consumption on liver enzymes. *Eur. J. Epidemiol.* 9, 293–297.

Castellanos F.X. and Rapoport J.L. (2002) Effects of caffeine on development and behaviour in infancy and childhood: a review of the published literature. *Food Chem. Toxicol.* 40 (9), 1235–1242.

Cavin C., Holzhäuser D., Scharf G. et al. (2002) Cafestol and kahweol, two coffee specific diterpenes with anticarcinogenic activity. *Food Chem. Toxicol.* 40, 1155–1163.

Checkoway H., Powers K., Smith-Wellers T., Franklin G.M., Longstreth W.T. and Swanson P.D. (2002) Parkinson's disease risks associated with cigarette smoking, alcohol consumption, and caffeine intake. *Am. J. Epidemiol.* 155, 732–738.

Chen J.F., Xu K., Petzer J.P., Staal R., Xu Y.H., Beilstein M. et al. (2001) Neuroprotection by caffeine and A_{2A} adenosine receptor inactivation in a model of Parkinson's disease. J. Neurosci. 21, 143–149.

Clausson B., Granath F., Ekbom A. et al. (2002) Effect of caffeine exposure during pregnancy on birth weight and gestational age. Am. J. Epidemiol. 155, 429–436.

Cnattingius S., Signorello L.B., Anneren G. et al. (2000) Caffeine intake and risk of first-trimester spontaneous abortion. N. Engl. J. Med. 343, 1839–1845.

Cole K., Costill D.L., Starling R.D., Goodpaster B.H., Trappe S.W. and Fink W.J. (1996) Effect of caffeine ingestion on perception of effort and subsequent work production. Intern. J. Sport Nutr. 6, 14–23.

Corrao G., Lepore A.R., Torchio P., Valenti M., Galatola G., D'Amicis A. et al. (1994) The effect of drinking coffee and smoking cigarettes on the risk of cirrhosis associated with alcohol consumption: a case-control study. Eur. J. Epidemiol. 10, 657–664.

Corrao G., Zambon A., Bagnardi V., D'Amicis A., Klatsky A. et al. (2001) Coffee, caffeine, and the risk of liver cirrhosis. Ann. Epidemiol. 11 (7), 458–465.

Corti R., Binggeli C., Sudano I. et al (2002) Coffee acutely increases sympathetic nerve activity and blood pressure independently of caffeine content. Circulation 106, 2935–2940.

Costa F.V. (2002) Non-pharmacological treatment of hypertension in women. J. Hypertens. 20 (Suppl. 2), 57–61.

Costill D.L., Dalsky G.P. and Fink W.J. (1978) Effects of caffeine ingestion on metabolism and exercise performance. Med. Sci. Sports 10, 155–158.

Cui Y., Winton M.I., Zhang Z.F., Rainey C., Marshall J., De Kernion J.B. and Eckert C.D. (2004) Dietary boron intake and prostate cancer risk. Oncol. Rep. 11, 887–892.

Curhan G.C., Willett W.C., Rimm E.B., Spiegelman D. and Stampfer M.J. (1996) Prospective study of beverage use and the risk of kidney stones. Am. J. Epidemiol. 143, 240–247.

Curhan G.C., Willett W.C., Speizer F.E. and Stampfer M.J. (1998) Beverage use and risk for kidney stones in women. Ann. Intern. Med. 128, 534–540.

Daglia M., Tarsi R., Papetti A., Grisoli P., Dacarro C., Pruzzo C. and Gazzani G. (2002) Antiadhesive effect of green and roasted coffee on Streptococcus mutans adhesive properties on saliva-coated hydroxyapatite beads. J. Agric. Food Chem. 50 (5), 1225–1229.

Daglia M., Papetti A., Gregotti C., Berte F. and Gazzani G. (2002) In vitro antioxidant and ex vivo protective activities of green and roasted coffee. J. Agric. Food Chem. 48, 1449–1454.

Dall'Igna O.P., Porciuncula L.O., Souza D.O., Cunha R.A. and Lara D.R. (2003) Neuroprotection by caffeine and adenosine A_{2A} receptor blockade of beta-amyloid neurotoxicity. Br. J. Pharmacol. 138, 1207–1209.

Daly P.A., Krieger D.R., Dulloo A.G., Young J.B. and Landsberg L. (1993) Ephedrine, caffeine and aspirin: safety and efficacy for treatment of human obesity. *Intern. J. Obes. Relat. Metab. Disord.* 17 (Suppl.), 73–78.

D'Amicis A. and Viani R. (1993) The consumption of coffee. In S. Garattini (ed.), *Caffeine, Coffee and Health*. New York: Raven Press, pp. 1–16.

D'Amicis A., Scaccini C., Tomassi G. *et al.* (1996) Italian style brewed coffee: effect on serum cholesterol in young men. *Int. J. Epidemiol.* 25, 513–520.

D'Avanzo B., La Vecchia C., Tognoni G. *et al.* (1993) Coffee consumption and the risk of acute myocardial infarction in Italian males. *Ann. Epidemiol.* 3, 595–600.

De Paulis T., Schmidt D.E., Bruchey A.K., Kirby M.T., McDonald M.P., Commers P., Lovinger D.M. and Martin P.R. (2002) Dicinnamoylquinides in roasted coffee inhibit the human adenosine transporter. *Eur. J. Pharmacol.* 242, 213–221.

De Paulis T. and Martin P.R. (2003) Effects of non-caffeine constituents in roasted coffee on the brain. In A. Nehlig (ed.), *Coffee, Tea, Chocolate and the Brain*. London: Taylor & Francis.

De Roos B., Caslake M.J., Stalenhoef A.F. *et al.* (2001) The coffee diterpene cafestol increases plasma triacylglycerol by increasing the production rate of large VLDL apolipoprotein B in healthy normolipidemic subjects. *Am. J. Clin. Nutr.* 73, 45–52.

Debrah K., Sherwin R.S., Murphy J. and Kerr D. (1996) Effect of caffeine on the perception of and physiological responses to hypoglycaemia in patients with IDDM. *Lancet* 347, 19–24.

Debry G. (1994) *Coffee and Health*. Paris: John Libbey Eurotext.

Del Castillo M.D., Ames J.M. and Gordon M.H. (2002) Effect of roasting on the antioxidant activity of coffee brews. *J. Agric. Food Chem.* 50, 3698–3703.

De Valck E. and Cluydts R. (2001) Slow-release caffeine as a countermeasure to driver sleepiness induced by partial sleep deprivation. *J. Sleep Res.* 10, 203–209.

Devasagayam T.P., Kamat J.O., Moham H. and Kesevan P.C. (1996) Caffeine as an antioxidant: inhibition of lipid peroxidation induced by reactive oxygen species. *Biochem. Biophys. Acta* 1282 (1), 63–70.

Dews P., O'Brien C. and Bergman J. (2002) Caffeine: behavioral effects of withdrawal and related issues. *Food Chem. Toxicol.* 40, 1257–1261.

Di Chiara G., Acquas E., Tanda G., Marrocu P. and Pisanu A. (2001) Effects of caffeine on dopamine and acetylcholine release on short term memory function: a brain microdialysis and spatial delayed alternation task study. *Proc. 19th ASIC Coll.*, CD-ROM.

Dulloo A.G., Geissler C.A., Horton T., Collins A. and Miller D.S. (1989) Normal caffeine consumption: influence on thermogenesis and daily energy expenditure in lean and post-obese human volunteers. *Am. J. Clin. Nutr.* 49, 44–50.

Ellis J. (1824) *The History of Coffee*. Café Press. Available online at: www.cafepress.pair.com.

El Yacoubi M., Ledent C., Parmentier M., Costentin J. and Vaugeois J.M. (2003) Caffeine reduces hypnotic effects of alcohol through adenosine A(2A) receptor blockade. *Neuropharmacol.* 45, 977–985.

Eskenazi B., Stapleton A.L., Kharrazi M. and Chee W.Y. (1999) Associations between maternal decaffeinated and caffeinated coffee consumption and foetal growth and gestational duration. *Epidemiol.* 10, 242–249.

Esposito F., Morisco F., Verde V., Ritieni A., Alezio A., Caporaso N. and Fogliano V. (2003) Moderate coffee consumption increases plasma glutathione but not homocysteine in healthy subjects. *Aliment. Pharmacol. Ther.* 17, 595–601.

FAO/WHO (2002) Health implications of acrylamide in food. Report of a Joint FAO/WHO Consultation. WHO, Geneva, 25–27 June 2002, p. 17.

Favero A., Franceschi S., La Vecchia C. *et al.* (1998) Meal frequency and coffee intake in colon cancer. *Nutr. Cancer* 30, 182–185.

Fenster L., Quale C., Hiatt R.A. *et al.* (1998) Rate of caffeine metabolism and risk of spontaneous abortion. *Am. J. Epidemiol.* 147, 503–510.

Fernandes O., Sabharwal M., Smiley T. *et al.* (1998) Moderate to heavy caffeine consumption during pregnancy and relationship to spontaneous abortion and abnormal fetal growth: a meta-analysis. *Reprod. Toxicol.* 12, 435–444.

Ferrauti A., Weber K. and Strüder H.K. (1997) Metabolic and ergogenic effects of carbohydrate and caffeine beverages in tennis. *J. Sports Med. Phys. Fitness* 37, 258–266.

Ford R.P., Schluter P.J., Mitchell E.A. *et al.* (1998) Heavy caffeine intake in pregnancy and sudden infant death syndrome – New Zealand Cot Death Study Group. *Arch. Dis. Child.* 78, 9–13.

Foukas L.C., Daniele N., Ktori C., Anderson K.E., Jensen J. and Shepherd P.R. (2002) Direct effects of caffeine and theophylline on P110delta and other phosphoinositiode 3-kinases. *J. Biol. Chem.* 277, 37124–37130.

Franceschi S., Schinella D., Bidoli E. *et al.* (1996) The influence of body size, smoking, and diet on bone density in pre- and postmenopausal women. *Epidemiology* 7, 411–414.

Fredholm B.B., Bättig K., Holmén J. *et al.* (1999) Actions of caffeine in the brain with special reference to factors that contribute to its widespread use. *Pharmacol. Rev.* 51, 83–133.

Gallus S., Tavani A., Negri E. and La Vecchia C. (2002a) Does coffee protect against liver cirrhosis? *Ann. Epidemiol.* 12 (3), 202–205.

Gallus S., Bertuzzi M., Tavani A. *et al.* (2002b) Does coffee protect against hepatocellular carcinoma? *Br. J. Cancer* 87, 956–959.

George K.C., Hebbar S.A., Kale S.P. and Kesavan P.C. (1999) Caffeine protects mice against whole-body lethal dose of gamma-irradiation. *J. Radiol. Prot.* 19, 171–176.

Giovannuccci E. (1998) Meta-analysis of coffee consumption and the risk of colorectal cancer. *Am. J. Epidemiol.* 147, 1043–1052.

Graham T.E. (2001) Caffeine, coffee and ephedrine: impact on exercise performance and metabolism. *Canad. J. Appl. Physiol.* 26, 103–119.

Gray J. (1998) Caffeine, coffee and health. *Nutr. Food Sci.* 6, 314–319.

Gross G., Jaccoud D. and Huggett A.C. (1997) Analysis of the content of the diterpenes cafestol and kahweol in coffee brews. *Food Chem. Toxicol.* 35, 547–554.

Grosso L.M., Rosenberg K.D., Belanger K., Saftlas A.F., Leaderer B. and Bracken M.B. (2001) Maternal caffeine intake and intrauterine growth retardation. *Epidemiology* 12 (4), 447–455.

Harris S.S. and Dawson-Hughes B. (1994) Caffeine and bone loss in postmenopausal women. *Am. J. Clin. Nutr.* 60, 573–578.

Hartter S., Nordmark A., Rose D.M., Bertilsson L., Tybring G. and Laine K. (2003) Effects of caffeine intake on the pharmacokinetics of melatonin, a probe drug for CYPIA2 activity. *Br. J. Clin Pharmacol.* 56, 679–682.

Henderson J.C., O'Connell F. and Fuller R.W. (1993) Decrease of histamine induced bronchoconstriction by caffeine in mild asthma. *Thorax* 48, 824–826.

Hernàn M.A., Takkouche B., Caamaño-Isorna F. and Gestal-Otero J.J. (2002) A meta-analysis of coffee drinking, cigarette smoking, and the risk of Parkinson's disease. *Ann. Neurol.* 52, 276–284.

Hertog M.G.L., Feskens E.L.M., Hollman O.C.H., Katan M.B. and Krombout D. (1993) Dietary antioxidant flavonoids and risk of coronary heart disease. *Lancet* 342, 1007–1011.

Hollands M.A., Arch J.R.S., Phil D. and Cawthorne M.A. (1981) A simple apparatus for comparative measurements of energy expenditure in human subjects: the thermic effect of caffeine. *Am. J. Clin. Nutr.* 34, 2291–2294.

Honig L.S. (2000) Relationship between caffeine intake and Parkinson disease. *JAMA* 284, 1378–1379.

Honjo S., Kono S., Coleman M.P. *et al.* (1999) Coffee drinking and serum gammaglutamyltransferase: an extended study of self-defense officials of Japan. *Ann. Epidemiol.* 9, 325–331.

Horne J.A. and Reyner L.A. (1996) Counteracting driver sleepiness: effects of napping, caffeine and placebo. *Psychophysiol.* 33, 306–309.

Horne J.A. and Reyner L.A. (1999) Vehicle accidents related to sleep: a review. *Occup. Environ. Med.* 56, 289–294.

Huber W., Scharf G., Nagel G. *et al.* (2002) Coffee and its chemopreventive components kahweol and cafestol increase the activity of O6-methylguanine-DNA methyltransferase in rat liver – comparison with phase II xenobiotic metabolism. *Mutat. Res.* 522, 57–68.

Huggett A.C. and Schilter B. (1995) Chemoprotective effects of coffee and its components cafestol and kahweol: effects on xenobiotic metabolising enzymes. *Proc. 16th ASIC Coll.*, pp. 65–72.

Huopio J., Kroger H., Honkanen R. *et al.* (2000) Risk factors for perimenopausal fractures: a prospective study. *Osteoporos. Int.* 11, 219–227.

IARC (1994) Monographs on the evaluation of carcinogen risk to humans: some industrial chemicals. No. 60, International Agency for Research on Cancer, Lyons.

ICO (2003) Coffee Statistics 27 – June 2002. London: International Coffee Organization; available online at: www.ico.org/frameset/traset.htm.

Inoue M., Tajina K., Hirose K. et al. (1998) Tea and coffee consumption and the risk of digestive tract cancers: data from a comparative case-referent study in Japan. Cancer Causes Control 9, 209–216.

IOC (2000) Appendix A: Prohibited classes of substances and prohibited methods. In Olympic Movement Anti-doping Code. Lausanne: International Olympic Committee, p. 15.

Jackman M., Wendling P., Friars D. and Graham T.E. (1996) Metabolic, catecholamine, and endurance responses to caffeine during intense exercise. J. Appl. Physiol. 81, 1658–1663.

James J.E. (1994) Does caffeine enhance or merely restore degraded psychomotor performance? Neuropsychobiol. 30 (2–3), 124–125.

Johnson K.C., Pan S. and Mao Y. (2002) Risk factor for male breast cancer in Canada, 1994–1998. Eur. J. Cancer Prev. 11, 253–263.

Johnson-Kozlow M., Kritz-Silverstein D., Barrett-Connor E. and Morton D. (2001) Coffee consumption and cognitive function among older adults. Am. J. Epidemiol. 156, 842–850.

Johnston K.L., Clifford M.N. and Morgan L.M. (2003) Coffee acutely modifies gastrointestinal hormone secretion and glucose tolerance in humans: glycemic effects of chlorogenic acid and caffeine. Am. J. Clin. Nutr. 78, 728–733.

Jussieu A. (1715) Histoire du café. Memoires de l'Academie Royale des Sciences, 4 May 1715, pp. 291–298.

Kato T., Takanasu S. and Kikugawa K. (1991) Loss of heterocyclic amine mutagens by insoluble hemicellulose fiber and high-molecular-weight soluble polyphenolics of coffee. Mutat. Res. 246, 169–178.

Kawachi I., Willett W.C., Colditz G.A., Stampfer M.J. and Speizer F.E. (1996) A prospective study of coffee drinking and suicide in women. Arch. Intern. Med. 156, 521–525.

Keijzers G.B., De Galan B.E., Tack C.J. and Smits P. (2002) Caffeine can decrease insulin sensitivity in humans. Diabetes Care 25, 364–369.

Kerr D., Sherwin R.S., Pavalkis F. et al. (1993) Effect of caffeine on the recognition of and responses to hypoglycemia in humans. Ann. Intern. Med. 119, 799–804.

Kivity S., Ben Aharon Y., Man A. and Topilsky M. (1990) The effect of caffeine on exercise-induced bronchoconstriction. Chest 97, 1083–1085.

Kiyohara C., Kono S. and Honjo S. (1999) Inverse association between coffee drinking and serum uric acid concentrations in middle-aged Japanese males. Br. J. Nutr. 82, 125–130.

Klag M.J., Wang N.Y., Meoni L.A. et al. (2002) Coffee intake and risk of hypertension: the Johns Hopkins precursors study. Arch. Intern. Med. 162, 657–662.

Klatsky A.L. and Armstrong M.A. (1992) Alcohol, smoking, coffee, and cirrhosis. Am. J. Epidemiol. 136, 1248–1257.

Knapp C.M., Foye M.M., Cottam N., Ciraulo D.A. and Kornetsky C. (2001) Adenosine agonists CGS 21680 and NECA inhibit the initiation of cocaine self-administration. *Pharmacol. Biochem. Behav.* 68, 797–803.

Kono S., Shinchi K., Imanishi K., Todoroki I. and Hatsuse K. (1994) Coffee and serum gamma-glutamyltransferase: a study of self-defence officials in Japan. *Am. J. Epidemiol.* 139, 723–727.

Kubik A., Zatioukal P., Tomasek L. *et al.* (2001) Diet and the risk of lung cancer among women. A hospital-based case-control study. *Neoplasma* 48, 262–266.

Lane J.D. (1997) Effects of brief caffeinated-beverage deprivation on mood, symptoms, and psychomotor performance. *Pharmacol. Biochem. Behav.* 58, 203–208.

La Vecchia C., D'Avanzo B., Negri E. *et al.* (1993) Decaffeinated coffee and acute myocardial infarction. A case-control study in Italian women. *Ann. Epidemiol.* 3, 601–604.

Leitzmann M.F., Stampfer M.J., Willett W.C. *et al.* (2002) Coffee intake is associated with lower risk of symptomatic gallstone disease in women. *Gastroenterol.* 123, 1823–1830.

Leitzmann M.F., Willett W.C., Rimm E.B. *et al.* (1999) A prospective study of coffee consumption and the risk of symptomatic gallstone disease in men. *JAMA,* 281, 2106–2112.

Lelo A., Miners J.O., Robson R. and Birkett D.J. (1986) Assessment of caffeine exposure: caffeine content of beverages, caffeine intake, and plasma concentrations of methylxanthines. *Clin. Pharmacol. Ther.* 39, 54–55.

Lekse J.M., Xia L., Stark J., Morrow J.D. and May J.M. (2001) Plant catechols prevent lipid peroxidation in human plasma and eythrocytes. *Mol. Cell. Biochem.* 226 (1–2), 89–95.

Leviton A. and Cowan L. (2002) A review of the literature relating caffeine consumption by women to their risk of reproductive hazards. *Food Chem. Toxicol.* 40, 1271–1310.

Libby P. (2002) Atherosclerosis: the new view. *Sci. Am.* June, 29–37.

Lieberman H.R. (1987) Positive behavioural effects of caffeine. *Proc. 12th ASIC Coll.,* pp. 102–107.

Lin Y., Tamakoshi A., Kawamura T. *et al.* (2002) Risk of pancreatic cancer in relation to alcohol drinking, coffee consumption and medical history: findings from the Japan collaborative cohort study for evaluation of cancer risk. *Int. J. Cancer* 99, 742–746.

Lindsay J., Laurin D., Verreault R. *et al.* (2002) Risk factors for Alzheimer's disease: a prospectic analysis from the Canadian study of health and aging. *Am. J. Epidemiol.* 156, 445–453.

Lindskog M., Svenningsson P., Pozzi L., Kim Y., Fienberg A.A., Bibb J.A., Fredholm B.B., Nairn A.C., Greengard P. and Fisone G. (2002) Involvement of DARPP-32 phosphorylation in the stimulant action of caffeine. *Nature* 418 (6899), 774–778.

Lloyd T., Rollings N., Eggli D. *et al.* (1997) Dietary caffeine intake and bone status of postmenopausal women. *Am. J. Clin. Nutr.* 65, 1826–1830.

Lloyd T., Rollings N., Kieselhorts K. *et al.* (1998) Dietary caffeine intake is not correlated with adolescent bone gain. *J. Am. Coll. Nutr.* 17, 454–457.

LMC (2002) LMC Commodity Bulletin – Coffee. Oxford: LMC International Ltd; available online at: www.lmc.co.uk/coffee/coffee_regs_set.html.

Lu Y.P., Lou Y.R., Xie J.G. *et al.* (2002) Topical applications of caffeine or ([–])-epigallocatechin gallate (EGCG) inhibit carcinogenesis and selectively increase apoptosis in UVB-induced skin tumors in mice. *Proc. Natl Acad. Sci.* 99 (19), 12455–12460.

MacIntosh B.R. and Wright B.M. (1995) Caffeine ingestion and performance of a 1,500 m swim. *Canad. J. Appl. Physiol.* 20, 168–177.

Mack W.J., Preston-Martin S., Bernstein L. and Qian D. (2002) Lifestyle and other risk factors for thyroid cancer in Los Angeles County females. *Ann. Epidemiol.* 12 (6), 395–401.

Maia L. and De Mendonça A. (2002) Does caffeine intake protect from Alzheimer's disease? *Eur. J. Neurol.* 9 (4), 377–382.

Martin P.R., De Paulis T. and Lovinger D.M. (2001) Non-caffeine dicinnamoylquinide constituents of roasted coffee inhibit the human adenosine transporter. *Proc. 19th ASIC Coll.*, CD-ROM.

Meilgaard M., Civille G.V. and Carr B.T. (1999) *Sensory Evaluation Techniques*, 3rd edn. Boca Raton, FL: CRC Press.

Michels K.B., Holmberg L., Bergkvist L. and Wolk A. (2002) Coffee, tea, and caffeine consumption and breast cancer incidence in a cohort of Swedish women. *Ann. Epidemiol.* 12, 21–26.

Miraglia M. and Brera C. (2002) Assessment of dietary intake of ochratoxin A by the population of EU Member States. Reports on tasks for scientific cooperation, January 2002. European Commission, Directorate-General Health and Consumer Protection, Brussels; available on-line at: http://europa.eu.int/comm/food/fs/scoop/3.2.7_en.pdf.

Montesinos M.C. (2000) Adenosine, acting via A_{2A} receptors, is a critical modulator of inflammation and angiogenesis. 3rd International Symposium of Nucleosides and Nucleotides, Madrid.

Morales F.J. and Babbel M.B. (2002) Melanoidins exert a weak antiradical activity in watery fluids. *J. Agric. Food Chem.* 50, 4657–4661.

Myers M.G. (1991) Caffeine and cardiac arrhythmias. *Ann. Intern. Med.* 114, 147–150.

Nardini M., Cirillo E., Natella F. and Scaccini C. (2002) Absorption of phenolic acids in humans after coffee consumption. *J. Agric. Food Chem.* 50, 5735–5741.

Natella F., Nardini M., Giannetti I., Dattilo C. and Scaccini C. (2002) Coffee drinking influences plasma antioxidant capacity in humans. *J. Agric. Food Chem.* 50 (21), 6211–6216.

NCA (2003) National Coffee Drinking Trends 2002. New York: National Coffee Association of USA; available online at www.ncausa.org/public/pages/index.cfm?pageid=201.

Nehlig A. (1999) Are we dependent on coffee and caffeine? A review on human and animal data. *Neurosci. Biobehav. Rev.* 23, 563–576.

Nehlig A. and Boyet S. (2000) Dose–response study of caffeine effects on cerebral functional activity with a specific focus on dependence. *Brain Res.* 6, 71–77.

Nguyen Van Tam D.P. and Smith A.P. (2001) Caffeine and human memory: a literature review and some data. *Proc. 19th ASIC Coll.*, CD-ROM.

Nishi M., Ohba S., Hirata K. and Miyake H. (1996) Dose–response relationship between coffee and the risk of pancreas cancer. *Jpn J. Oncol.* 26, 42–48.

Nurminen M.L., Niitynen L., Korpela R. and Vapaatalo H. (1999) Coffee, caffeine and blood pressure: a critical review. *Eur. J. Clin. Nutr.* 53, 831–839.

Olsen J. (1991) Cigarette smoking, tea and coffee drinking, and subfecundity. *Am. J. Epidemiol.* 133, 734–739.

Olthof M.R., Hollman C.H. and Katan M.B. (2001) Chlorogenic acid and caffeic acid are absorbed in humans. *J. Nutr.* 131, 66–71.

Pagano R., Negri E., Decarli A. and La Vecchia C. (1988) Coffee drinking and prevalence of bronchial asthma. *Chest* 94, 386–389.

Pasman W.J., Van Baak M.A., Jeukendrup A.E. and De Haan A. (1995) The effect of different doses of caffeine on endurance performance time. *Int. J. Sports Med.* 16, 225–230.

Pehl C., Pfeiffer A., Wendl B. and Kaess H. (1997) The effect of decaffeinated coffee on gastro-oesophageal reflux in patients with reflux disease. *Aliment. Pharmacol. Ther.* 11, 483–486.

Peters A. (1991) Brewing makes the difference. *Proc. 14th ASIC Symp.*, pp. 97–106.

Petracco M. (1989) Physico-chemical and structural characterisation of espresso coffee brew. *Proc. 13th ASIC Symp.*, pp. 246–261.

Petracco M. (2001) Beverage preparation: brewing trends for the new millennium. In R.J. Clarke and O.G. Vitzthum (eds), *Coffee: Recent Advances*. Oxford: Blackwell Science, pp. 140–164.

Plestina R. (1996) Nephrotoxicity of ochratoxin A. *Food Addit. Contam.* 13 Suppl., pp. 19–50.

Poikolainen K. and Vartiainen E. (1997) Poikolainen K. and Vartiainen E. (1997) Determinants of gamma-glutamyltransferase: positive interaction with alcohol and body mass index, negative association with coffee. *Am. J. Epidemiol.* 146, 1019–1024.

Ragonese P., Salemi G., Morgante L. *et al.* (2003) A case–control study on cigarette, alcohol, and coffee consumption preceding Parkinson's disease. *Neuroepidemiol.* 22, 297–304.

Rapuri P.B., Gallagher J.C., Kinyamu H.K. and Ryschon K.L. (2001) Caffeine intake increases the rate of bone loss in elderly women and interacts with vitamin D receptors genotypes. *Am. J. Clin. Nutr.* 74, 694–700.

Reyner A. and Horne J.A. (1997) Suppression of sleepiness in drivers: combination of caffeine with a short nap. *Psychophysiol.* 34, 721–725.

Reyner L.A. and Horne J.A. (1998) Evaluation of 'in-car' countermeasures to sleepiness: cold air and radio. *Sleep* 21, 46–51.

Reyner L.A. and Horne J.A. (2000) Early morning driver sleepiness: effectiveness of 200 mg caffeine. *Psychophysiology* 37, 251–256.

Richelle M., Tavazzi I. and Offord E. (2001) Comparison of the antioxidant activity of commonly consumed polyphenolic beverages (coffee, cocoa, and tea) prepared per cup serving. *J. Agric. Food Chem.* 49, 3438–3442.

Riedel W., Hogervorst E. and Jolles J. (1995a) Cognition enhancers and aging. In J. Jolles, P.J. Houx, M.P.J. Van Boxtel and R.W.H.M. Ponds (eds), *The Maastricht Aging Study – Determinants of Cognitive Aging.* Maastricht: Neuropsych Publishers, pp. 149–156.

Riedel W., Hogervorst E, Leboux R. *et al.* (1995b) Caffeine attenuates scopolamine-induced memory impairment in humans. *Psychopharmacol.* 122 (2), 158–168.

Rosengren A., Dotevall A., Wilhelmsen L., Thelle D. and Johansson S. (2004) Coffee and incidence of diabetes in Swedish women: a prospective 18-year follow-up study. *J. Intern. Med.* 255, 89–95.

Rosmarin P.C. (1989) Coffee and coronary heart disease: a review. *Prog. Cardiovasc. Dis.* 32, 239–245.

Ross G.W. and Petrovitch H. (2001) Current evidence for neuroprotective effects of nicotine and caffeine against Parkinson's disease. *Drugs Aging* 18, 797–806.

Ross G.W., Abbott R.D., Petrovitch H. *et al.* (2000) Association of coffee and caffeine intake with the risk of Parkinson disease. *JAMA* 283, 2674–2679.

Sala M., Cordier S., Chang-Claude J. *et al.* (2000) Coffee consumption and bladder cancer in non-smokers: a pooled analysis of case-control studies in European countries. *Cancer Causes Control* 11, 925–931.

Salazar-Martinez E., Willett W.C., Ascherio A., Manson J.E., Leitzmann M.F., Stamfer M.J. and Hu F.B. (2004) Coffee consumption and risk for type 2 diabetes mellitus. *Ann. Intern. Med.* 140, 1–8.

Scharf G., Prustomersky S., Huber W.W. *et al.* (2001) Elevation of glutathione levels by coffee components and its potential mechanisms. *Adv. Exper. Med. Biol.* 500, 535–539.

Schilter B., Holzhäuser D. and Cavin C. (2001) Health benefits of coffee. *Proc. 19th ASIC Coll.*, CD-ROM.

Schnackenberg R.C. (1973) Caffeine as a substitute for schedule II stimulants in hyperkinetic children. *Am. J. Psychiat.* 130, 796–798.

Schwartz G.G. (2002) Hypothesis: does ochratoxin A cause testicular cancer? *Cancer Causes Control* 13, 91–100.

Schwartz J. and Weiss S.T. (1992) Caffeine intake and asthma symptoms. *Ann. Epidemiol.* 2, 627–635.

Schwarzchild M.A., Chen J.F. and Ascherio A. (2002) Caffeinated clues and the promise of adenosine A(2A) antagonists in PD. *Neurology* 58, 1154–1160.

Sharp D.S. and Benowitz, N.L. (1990) Pharmacoepidemiology of the effect of caffeine on blood pressure. *Clin. Pharmacol. Ther.* 1990, 47 (1), 57–60.

Shearer J., Farah A., De Paulis T., Bracy D.P., Pencek R.R., Graham T.E. and Wasserman D.H. (2003) Quinides of roasted coffee enhance insulin action in conscious rats. *J. Nutr.* 133, 3529–3532.

Shin H.Y., Lee C.S., Chae H.J. *et al.* (2000) Inhibitory effects of anaphylactic shock by caffeine in rats. *Int. J. Immunopharmacol.* 22, 411–418.

Singhara A., Macku C. and Shibamoto T. (1998) Antioxidative activity of brewed coffee extracts. In T. Shibamoto, J. Terao and T. Osawa (eds), Functional Foods for Disease Prevention II: Medicinal Plants and Other Foods. ACS Symposium Series, 701. Washington, DC: American Chemical Society, pp. 101–109.

Smith A.P. (1994) Caffeine and psychomotor performance: a reply to James. *Addiction* 90 (9), 1261–1265.

Smith A.P. (2002) Effects of caffeine on human behaviour. *Food Chem. Toxicol.* 40 (9), 1243–1255.

Smith A.P. and Brice C. (2000) Behavioural effects of caffeine. In T.H. Parliment, Chi-Tang Ho and P. Schiberk (eds), Caffeinated Beverages: Health Benefits, Physiological Effects and Chemistry. ACS Symposium Series, 754. Washington, DC: American Chemical Society, pp. 30–36.

Smith A.P., Brockman P., Flynn R. *et al.* (1993) Investigation of the effects of coffee on alertness and performance during the day and night. *Neuropsychobiol.* 27, 217–223.

Sobel R.K. (2000) A cappuccino a day . . . Caffeine may ward off Parkinson's disease. *US News and World Report* 5 June, p. 63.

Somoza V., Lindenmeier M., Wenzel E., Frank O., Erbesdobler H.F. and Hofmann T. (2003) Activity-guided identification of a chemopreventive compound in coffee beverage using in vitro and in vivo techniques. *J. Agric. Food Chem.* 51, 6861–6869.

Spiller G. (ed.) (1998) *Caffeine*. Boca Raton, FL: CRC Press.

Stavric B. (1992) An update on research with coffee/caffeine (1989-1990). *Food Chem. Toxicol.* 30, 553–555.

Stein M.A., Krasowski M., Leventhal B.L. *et al.* (1996) Behavioural and cognitive effects of methylxanthines: a meta-analysis of theophylline and caffeine. *Arch. Pediatr. Adolesc. Med.* 150, 284–288.

Stookey J.D. (1999) The diuretic effects of alcohol and caffeine and total water intake misclassification. *Eur. J. Epidemiol.* 15 (2), 181–188.

Tanskanen A., Tuomilehto J., Viinamaki H. *et al.* (2000) Heavy coffee drinking and the risk of suicide. *Eur. J. Epidemiol.* 16, 789–791.

Tareke E., Rydberg P., Karlsson P., Eriksson S. and Törnqvist M. (2002) Analysis of acrylamide, a carcinogen formed in heated foodstuffs. *J. Agric. Food Chem.* 50, 4988–5006.

Tavani A. and La Vecchia C. (2000) Coffee and cancer: a review of epidemiological studies, 1990–1999. *Eur. J. Cancer Prev.* 9, 241–256.

Tavani A., Gallus S., Dal Maso L. *et al.* (2001) Coffee and alcohol intake and risk of ovarian cancer: an Italian case-control study. *Nutr. Cancer* 39, 29–34.

Tverdal A. and Skurtveit S. (2003) Coffee intake and mortality from liver cirrhosis. *Ann. Epidemiol.* 13 (6), 419–423.

Urgert R. and Katan M.B. (1997) The cholesterol-raising factor from coffee beans. *Annu. Rev. Nutr.* 17, 305–324.

Van Boxtel M.P.J. and Jolles J. (1999) Habitual coffee consumption and information processing efficiency. *Proc. 18th ASIC Coll.*, 209–212.

Van Dam R.M. and Feskens E.J.M. (2002) Coffee consumption and risk of type 2 diabetes mellitus. *Lancet* 360, 1477–1478.

Van Dusseldorp M. and Katan M.B. (1990) Headache caused by caffeine withdrawal among moderate coffee drinkers switched from ordinary to decaffeinated coffee: a 12 week double blind trial. *BMJ* 300, 1558–1559.

Varani K., Merighi S., Ongini E. *et al.* (1999) Caffeine alters A_{2A} adenosine receptors and their function in human platelets. *Circulation* 99 (19), 2499–2502.

Varani K., Rigamonti D., Sipione S., Camurri A., Borea P.A., Cattabeni F., Abbracchio M.P. and Cattaneo E. (2001) Aberrant amplification of A_2A receptor signaling in striatal cells expressing mutant huntingtin. *FASEB J* 15, 1245–1247.

Watson J.M., Jenkins E.J., Hamilton P. *et al.* (2000) Influence of caffeine on the frequency and perception of hypoglycemia in free-living patients with type 1 diabetes. *Diabetes Care* 23, 455–459.

Weber J.G., Klindworth J.T., Arnold J.J., Danielson D.R. and Ereth M.H. (1997) Prophylactic intravenous administration of caffeine and recivery after ambulatory surgical procedures. *Mayo Clinic Proc.* 72, 621–626.

Weinberg B.A. and Bealer B.K. (2001) *The World of Caffeine*. New York: Routledge.

Wen W., Shu X.O., Jacobs D.R. Jr. and Brown J.E. (2001) The association of maternal caffeine consumption and nausea with spontaneous abortion. *Epidemiology* 12, 38–42.

Weusten-van der Wouw M.P., Katan M.B., Viani R. *et al.* (1994) Identity of the cholesterol-raising factor from boiled coffee and its effects on liver function enzymes. *J. Lipid Res.* 35, 721–733.

WHO (1994) *The ICD-10 Classification of Mental and Behavioural Disorders*. Geneva: World Heath Organization.

Wilcox A., Weinberg C. and Baird D. (1988) Caffeinated beverages and decreased fertility. *Lancet* ii, 1453–1456.

Wynne K.N., Familari M., Boublik J.H., Drummer O.H., Rae I.D. and Funder J.W. (1987) Isolation of opiate receptor ligands in coffee. *Clin. Exper. Pharmacol. Physiol.* 14, 785–790.

Yagasaki K., Okauchi R. and Miura Y. (2002) Bioavailabilities and inhibitory actions of trigonelline, chlorogenic acid and related compounds against hepatoma cell invasion in culture and their mode of action. *Animal Cell Technol: Basic and Applied Aspects* 12, 421–425.

Yoshida T., Sakane N., Umekawa T. and Kondo M. (1994) Relationship between basal metabolic rate, thermogenic response to caffeine, and body weight loss following combined low calorie and exercise treatment in obese women. *Intern. J. Obes. Relat. Metabol. Disord.* 18 (5), 345–350.

Zeeger M.P.A., Tan F.E.S., Goldbohm R.A. and van der Brandt P. (2001) Are coffee and tea consumption associated with urinary tract cancer risk? A systematic review and meta-analysis. *Intern. J. Epidemiol.* 30, 353–362.

Zepnik H., Pahler A., Schauer U. and Dekant W. (2001) Ochratoxin A-induced tumor formation: is there a role of reactive ochratoxin A metabolites? *Toxicol. Sci*, 59 (1), 59–67.

Zepnik H., Völkel W. and Dekant W. (2003) Metabolism and toxicokinetics of the mycotoxin ochratoxin A in F344 rats. *25th Mykotoxin-Workshop* 19–21 May 2003, Giessen.

Closing remarks

E. Illy

This book reminds me of a big tree with its beautiful foliage dancing in the wind, representing the knowledge selected and made accessible by the experts, and its invisible but equally imposing root system, constituted by the broad literature which allows whosoever is interested in a deeper vision to easily satisfy their curiosity.

This 'tree of knowledge' gives the reader a sense of the complexity of all that concerns coffee; from genetics, to interaction with the environment, processing of the cherries, to the transformation in the roaster and, finally, to the delicious cup that delivers such a wonderful message to our senses, followed by a mild stimulation of our brain and activation of the centres of alertness, relaxation and well being, helping us to live better and even possibly to live longer. This complexity is one of coffee's major attractions and the source of an interest that can sometimes evolve into a real love for the product and the culture that surrounds its consumption.

The authors, the editors, all the people involved in the realization of this work, have a common feeling. They love coffee and hope that their efforts will contribute to a deeper understanding of this extraordinary product and a profound respect for all those involved in the production, industrialization and, finally, the preparation of a perfect cup of coffee – the consumption of which is an unforgettable experience.

Meanwhile, the tree will continue to grow, and already we look forward to engaging your attention with a future edition.

Index

Page numbers in *italic* refer to Tables and Figures.

Printed and bound by CPI Group (UK) Ltd, Croydon, CR0 4YY

08/05/2025

01864818-0001